THE
CONTROL OF SULPHIDES
IN
SEWERAGE SYSTEMS

This Manual was edited and prepared for publication by D. K. B. THISTLETHWAYTE, from an original draft by the late W. J. DAVY, for the Technological Standing Committee on Hydrogen Sulphide Corrosion in Sewerage Works, set up by the Biennial Conference of the Senior Technical Officers of the Principal Water Supply and Sewerage Authorities in Australia.

THE CONTROL OF SULPHIDES IN SEWERAGE SYSTEMS

Edited by

D. K. B. THISTLETHWAYTE

B.Sc., F.R.A.C.I., A.S.T.C.(Chem.Eng.),
M.I.E.(Aust.), M.Inst.W.P.C.

Chief Chemist (Research), Metropolitan Water Sewerage and Drainage Board, Sydney; Research Consultant in Public Health Engineering, University of New South Wales.

ann arbor science PUBLISHERS INC.
POST OFFICE BOX 1425 • ANN ARBOR, MICHIGAN 48106

ANN ARBOR SCIENCE PUBLISHERS INC.
P.O. BOX No. 1425
ANN ARBOR, MICHIGAN 48106 USA

First published in Australia 1972 by
Butterworths Pty. Ltd.

ISBN 0-250-40003-0

Library of Congress Catalog Card No. 72-83514

WHOLLY SET UP AND PRINTED IN AUSTRALIA BY
HOGBIN, POOLE (PRINTERS) PTY. LTD., SYDNEY

H_2S TECHNOLOGICAL STANDING COMMITTEE

AUTHORITY	REPRESENTATIVE		
Brisbane City Council, Queensland	Michele, Dr. L., M.Sc., Ph.D., A.R.A.C.I.		1951-1962
	Pope, K. G. S., M.Sc., F.R.A.C.I.		1962-1966
	Sharp, C. F., B.Sc., B.Econ., Dip.Pub.Adm.		1966-1971
Department of Local Government, Queensland	King-Scott, R. D., F.I.E. (Aust.), M.I.C.E., F.R.S.H.		1954-1957
			1964-1971
	Simmonds, M. A., B.E. (Chem.), M.I.E. (Aust.), A.R.A.C.I.		1957-1961
Department of Works & Local Government, New South Wales	Hulton, H. R. B., B.Sc., P.T.S.C., M.I.E. (Aust.)		1951-1971
Engineering and Water Supply Department, South Australia	Davis, J. A., B.E. (Civ.), F.S.A.S.M., M.I.E. (Aust.)		1951-1968
	Johnston, J. J., A.S.A.S.M., A.R.A.C.I.		1951-1971
	Lewis, K. W., B.E. (Civ.), D.I.C. (Lond.), F.I.E. (Aust.)		1968-1971
Hunter District Water Board, Newcastle, New South Wales	Callen, I. P. A., M.Sc.		1951-1965
	Marriott, P. J., A.S.T.C.		1965-1971
Melbourne and Metropolitan Board of Works, Victoria	Benson, L., Dip.C.E., M.I.E. (Aust.)	Techn. Sec.	1968-1971
	Bird, A. W., B.C.E., M.I.E. (Aust.)	Techn. Sec.	1956-1968
		Chairman	1968-1971
	Davy, W. J., B.C.E., M.I.E. (Aust.)	Chairman	1951-1960
		(as Consultant)	1960-1963
	McIntosh, J. A., B.C.E., M.I.E. (Aust.)	Chairman	1960-1967
	McKenzie, J. P., B.C.E., Dip.C.E., F.I.E. (Aust.)	Techn. Sec.	1951-1956
	Mitchell, I. G., B.Sc., A.R.A.C.I.		1962-1971
	Parker, C. D., B.Sc., Dip.Bact. (I.C.Lond.), F.R.A.C.I.		1951-1962
Metropolitan Water Sewerage and Drainage Board, Sydney, New South Wales	Aiken, D. G., B.E. (Civ.), M.I.E. (Aust.)		1963-1971
	O'Brien, T. J. T., B.E. (Civ.), B.Sc.		1962-1963
	Thistlethwayte, D. K. B., B.Sc., F.R.A.C.I., A.S.T.C. (Chem.Eng.), M.I.E. (Aust.), M.Inst. W.P.C.		1951-1971
	Venables, F. de L., B.E. (Civil), M.I.E. (Aust.)		1951-1962
Metropolitan Water Supply, Sewerage and Drainage Board, Perth, Western Australia	Hood, J. C., B.E.M., A.R.A.C.I.		1951-1956
	Montgomery, D. D. T., B.E.		1959-1971
	Morison, F. W. L., B.E., M.I.E. (Aust.), A.M.I.C.E.		1951-1956
	Müller, W. J., Dipl.Ing. (Berlin), Dr.-Ing. (Stuttgart), M.I.E. (Aust.)		1956-1959
State Rivers and Water Supply Commission, Victoria	Evans, R. W. G., Dip. C.E., M.I.E. (Aust.)		1969-1971
	Olsen, P. E., B.E., M.I.E. (Aust.)		1954-1961
	Turner, R. B., B.C.E., M.I.E. (Aust.)		1961-1969

PREFACE

Hydrogen sulphide corrosion has been experienced in many parts of the world, including the sewerage systems of all the principal Sewerage Authorities in Australia.

In 1939, due to the paucity of knowledge on the subject, the Melbourne and Metropolitan Board of Works initiated a special programme of laboratory and field work, in the course of which C. D. Parker and co-workers isolated and described several species of sulphur-oxidizing bacteria in samples taken from corroded concrete sewers, and elucidated the principal mechanisms involved in the corrosion process. Their work also included studies of slime growth and the oxidation of sulphide on sewer walls. These investigations, coupled with reports from overseas, led to a greater understanding of the factors involved and gave promise for control of sulphide problems by better sewerage engineering practice.

The problem in Australia was considered sufficiently important to justify special co-operative investigations, and in 1951 the principal Sewerage Authorities in Australia established a representative Standing Committee of technical officers to collate knowledge and experience, and to co-ordinate further investigations of the problem. The Committee comprised bacteriologists, chemists and engineers engaged in sewerage design or operation or in specific research into hydrogen sulphide problems. The aim was to produce a Code of Practice or Manual for sulphide control.

In 1954, the Committee produced an "Interim Manual of Practice for Minimizing Hydrogen Sulphide Corrosion in Sewage Works". Its purpose was to present in concise form a statement of knowledge then known to the Committee and to direct attention to those gaps in knowledge which could not be satisfactorily filled without further investigation. Following this an Australia-wide series of investigations were undertaken to determine the rates of generation of hydrogen sulphide in pumping systems, detention tanks and inverted syphons. The investigations were carried out by the respective participating Authorities under the direction of a special Sub-Committee, and was co-ordinated by an officer seconded by one of the Authorities. The rates of generation of sulphide under various conditions of sewage temperature and composition were found to be predicted satisfactorily by an empirical formula derived from the data gathered. The investigations led also to an empirical formula for estimating the rate of build-up of hydrogen sulphide due to sewer slimes. This was a major step towards the rationalisation of sewerage design procedures.

The Chairman of the Standing Committee, until his retirement from the Melbourne and Metropolitan Board of Works in 1961, was the late Mr. W. J. Davy, who following his retirement was retained by the Authorities to prepare the first draft of this new Manual. For this he developed the original concepts and fundamental equations relating to gravity sewers which are presented herein, and linked them with the empirical equations for sulphide generation rates. His work thus provided the basis of design for sulphide control. Without the personal contributions of Mr. Davy this Manual could not have been written in its present form.

The Committee hopes that the Manual will prove not only a substantial contribution to sewerage design practice, but also will provide a better basis for evaluating future experience, and lead to the progressive development of more economical and reliable sulphide control. The Committee wishes to emphasise, therefore, the need for further research in order that fundamental concepts may be adjusted and design procedures refined. Nevertheless the Committee records here its conviction that the application of the principles and data contained in the Manual generally will lead to more satisfactory solutions to the numerous problems confronting those responsible for the design and operation of sewerage systems.

It should be understood that the particular statements and judgments incorporated herein are endorsed by the Committee as a whole. They have not necessarily been endorsed by all or any of the participating Authorities.

THE EDITOR

Sydney, Australia
August, 1971

EDITORIAL

In the end some one member of the members of the Australian Standing Committee for Hydrogen Sulphide Corrosion in Sewage Works had to be made responsible for assembling and editing this Manual. It so happened that the job was allotted to me, and my first purpose in these few words is to express my deep appreciation of the honour and trust accorded to me by my colleagues. As the actual publication comes into sight, I am fearful of the many ways, albeit some may be of lesser import, in which I may have failed to justify this trust; and so acknowledge that any errors and omissions are to be attributed largely to me rather than to my colleagues individually and collectively.

The Committee's Preface makes its own acknowledgments and I am of course very much a party to that, especially the mention of our first chairman, our beloved and respected "Bill" Davy. It is then for me to acknowledge especially the contribution by our chairman of this stage, Alan Bird, formerly Technical Secretary to the Committee, who later took Bill's place also as Engineer for Major Sewerage Design of the Melbourne and Metropolitan Board of Works, and his staff; the painstaking help and support of the present Technical Secretary, Lawrie Benson; that of John Lovell, Jr. of the Metropolitan Water Sewerage and Drainage Board of Sydney, who corrected the whole of the mathematics of the Manual with characteristic thoroughness, previously having helped me and the Committee in many other ways; and members of the same Board's Survey Drafting staff especially Messrs. Rogers, Ferguson and colleagues. Among those who should be named also are the late Frank Rossback also of the Sydney Board, and Messrs. Cikalov and Robertson of the Melbourne Board's Design staff, and their colleagues. Of course there are others, too many for detailing here, who I trust will accept thanks anonymously for their contributions. I must thank Messrs. Butterworth's Aust. also for extending more than the common courtesies of good business practice, including Frank Judson now retired. Finally, the faithful interest, patience and support of my wife, Mollie, must come close to a fair claim for honours with these other prime factors which have brought the work to publication.

D. K. B. THISTLETHWAYTE
Editor

Sydney

CONTENTS

LIST OF FIGURES

PART ONE

DIGEST AND DETAILED DISCUSSIONS

Chapter 1

DIGEST OF THE SULPHIDE PROBLEM

Introductory

The problems of deterioration in sewerage systems include all the normal types of problems associated with the deterioration of metals, mortars, concrete and timbers, but involve in addition problems arising from the special nature of water-borne wastes flowing in sewers. These include problems caused by the activities of biological organisms.

This Manual deals in particular with one of the most destructive of these, viz., the corrosion and deterioration which results from sulphides in the wastes, the evolution of gaseous sulphide and its oxidation to form an acidic sulphated environment on the surfaces above the sewage. The Manual describes this sequence setting out first the main factors controlling sulphide concentration levels, then those which operate in connexion with sulphide evolution and oxidation; and finally propounding ways in which these factors may be controlled.

Included also is a discussion of related aspects of sewerage ventilation, both respecting odours and the possible development of irrespirable or toxic conditions.

These discussions and recommended procedures are believed to represent the best information presently available. This information has been assembled on the basis of practical experience in Australia as well as published data from overseas sources and on the basis, also, of the results from co-operative and individual investigations and research programmes conducted in Australia.

FIG. 1.1 Corroded roof of Digestion Tank which collapsed as the attendant stepped onto it, resulting in his death.

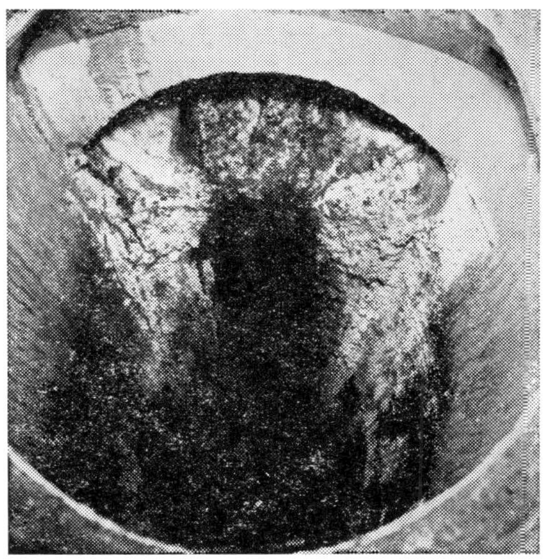

FIG. 1.2 Pasty corrosion and erosion of corrosion products in ventilation and access manhole.

Typical examples of hydrogen sulphide corrosion

The corrosion discussed in this Manual is that which occurs to the surface of concrete, mortar, some metals and other structural materials in sewage works when they are exposed to an atmosphere containing hydrogen sulphide. Characteristic features of such a corroding concrete or mortar surface are high sulphate and sulphuric acid concentrations with pH values even as low as pH 1.0, or less.

The following illustrations are examples of the damage caused to sewerage structures as a result of hydrogen sulphide attack.

These typical examples of the incidence of hydrogen sulphide corrosion under Australian conditions serve to underline the gravity of the problem for sewerage authorities. Expressed in general terms, a set of working rules is required. This requires that the principal factors operating be recognized and their quantitative effects elucidated.

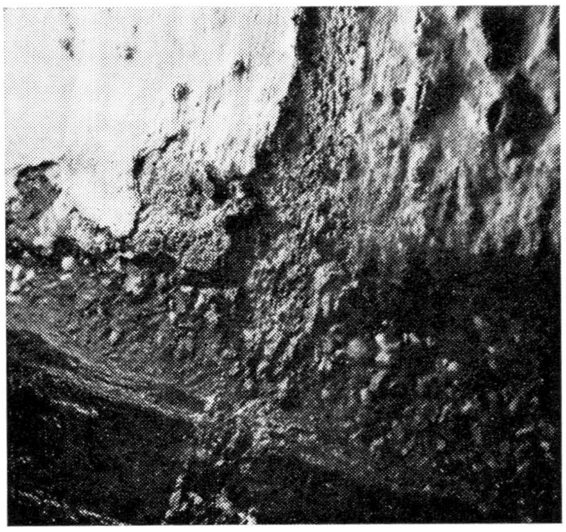

FIG. 1.3 Appearance of 36in. diameter concrete pipe after three years' exposure to severe sulphide corrosion.

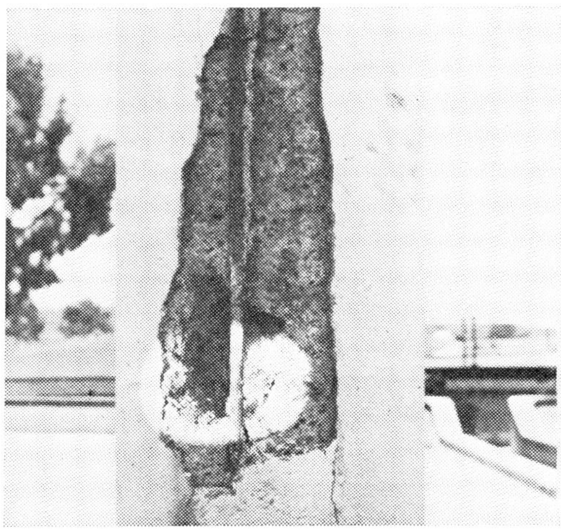

FIG. 1.4 Corrosion of galvanized steel vent stack.

FIG. 1.5 "Pock-marking", scaling, and paste formation.

Train of conditions and processes resulting in hydrogen sulphide corrosion

Briefly, the sequence required for hydrogen sulphide corrosion to take place is as follows:—

(1) Sulphides must be present, dissolved in the sewage. The dissolved sulphides may come from sulphide-bearing wastes, or may be generated by anaerobic bacteria within the sewers.

(2) Significant amounts of hydrogen sulphide gas must escape from the sewage into the sewer air.

(3) Some of this hydrogen sulphide gas must be transferred from the sewer air to the surfaces above sewage level.

(4) The sulphide thus reaching the surfaces not washed continuously by the sewage must next be oxidised with formation of sulphuric acid. This oxidation takes place principally as a result of bacterial action.

(5) The final stage in the sequence is the chemical attack of the concrete or mortar progressively by the sulphuric acid and by the sulphates formed.

Each of these conditions and processes is an essential step in the sequence and consequently hydrogen sulphide corrosion can be minimized or prevented by controlling any one or combination of them. The inter-relationships of these sequences are discussed at length in the following Chapters but are summarized here in the following.

The sequence is illustrated by the outline diagram behind the front cover.

Sewer and sewage conditions in relation to sulphide problems

Before laying down rules for the control of the sulphide problems which may be encountered in a sewerage system, it is desirable to have some overall perspective of those factors which influence the intensity of these problems and the concentrations of sulphides in the sewage as they vary with time and location in the system. Sulphide concentration in the sewage is by no means the only criterion determining the intensity of the sulphide problem at any particular place in the system, but there can be no such problem if there are no sulphides.

Conditions conducive to the development of the sulphide problem:—The conditions listed in the following are related more particularly to the sulphide corrosion problem. Aspects of odour nuisance and safety of personnel are correlated.

(a) *Conditions which commonly result in higher concentrations of molecular hydrogen sulphide in sewage* (Step No. 1, see pp. 7-9, 33 et seq.).

(1) Admixture in the sewage of a large proportion of trade wastes with substantial sulphide or organic sulphur contents (pp. 8, 41).

FIG. 1.6 Concrete sewer pipes corroded internally above sewage-level and undamaged below it.

(2) Relatively high sulphate content in sewage. The higher the sulphate is the higher is the potential sulphide concentration which can be generated under conditions favouring septicity (Fig. 8.9, p. 91).

(3) Low pH value of sewage—the lower the pH value is the greater the proportion of molecular hydrogen sulphide (pp. 7, 9, Fig. 5.1, p. 35).

(4) A high rate of oxygen demand which rapidly exhausts available dissolved oxygen. Also strong sewage usually contains more surface-active material which reduces surface rate of re-aeration (pp. 27 et seq.).

(5) High sewage temperature accelerates biological activity, so that oxygen is consumed more rapidly; likewise hydrogen sulphide generation starts sooner and then proceeds faster; while oxygen solubility and re-aeration rates are also lower (Fig. 8.7, p. 90).

(6) Holding up the sewage flow under conditions favouring septicity, slime growth and siltation. Examples are long sewers with flat gradients, long pressure mains, large wet wells, and so on.

(7) Low sewage velocity favours oxygen depletion and hence sulphide generation commences earlier. It decreases the rate of re-aeration, i.e. oxygen absorption at the surface of the flowing sewage (pp. 21 et seq.).

(b) *Factors tending to increase the emission of hydrogen sulphide from sewage* (Step No. 2).

(1) The higher the concentration of molecular hydrogen sulphide in the sewage, the higher is the rate of emission (pp. 34 et seq.).

(2) High sewage velocity or shock turbulence encourage emission (p. 36).

(3) The higher the relative velocity and turbulence in the airway above the sewage, the higher is the emission (pp. 37 et seq.).

(4) The cleaner the surface of the sewage with respect to oil films, surface active agents, etc., the higher the emission (pp. 35, 36).

(c) *Factors tending to increase the rate of transfer of hydrogen sulphide from the sewer atmosphere to structural surfaces above sewage level* (Step No. 3, p. 67).

(1) The higher the concentration of hydrogen sulphide in the sewer air, the higher the rate of transfer (p. 69).

(2) Higher pH of wall moisture increases the transfer rate because the gas then is more soluble. Sulphide may accumulate in this way before bacterial oxidation actually sets in.

(3) Increased turbulence in the airway, due to high ventilating air velocity or wall roughness or projections, increases the rate of transfer.

(d) *Conditions tending to promote the conversion of hydrogen sulphide to sulphuric acid on the structural surfaces* (Step No. 4, pp. 65 et seq.).

(1) The presence of sulphur or sulphide on the moist sewer surface, and of carbon dioxide, oxygen, ammonia, phosphate, etc., necessary for the proliferation of the sulphuric acid-forming bacteria.

(2) A suitable environment for the sulphur bacteria to commence their activities. In the case of concrete, this may develop initially as a result of carbonation of the originally highly alkaline concrete surface.

(3) Adequate moisture, normally requiring high relative humidity of the sewer air (p. 58).

(e) *Conditions tending to promote attack on concrete* (Step No. 5, pp. 65 et seq.).

(1) Permeable and lean-mix concrete is more susceptible to attack than dense concrete. Less acid is required to cause disintegration of the same volume of lean concrete than is needed with a richer mix. Also,

it has less structural strength and commonly it is the more porous. It is to be noted that sharp edges or arrises are subject to attack from two directions and moreover cause increased turbulence.

(2) Condensed moisture running down sewer walls tends to concentrate acid conditions initially and cause the scouring of channels. Acid also tends to accumulate just above sewage level.

Control of sulphide problems

The means available to prevent the development of corrosion conditions or to bring such conditions under control, are related to the foregoing. A brief list of control measures follows. They are presented separately as preventative measures in the design of systems to be constructed (new works), as corrective measures for existing systems and as methods of protection, either for existing or new works.

(a) *Control of conditions by Design in new systems (preventative measures).*

(1) Design sewer sizes and grades so that oxygen intake rates balance or exceed oxygen consumption rates, also to control slime growths (pp. 78 et seq.).

(2) Design junctions and drops with regard for turbulence effects (pp. 94 et seq.).

(3) Design pumping systems to avoid significant build-up and to control oxygen depletion (p. 132).

(4) Control or regulate sulphide admission in industrial and trade waste discharges (pp. 84, 115).

(b) *Control of sulphide conditions in existing systems (corrective measures).*

"Sulphide Conditions" Surveys should first be made so as to indicate where the most significant sulphide concentration increments occur. The possibilities for control then may be usefully investigated.

(1) Regulate or control sulphide-bearing discharges.

(2) Regulate or control sulphate-bearing discharges so that their effects if any become insignificant.

(3) Operate pumping systems to minimize their contributions to sulphide concentration levels (rearrange or reconstruct if necessary).

(4) Operate, adjust or reconstruct gravity sewer systems to minimize sulphide generation.

(5) Operate, adjust, instal or reconstruct ventilating systems so as to minimize sulphide problems of corrosion, odour, nuisance, toxicity hazards etc. (Chapter 6).

(c) *Protective (curative) measures*

Measures aimed at preventing the *effects* of corrosion fall into three categories:

(1) by complete or partial "insulation" of the structures from the corrosive environment—e.g. by encasing the exposed concrete surfaces in acid resisting plastic coatings.

(2) by reducing the time-intensity effects of corrosion, e.g. by using acid-resistant cements or mortars, free lime control by incorporation of pozzolans, etc. (pp. 71 et seq.).

(3) by incorporating additional thicknesses of concrete to allow for destruction by corrosion, designing to facilitate removal of deterioration and replacement from time to time.

Materials of construction

The effects of sulphides and consequently the corrosion sequence differs for the different types of materials used in sewerage systems. Galvanized coatings may extend the life of mild steel exposed to sewer atmospheres where it is possible to avoid acidic drainage, for example, under water. Where biological growth is a factor, the supply of nutrients and trace elements plays a vital part. The choice of special materials of construction consequently involves many questions, many of them specific for a particular application. Lengthy field trials may be essential.

Most of these questions are beyond the scope of this Manual, which is directed primarily towards concrete construction.

Economics of sulphide corrosion

While there is no doubt that prevention of sulphide problems is always desirable, in some cases at least the cost of ensuring against all or any such problems may be rather too high. The possibilities will be seen to include the following:

(a) design to prevent any possibility of sulphide corrosion;

(b) accept that some damage may occur from time to time along some sections of the sewerage system and set aside the appropriate moneys required to provide for repairs or replacement as partial or near failure approaches; other alternatives are

(c) incorporate replaceable wastage in those parts of the sewer structures likely to suffer damage, or

(d) incorporate relatively costly corrosion resistant linings or other materials in sections likely to suffer damage, paying for these additional costs out of overall savings.

These possibilities may be complicated where factors such as odour nuisance or special ventilation requirements are brought into account.

For discussion and mathematical treatment of the related engineering economics see pp. 77, 78.

CHAPTER 2

SOURCES AND NATURE OF SULPHIDES IN SEWERS

Introductory

As stated in Chapter 1, sulphide corrosion cannot be initiated unless hydrogen sulphide be present in the airway above the sewage. The nature and origins of sulphides therefore are the basic considerations of this chapter.

The nature of sulphide compounds found in sewage

When the total sulphide content of sewage is measured, the types of compounds taken into account include (i) Sulphides in solution as sulphide ion (symbol $S^=$), as hydrosulphide ion (HS^-), and as molecular hydrogen sulphide (H_2S), and (ii) insoluble compounds, such as iron sulphide (FeS) or zinc sulphide (ZnS), etc.

When molecular hydrogen sulphide dissolves in water, the H_2S dissociates in accordance with reversible ionization reactions, as follows:

$$H_2S \rightleftharpoons H^+ + HS^- \rightleftharpoons 2H^+ + S^= \qquad [2.1]$$

These relationships indicate that an increase in the hydrogen ion concentration (i.e. a decrease in pH) results in a greater proportion of undissociated H_2S.

Any dissolved sulphide present usually is partly in the undissociated molecular form (H_2S) and partly in dissociated form. The distinction between these forms is of the utmost importance for the train of processes leading to the development of sulphide problems since it is only as H_2S molecules that hydrogen sulphide can escape to the sewer atmosphere and play its part in the corrosion process.

There is always a balance between the proportions of these dissolved sulphide forms, the proportions depending upon the pH and temperature of the sewage, and also on the contents if any of metallic ions such as of iron or zinc. Of the sulphide sulphur present as $S^=$, HS^-, H_2S, the proportion of molecular H_2S depends on pH in the sewage in such a way that the H_2S (molecular) concentration is increased with decrease in the pH value (see Fig. 2.1). For example, a reduction in pH (i.e. an increase in the degree of acidity) from pH 7.0 to 6.0, results in the proportion of molecular hydrogen sulphide being doubled, whilst an increase of pH from 7.0 to 8.0 would reduce the proportional molecular sulphide concentration to about one-fifth of its original value.

The discharge of acid wastes to sewers can have bad effects in causing increased emission rates of hydrogen sulphide. In some cases, the sudden discharge of acid wastes into sewage containing sulphides, and the consequent release of hydrogen sulphide from the sewage, has produced such high local concentrations of sulphide in the sewer atmosphere as to cause the deaths of maintenance workmen.

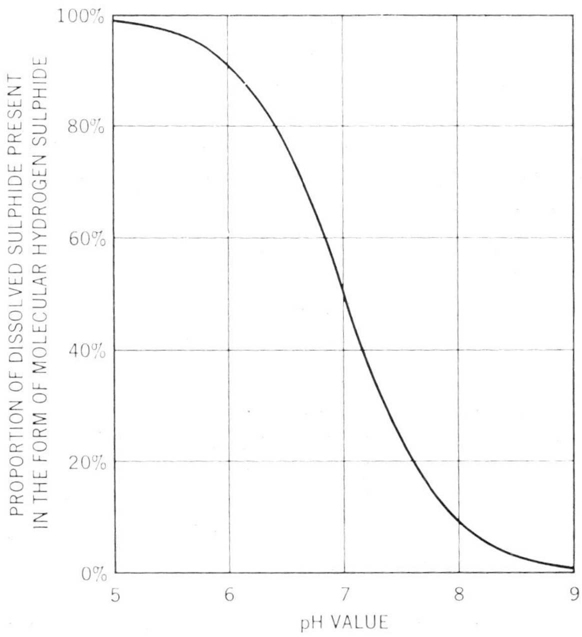

FIG. 2.1 Relative concentrations of dissolved molecular hydrogen sulphide.

The proportion of molecular sulphide present is decreased also by an increase in temperature. Increasing the temperature by 1°C. has the same effects as increasing the pH by 0.15(*1*). The equivalent figure for 1°F. is 0.008 pH units.

Sewage normally contains both dissolved and insoluble sulphides and also significant concentrations of iron amounting usually to several mg/1 or more. The iron is contributed variously by the water supply, and by domestic and industrial wastes.

The black colour of old sewage is due to the presence of iron sulphides in suspension; also the submerged sewer surfaces behind the wall slime are usually black. This latter phenomenon may well be due to the fact that sulphide concentrations within wall slimes are very high (several hundred p.p.m.) and this high concentration may increase the reaction of sulphide and iron compounds in the slimes and concrete. Observations generally indicate that with domestic sewage the first 0.25 mg/1 (at least) of sulphide released are converted to the insoluble form (i.e. within the sewage).

Some of the sulphide sulphur generated in the wall slimes diffuses into the sewage, but must be in a soluble form ($S^{=}$, HS^{-}, or H_2S; most sulphide sulphur diffusing from the slimes is actually in the form of molecular H_2S). Some is held in the slimes as black FeS, and some may become oxidised to elemental sulphur at the slime surface if the redox potential of the flowing sewage—as for example with fresh sewage—is high enough. Sulphide sulphur brought in with industrial or other waste discharges—or groundwater infiltration, etc.—likewise may be initially in a soluble form, or may be wholly or partly insoluble. In any case, sulphides initially either soluble or insoluble may be rendered respectively either insoluble or soluble on admixture with the flowing sewage, e.g. because of the presence of metallic ions such as Fe^{++}, Zn^{++} or Cu^{++}, etc., alternatively because of pH conditions.

The soluble sulphides may:—

(1) *be wholly or partly oxidized in the sewage* to stable compounds either by the dissolved oxygen present (aerobic conditions) or even under anaerobic conditions if the oxidizing potential of the sewage remains high enough, e.g. where the sewage has become stale (p. 94) but not yet septic. The portion of the sulphides so oxidized plays no further part in producing sulphide problems (unless it be subsequently reduced again to sulphide somewhere downstream);

or (2) *be wholly or partly converted to insoluble sulphides* by interaction with other substances such as metallic compounds present in the sewage. Insoluble sulphides, unable to escape into the sewer air as gases, cannot contribute to the sulphide corrosion or odour problems as does soluble sulphide.

Although other contributory factors are present in sulphide corrosion, the concentration of sulphide in sewage (or more correctly the concentration of dissolved free hydrogen sulphide) is the important primary factor in determining the intensity of corrosion. The presence of free hydrogen sulphide in sewage is a necessary prerequisite for the existence of the problem. In this Manual the concentration of sulphide sulphur (in whatever form) is expressed as the mass of sulphide sulphur in milligrams per litre of sewage (nearly enough, parts per million by weight). While the concentration of total sulphides may be useful for a preliminary appraisal of a sulphide problem, nevertheless only that fraction of the total sulphides present as soluble sulphide can affect the corrosion process at a particular time and place.

Whatever the site and the mechanism may be, while it is known that the percentages of total sulphide occurring in the insoluble form vary considerably in normal sewage, this percentage appears most frequently to range between 40% and 60%, and it is suggested that a safe figure to adopt for most circumstances would be 40%.

Sources of sulphide in sewers

Sulphides if present in sewage may originate in either or both of the following ways:

(1) By the direct discharge into the sewers of sulphides as such, for example in various industrial wastes.

(2) By the generation in the sewers of sulphides through the decomposition of sulphur compounds normally present in the sewage (such as inorganic sulphates or organic sulphur compounds).

Direct discharges of sulphide

Certain industries are likely to discharge relatively large amounts of sulphide sulphur, including, for example, abattoirs and fell-mongering works, tanneries, coal-gas works and oil refineries, sulphur dye works, viscose factories, and many others. Where any particular area is under investigation, it is necessary to examine the situation in detail, but experienced chemists can assess the possibilities adequately and rapidly by inspection, only a few special cases then requiring detailed attention. Some industries discharge wastes carrying sulphide concentrations measured in hundreds or even thousands of mg/l of waste, so that even comparatively very small volumes may contribute substantially to the sulphide content of the flowing sewage, especially in the local sewers just downstream.

Generation of sulphides in the sewerage systems

Bacterial metabolism within the flowing sewage or within slimes and silt deposits commonly results in some sulphide formation and frequently is the most important source even in industrial areas. Sulphur must be available in some form ranging from elemental sulphur (rarely important) through compounds of sulphur in oxidation states varying up to inorganic sulphates, organic sulphates, or sulphonates, but including also proteins, thioamino acids, etc. In cases where the sulphur already is in the reduced state sulphide ion may be released perhaps by some metabolic processes of decomposition, or by some substitution reaction.

The possibilities of such sources of sulphide require attention to many of the physical circumstances which may be connected with sewerage areas in addition to domestic and industrial waste discharges.

Septic tanks usually contain significant amounts of sulphide, hence dumping of sludges or effluents from such tanks may be of importance. Likewise any detention tank, pond, sedimentation apparatus, or even screen pits, should be taken into account.

The sulphate present in sewage is usually derived principally from the mineral content of the town water supply or from saline groundwater infiltration, but it may occur also as a result of saline industrial wastes or from sea-water used by industry and discharged to the sewer. The sulphate content, expressed as sulphur, may vary from only a few parts per million to hundreds of parts per million, and the limiting amount of hydrogen sulphide which can be formed therefrom in the sewage varies accordingly.

Organic sulphur compounds are present in excreta, domestic sewage containing usually from 1 to 3 p.p.m. of organic sulphur. Organic sulphur occurs in much higher concentrations in certain trade wastes, for example, meat industry wastes in which organic sulphur concentrations may amount to many tens of mg/1 or even hundreds. Some chemical industry wastes may contain even more pro rata.

According to Summer (*ref. 2, p. 301*) excreted wastes from human beings contain, on the average, about 2.6 gms. of sulphates as SO_4 per head per day. This would correspond to concentrations of sulphate sulphur of from 3.8 to 6.4 mg/1 on the basis of sewage flows of 30 to 50 gallons/head/day.

Regarding town water supplies, the sulphate content may be derived not only from the natural content of the raw water, but may be increased by the use of sulphate-containing coagulants.

Since all such sulphur in oxidized or reduced forms, organic or inorganic, represents a potential for sulphide ion formation, an assessment of these sources should be made as part of any sulphide corrosion investigation, or as a preliminary for all sewer design. Generally, for domestic sewerage, the main source will be sulphate. In industrial areas, the case may well be different.

Sites of generation.—Over most of the underwater surface of sewage works there develops naturally a biological slime varying in effective thickness from a few hundredths of an inch to half an inch or more. Evidence has been advanced to show that normally this slime is the site where most of the sulphide is generated(*3-9*). For example, it has been observed that when these slimes are removed, either by mechanical cleaning or chemical means, sulphide generation then either is greatly reduced or ceases entirely and the generation rate does not return to normal until some weeks later when sufficient slime has been re-established. Another site of possibly significant sulphide production is in silt deposits. Also, where the sewage has become septic, significant generation may also occur in the body of flowing sewage.

The processes involved and effective rates of generation are considered in detail in Chapter 3.

Chapter 3

GENERATION OF SULPHIDES IN SEWERAGE SYSTEMS

Introductory

The importance, as a primary source of sulphides, of that generated in the sewerage system by biochemical action from sulphate or organic sulphur compounds has been emphasized. Two major aspects involved are respectively (1) the sites of generation and (2) the rates of generation.

Sites of generation

The main factor determining the amount of sulphide which builds up in the sewage is the number of sulphate-reducing organisms active from place to place. Slimes which form on sewer walls and other parts of sewerage systems normally provide sites favourable for the development of these organisms in very large concentrations.

Occurrence of slimes

In 1959 Reid and Tien-Sheng Yang(*10*) published results of slime growth experiments in experimental sewers using four parallel semi-circular channels, each constructed from different commercial pipe. Settled sewage from a common source flowed along all channels. As a result of their experiments it was found that:—

(1) The weight of slime per unit length (i.e. average thickness of slime) progressively increased up to a period of about twelve weeks, after which little or no further increase occurred. The rate of growth of slime was high at first and then tapered off. Two typical cases are shown in the following Fig. 3.1.

(2) Whilst the sewage itself contained no sulphide the wall slimes contained large concentrations. This suggests that slimes are a major factor in sulphide generation.

(3) The average thickness of slime was found to vary with the velocity of the flowing sewage, the thickness being greater at low velocities.

(4) As the slime developed, the carrying capacity of the channel decreased. This effect was marked at low velocities.

These investigators suggested that the final thickness of slime reached under any set of conditions is related to the tractive force of sewage on the slime surface.

Parker, Alford, Pope, Brzozek and Beer(*7*) investigated the occurrence of slimes on the submerged surfaces of gravity sewers. By specially developed methods they were able to measure the thickness of slimes at various points on the submerged surface with reasonable accuracy. This survey of slime thickness covered sewers with a wide range of size, grade, velocity and proportional depth of flow, and disclosed the following:—

(i) The slime thickness varied considerably from place to place along the sewer and also from point to point around the wetted perimeter of the pipe.

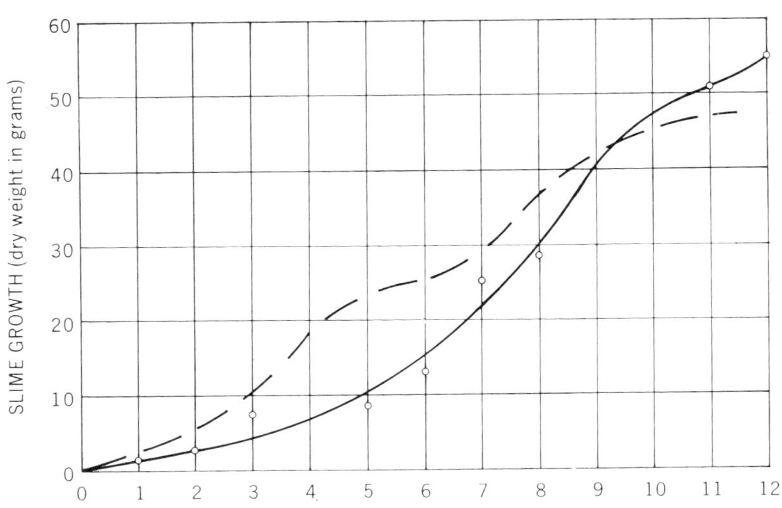

FIG. 3.1 Rates of growth of slimes in sewers.

(ii) At any one site the thickness varied more or less in a systematic way, being generally much greater at, or just below, daily low water level. From this point of maximum thickness the thickness gradually decreased to a minimum at the invert. This variation is shown diagrammatically in Fig. 3.2. The thickness at or near low water level may reach several inches, whereas at invert the depth of slime may be as shallow as only 1/40 inch, or in some instances entirely absent, near the invert.

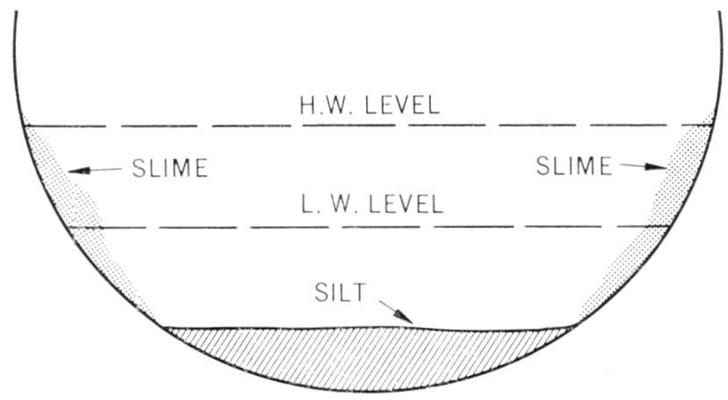

FIG. 3.2 Distribution and variations in thickness of sewer slimes.

(iii) The thickness was also small at high water level gradually increasing to the maximum at, or just below, low water level.

(iv) The thickness varied with sewage velocity, being greater at lower velocities.

(v) The thickness was greater on the outside of bends (except where critical values of wall shear are reached).

(vi) The thickest slimes were observed on outfall sewers, where the sewage usually is older.

(vii) When sewage was very fresh there was little slime growth.

(viii) Practically all slimes are black underneath and grey on the surface, the grey being of lighter colour for fresher sewages.

The absence of slimes presumably indicates conditions of little or no sulphide generation. It is important therefore to establish under what conditions slime is absent. As mentioned above, Reid and Tien-Sheng Yang suggested that slime thickness is related to the tractive force of the sewage on the slime; this force corresponding to the boundary shear between sewer wall and sewage.

Neglecting any shearing forces between the water surface and air above, the mean value of the wall shear around the wetted perimeter can be calculated using established criteria, as follows:

$$\tau_{om} = 62.32 \times S \times m \qquad [3.1]$$

where τ_{om} = mean value of wall shear around wetted perimeter (lb/sq. ft.)
S = slope of sewer channel
m = hydraulic radius (ft.)

The wall shear stresses vary as indicated in Fig. 3.3. The areas subtending the divided angle $(\theta_2 - \theta_1)$, include the wall areas covered by slimes. Within the areas subtending the angle θ_1, the wall shear is too high for slimes to develop.

As noted above, the surveys of slime thickness variations reported by Parker and his co-workers indicated that any areas on which slimes did not develop were always on the lower portion of the submerged arc.

An analysis of the cases where significant slime growth was absent indicated that there was a critical value of wall shear above which sulphide producing slimes do not develop. For gravity sewers, conveying sewage with normal grit loadings, this value appears to be above about 0.06 lb/sq. ft. Available rising main data indicate that sulphide generation ceases at values somewhat below 0.078 lb/sq. ft. Current practice for rising main design, however, has been such that the value of wall shear would seldom reach a value of 0.078 lb/sq. ft., for any significant portion of the day. Furthermore, recent tests on a 6 in. diameter test pipeline, conveying partially sedimented sewage, indicated a critical shear stress of approximately 0.07 lb/sq. ft. (*11*).

The determination of the exact value of wall shear necessary to prevent significant slime development on submerged sewer surfaces requires further detailed experimental investigation. Meanwhile it is recommended that a value of wall shear of 0.07 lb/sq. ft. be adopted as the critical value above which significant wall slimes will not develop for gravity sewers, and not less than 0.08 lb/sq. ft. for pressure mains and rising mains.

For straight uniform pressure mains of circular cross-section the value of wall shear is the same at all points around the periphery and can be calculated from the equation:—

$$\tau_{om} = 62.32 \times S \times \frac{D}{4} = 15.58 \times S \times D \text{ lb/sq. ft.} \qquad [3.2]$$

where τ_{om} = boundary (or wall) shear (lb/sq. ft.)
S = slope of hydraulic gradient (dimensionless)
and D = internal diameter of pipe (ft.)

The above equation applies at a temperature of 68°F. (20°C.); the term 62.32 being the density of water at this temperature.

For gravity sewers likewise if the product of slope and hydraulic radius be sufficiently high, then theoretically, no slime will develop at any point around the wetted perimeter. On the contrary, if the value be low enough, slime will develop around the whole of the wetted surface. Between these two limits slime will be present on some part or parts of the submerged surfaces.

The concept of "zero" slime growth, under high average shear stress conditions, is in fact, however,

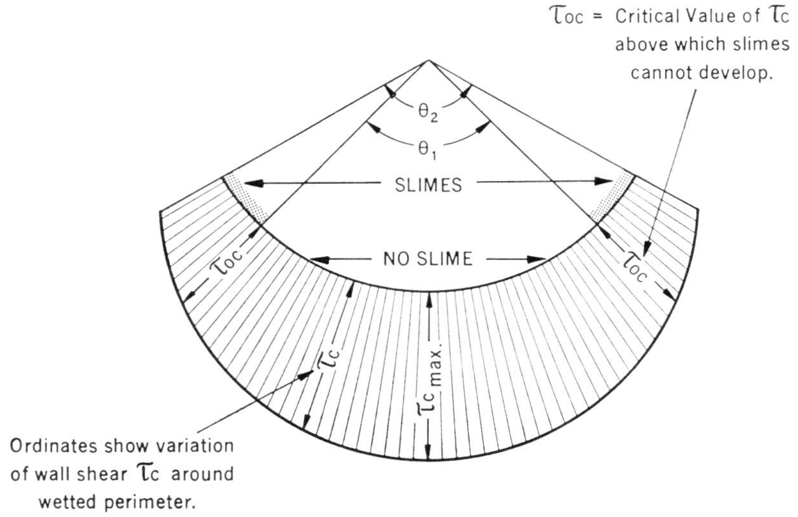

$F_s = \frac{\theta_2 - \theta_1}{\theta_2}$ = Proportion of submerged surface on which slimes develop.

FIG. 3.3 Slimes in sewers related to wall shear values.

somewhat of a misnomer. Actual shear stress conditions do tend to taper off quite rapidly near the limits of the wetted perimeter and usually are low enough to allow some slime growths, say from 5% to 15%, or even more of the wetted surface area. Hence, in order to allow for this, and also for construction variances in diameter and alignment, variations in pipe jointing details, diurnal variations in sewage levels and the like, an active slime growth of 15% of the wetted surface area has been adopted as a design basis (p. 111).

Figure 3.3 shows in a general way how the magnitude of wall shear stress varies around the wetted perimeter from a maximum at the invert to a minimum at the water surface level.

At present no fully proven method appears to have been developed which will enable the value of wall shear stress over any particular area to be accurately calculated. In this regard work by Replogle(*12*) appears to offer promise and it is expected his technique will shortly be extended in its application(*13*). However, for the purposes of this Manual the best practical formulation is one based on a segmental concept.

In applying this concept the flow is considered as a series of isovels or lines of equal velocity. From the isovel distribution a series of orthogonals may be constructed from the point of maximum velocity. If turbulent velocity fluctuations are ignored, there can be no momentum transfer across the orthogonals and hence it is reasonable to assume these are lines of zero shear. From this the average wall shear stress between any two of these orthogonals can be computed and the distribution of wall shear stress around the wetted perimeter established.

Strictly speaking such a concept is only applicable to cases of laminar flow. However, the relatively low level of turbulence encountered in most gravity sewers does not seriously invalidate application of the concept which tends to be slightly on the conservative side in estimating shear stress distributions.

Design Chart 10 presented in Chapter 10 is based on the segmental approach utilizing the results of research work undertaken at the University of Melbourne(*11*).

In gravity sewers the slime development problem tends to be further complicated by the scouring action of grit and the fact that the depth of flow varies at different times of the day and consequently wall shear stress values at any point on the wetted surface also vary. It would be expected that slime growths would tend to build up wherever, and during periods when wall shear stress was below critical levels, and tend to be reduced when wall shear stresses were greater.

For practical design purposes the value of the fraction "F_s" of the wetted surface on which slimes are maintained is important and approximate values of this fraction F_s can be assessed when the various influencing factors are taken into account (see Design Charts 1(a), 1(b), 10, pp. 109, 124). However, the problem is further complicated because it has not been conclusively established what minimum thickness of slime is necessary for significant sulphide generation. Present knowledge indicates this thickness to be about 1 mm. at least.

Structure and organization of slime

Reid and Tien-Sheng Yang(*10*) suggest that the slime layer contains a primary and secondary layer.

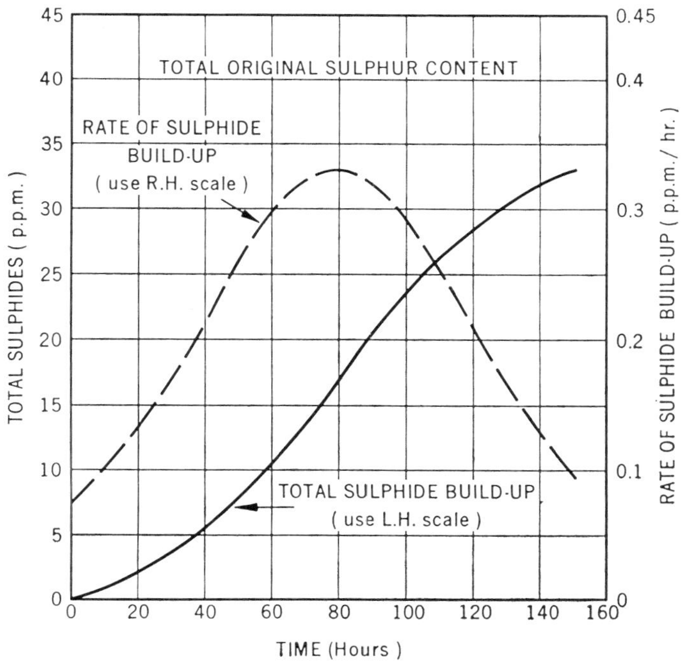

FIG. 3.4 Rate of sulphide build-up in stagnant sewage.

The primary layer is that which adheres to and fills the roughened surface of the channel wall. The secondary slime layer is that which develops on the primary layer. Since the cohesive forces between the secondary and primary slime layers are less than those holding the primary layer to the channel wall, the secondary layer sometimes sloughs off. They suggest that sulphide generation takes place in any portion of the slime where the redox potential is sufficiently low.

Parker and his co-workers(*7*) on the basis of their field and laboratory studies, make the following statement:—

The slime contains a mixed flora, including gram + ve cocci and rods and gram — ve rods. If the environmental conditions are aerobic then cocci predominate, until such time as actively proteolytic rod-shaped organisms such as *Proteus vulgaris, E. coli,* and *B. subtilis* proliferate and reduce the redox potential to a level low enough. This permits the production of hydrogen sulphide from the organic sulphide compounds present. The formation of hydrogen sulphide further lowers the redox potential and in turn permits the growth and activity of *Desulphovibrio desulphuricans*. This organism reduces sulphate to hydrogen sulphide at a redox potential of about —250 millivolts. The accumulation of hydrogen sulphide and its reaction with iron present in sewage causes blackening of the under-layers of slime. Under these conditions of low redox potential facultative aerobes and strict anaerobes (mainly rod-shaped organisms) will have the survival advantage, and provided the redox potential remains at a low level (as in stale or septic sewage) they will tend to persist. Stable sewer slimes are thus the slimes which develop under sewage of a constant low redox potential. Reversal of the above sequence, or its arrest at the first stage of anaerobic slime with cocci predominating, will occur only when the conditions of oxygenation in the sewage are such as to prevent the redox potential from being lowered enough, or to cause it to rise enough. The aerobic type of slime, which grows more quickly, tends to be a thicker slime but is subject to rapid fluctuation in thickness due to sloughing off and subsequent re-growth.

Viehl(*5*) gives populations of sulphate-reducing bacteria in slimes of both gravity sewers and pressure mains as reaching about 10^8 organisms per cc. Ferguson-Wood(*14*) quotes similar population densities for muds especially from estuaries. Raw sludge from sedimentation tanks contain relatively much fewer of these types of bacteria, say about 100 times less, while well-digested sludges from Imhoff tanks contained about 1000 times less and septic deposits from rivers about the same, or say 10^5 to 10^6 per cc.

Generation in the body of sewage

Experience indicates that some sulphide is generated in the body of sewage, especially when the sewage is already septic. Laboratory H_2S data are presented in Fig. 3.4 for stagnant test conditions in a sewage containing initially about 40 m.g./1 total equivalent sulphur (mainly sulphate). It seems likely that generation in flowing sewage would be more significant than would be the case for quiescent sewage held in bottles. Viehl's bacterial counts for sulphate-reducing organisms (see above) suggest however that normal sewer conditions are not favourable for rapid

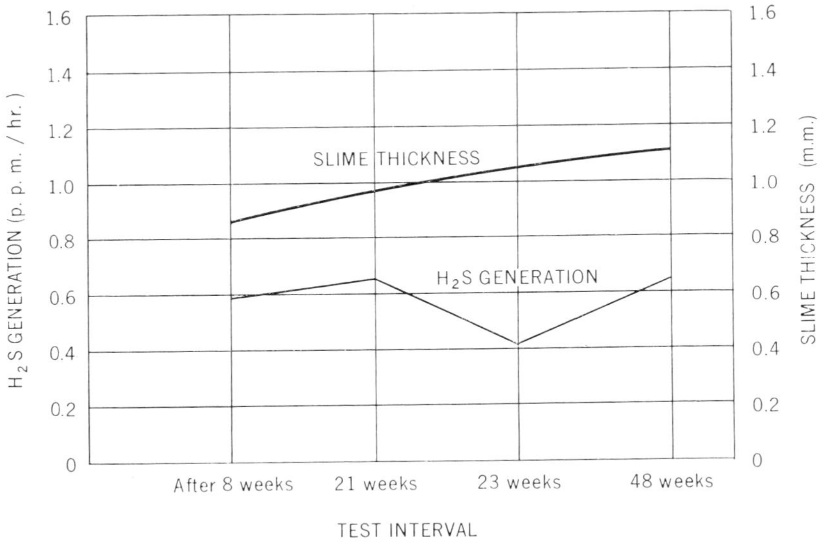

FIG. 3.5 Slime thickness and rates of sulphide generation.

sulphide production within the body of the flowing sewage.

Fig. 3.4 shows that the rate of generation of sulphide in stagnant sewage, placed originally in clear glass bottles and stored in the laboratory without visible slime formation, first increases exponentially, but then falls off, presumably due to a marked reduction in sulphate levels, and probably also in the supply of food and nutrients.

Rates of generation of sulphides in sewerage systems

The significance of sulphide generation in sewerage systems already has been stressed. Reference has been made in particular to sewer and pumping main wall slimes as important sites of generation. Conditions in gravity sewers differ to some extent from those in pumping mains, and in the following detailed discussions they are dealt with as separate cases, although there are many common features and some overlapping results accordingly.

Where the sewage is septic, there is no evidence indicating any material difference in effective generating rates for slimes in gravity sewers (below minimum flow level) and for slimes in pressure mains. In both instances the slime layers which develop include an anaerobic zone adjacent to the sewer wall within which are developed the very low redox values necessary for proliferation of sulphate-reducing organisms. That the generation rates per unit area of active slime are much the same for pressure mains as for gravity sewers is supported also by the fact that sulphide build-up in gravity sewers, as predicted from observations of build-up in pressure mains, agrees closely with the actual build-up observed. For gravity sewers, work by Parker and his co-workers(7) indicates that with septic sewage some little decrease occurs in the sulphide-generating ability of the slimes just below the free sewage surface where there is some degree of aeration of sewage. This suggests also that with fresh sewage in which some free dissolved oxygen is maintained, sulphide release from slimes may be substantially reduced or even prevented.

Rate of generation in slime in pressure mains

Parker and his co-workers(7) developed techniques for the quantitative study of the comparative sulphide generating ability of slimes from various sewers. The tests were operated under quiescent laboratory conditions. Under those conditions it was established that:

(a) The rate of sulphide generation does not vary greatly with different thicknesses of slime. Some indication of variation in generation rate with thickness of slime is shown in Fig. 3.5.

It would appear that, as a first approximation in formulating rates of sulphide build-up, the thicknesses of slimes may be neglected.

(b) The composition and condition of the sewage considerably influence the generation activity. Generally, slimes developed under stronger and more septic sewage will generate sulphides more rapidly than those developed under weaker and fresher sewage.

(c) The generation activity is not influenced much by sewage pH.

Note that Parker's generation rates are those applicable to slime submerged in quiescent sewage. In active sewers, slimes are almost always submerged by sewage in a state of turbulent flow. The rates at which nutrients can be transferred from the sewage to the slime complex are considerably greater in turbulent sewage, but even in turbulent flow there is a thin layer of sewage adjoining the submerged surface and in which mass transfer takes place virtually by molecular diffusion. Outside this layer the transfer of dissolved substances takes place by eddy diffusion. Compared with molecular diffusion rates, the rate of eddy diffusion (convection transfer) is so great that the body of sewage outside the laminar layer may be regarded as containing uniform concentrations of nutrient substances. The laminar layer thus constitutes a diffusion barrier to the transfer of these substances. The rate of transfer is proportional to the concentration gradient across the layer and since this gradient depends in turn on the thickness of the laminar layer, rate of nutrient transfer depends on the layer thickness, which depends in turn on the hydraulic conditions.

For a circular pipe running full, the thickness of the laminar layer is inversely proportional to the mean sewage velocity and consequently it might be anticipated that the rate of nutrient supply to the slime complex, and hence the overall rate of sulphide generation, increases with increased sewage velocity.

From Rouse's *Elementary Mechanics of Fluids* we take

$$\delta^1 = \frac{11.6 \times \nu}{\sqrt{\tau_0} \times (g/\rho)} \qquad [3.3]$$

where δ^1 = thickness of laminar layer in feet
ν = kinematic viscosity, $ft^2\ sec^{-1}$
ρ/g = lb. mass per ft^3
τ_0 = wall shear value, lb. $ft^{-1}\ sec^{-2}$.

Subs'g $\tau_0 = \frac{\rho}{g} \times \frac{f}{8} \times V_s^2$ (also from Rouse)

where f = wall roughness factor (dimensionless)
V_s = mean (sewage) velocity, ft. $sec.^{-1}$

$$\delta^1 = \frac{11.6 \times \nu}{V \times \sqrt{\frac{f}{8}}} \qquad [3.4]$$

Taking V_s at 2 fps and f at 0.024, and the temp. at 68°F $\nu = 1.08 \times 10^{-5}\ ft^2\ sec^{-1}$), δ then is about 0.014 inches.

It will be noted that the thickness of the laminar layer decreases with decreasing values of kinematic viscosity which in turn decreases with rising temperatures. It would therefore be expected that sulphide generation would increase at higher temperatures not only because of greater bacterial activity but also because of readier availability of nutrient requirements.

The rates of transfer of nutrients through the film must be directly proportional to their respective concentration gradients. It may be taken that domestic sewage composition can be represented generally in terms of BOD, so that the relative concentrations are measured more or less by applying suitable proportionality constants. For sulphide generation, however, the rate of supply of sulphate, which is not usually related to BOD levels, must also be a ruling factor.

Taking $\frac{d}{dt}(N_s)$ as representing the overall rate of supply of essential nutrient molecules to the sulphide generative slime bacteria, we may write as a fresh approximation

$$\frac{d}{dt}(N_s) \propto [BOD_5]^a \times [SO_4]^b \times D_L$$

where $[BOD_5]$ and $[SO_4]$ are concentrations in the sewage of BOD_5 and sulphate, respectively, in units of lb. oxygen and lb. SO_4 per $ft.^3$ and D_L is a general measure of molar diffusion rates, $ft.^2/hr$.

Concentration gradients however must depend on laminar film thickness.

$$\text{Hence} \quad \frac{d}{dt}(N_s) \propto \frac{[BOD_5]^a \times [SO_4]^b \times D_L}{\delta^1}$$

and from [3.4]

$$\frac{d}{dt}(N_s) \propto \frac{[BOD_5]^a \times [SO_4]^b \times D_L \times V_s \times \sqrt{\frac{f}{8}}}{11.6 \times \nu}$$

again, taking f = 0.024, temp. 20°C. (68°F.)

$$\frac{d}{dt}(N_s) \propto 438 \times V_s \times [BOD_5]^a \times [SO_4]^b \qquad [3.5]$$

Using units for N_s of lb/1000 ft^2/hr., and concentrations in units of mg/1, we have

$$\frac{d}{dt}(N_s) \propto \frac{4.38 \times 10^5 \times V_s \times [BOD_5]^a \times [SO_4]^b}{16{,}000^{(a+b)}} \qquad [3.6]$$

Note that a and b are exponential indices which by dimensional analysis should be such that a + b = 1. Experience indicates that the effect of sulphate is of less significance than the effect of factors measured by BOD relationships.

If the above holds, then as a rough approximation the rate of sulphide production by slimes "G_s" in lb./(1000 sq. ft.)/(hr.) might be given by the following equation:

$$G_s = \frac{4.38 \times 10^5}{16{,}000^{(a+b)}} D_L \times R_C \times V_s \times [BOD_5]^a[SO_4]^b \text{ at } 20°C. \ (68°F.) \qquad [3.7]$$

where R_C is the proportionality constant for $[BOD_5]^a \times R_C$ = nutrient concentration factor. Concentrations are expressed in terms of mg/1, or ppm.

For temperatures other than 20°C. (68°F.) it is necessary to apply a temperature adjustment factor based on the fact that both the bacterial activity and the diffusion coefficient increase with increasing temperature.

Based on the simplifying assumptions that the sulphide generation occurs only in the wall slimes, and that the slimes and their activity are uniform over the whole area of the pipelines, practical observations on pressure mains (i.e. in sewer mains running full) both in Australia and other countries, have indicated generation rates ranging from 0.05 to about 0.22 lb of sulphide sulphur/(1000 sq. ft.)/(hr.). Carefully co-ordinated tests made in Australia as a joint venture by the larger sewerage authorities(6) have shown that the rate of generation in rising mains from sewage pumping stations in which the pumps operate continuously is given nearly enough by the following empirical formula:—

$$G_S = 32.2 \times 10^{-6} \times V_S \times [BOD_5]^{0.8} \times [SO_4]^{0.4} \times 1.139^{(t°C - 20)} \qquad [3.8]$$

where G_S is slime generation rate in lb. of sulphide sulphur/1000 sq. ft. of slime/hr.
V_S is mean velocity of sewage flow in ft./sec.
$[BOD_5]$ is 5 day bio-chemical oxygen demand in mg. $O_2/1$.
$[SO_4]$ is sulphate expressed as mg.$SO_4/1$.
and t°C is temperature in degrees Centigrade.

The above equation [3.8] was derived from data obtained from investigations into sulphide build-up in pressure ("rising") mains of diameters varying from 1 ft. to 4 ft. and with average stream velocities of from 1 f.p.s. to 3 f.p.s. during pumping. Sewage temperatures varied from 19°C. to 27°C. (67°F. to 81°F.). The various sewages pumped showed BOD_5 values within an overall range of 90 mg/1 to 800 mg/1 and sulphate contents 42 mg/1 to 660 mg/1. Wall shear values ranged between 0.004 and 0.077 lb/sq. ft.

It should be noted that at very low velocities the value of G_S probably would be somewhat higher than is indicated by the equation, and that the equation cannot be used where V = 0. The empirical equation derived for pumping stations operated intermittently takes the form

$$G_S \propto [BOD_5]^{0.8} \times [SO_4]^{0.4} \times 1.139^{(t^\circ C - 20)} \times \frac{1.64^{\bar{V}_S}}{\bar{V}_S} \qquad [3.9]$$

where $\bar{V}$ is the effective overall mean velocity reckoned over the whole cycle of operation.

Equation [3.8] is similar in form to that of equation [3.7], suggesting that the argument leading to [3.7] was broadly verified by the experimental studies referred to; and [3.8] and [3.9] are used as a principal basis for sulphide control by sewerage design as outlined in Chapters 10 and 11.

Generation of sulphide in pressure mains, siphons, etc.

Equation [3.8] was derived empirically from the results obtained from a series of specific co-operative investigations carried out by major sewerage authorities in Australia and is the basic equation for estimating sulphide build-up wherever referred to in this book. It was not feasible to examine the wall slimes in all of the pressure mains investigated but from general observations it was concluded that the slimes were generally similar in structure to those found on the walls of gravity sewers, and that the relationships respecting thickness and hydraulic conditions are similar to those discussed above for both pressure mains and gravity sewers. The all-important difference between pipes running full, compared with pipes in which there is always a free air surface, as in gravity sewers, is that there is no opportunity for oxygen uptake in pressure mains (unless accidentally or deliberately aerated during pumping. This means also that there is commonly no loss of sulphide by oxidation during passage through rising mains.

The regression equation [3.8] (p. 15) should be valid at least for all cases where wall shear lies below about 0.08 (p. 139) for continuous pumping. However, it was observed that the equation over-estimated sulphide generation when the sewage entering the pumps still contained dissolved oxygen, and that it is not directly applicable for zero or for very low velocities.

The significance of this equation broadly stated rests on the following considerations:

(i) Once equilibrium has become established for overall sewage flows—slime thickness conditions, the rates at which sulphide is generated and released into the sewage depend directly on the rates of supply of sulphate sulphur into the slime layers. This sulphate is reduced rapidly, at ordinary concentration levels practically completely.

(ii) Wall slime activity depends also on the supply of nutrients to the zoögleal mass, including organic matter, nitrogen, phosphorus, etc. It is to be expected therefore that the rate of sulphide generation will be the greater, the "stronger" the sewage.

(iii) Temperature increase likewise increases slime activity.

It follows also from the equation that the rate of generation of sulphides will be greater while sewage of some particular composition is passing through the main (i.e. during pumping) than if the same sewage were stagnant in the main. Practical observations show that the rate of release of sulphide from the slimes into stagnant sewage (i.e. when pumps are stopped but other conditions and factors remain unchanged) is about one-half of the rate during pumping with $V_S = 1$.

It is important to realise that the local concentrations of sulphide, measured in mg/l, will increase as much or even more during stagnation, because although the rates of generation and release are lower, the volume of sewage passing through the main will be zero; hence the sulphide builds up continuously in the *same* sewage.

Thus there is usually little if any advantage in operating pumping stations intermittently, so far as sulphide problems are in question, unless pumping rates are high enough to reduce slime formation and growth (e.g. by high rates of wall shear).

Consider what happens when the flow in a rising main is suddenly stopped, as when pumping stops with intermittent pumping systems. The following changes result.

First the flow condition changes from steady turbulent flow to a quiescent state following a short period of oscillations with local eddy currents. A reduced rate of sulphide generation follows due to the slowing down of nutrient supply to the slimes because the eddy diffusion conditions of turbulent flow are changed during quiescence to molecular diffusion. These changes of slime generation rates would take place over a time period mostly determined by sewage nutrient and sulphate concentration levels. The change from turbulent to quiescent conditions would also reduce the movement of sulphide-producing organisms and thus slow down any generation in the body of the sewage.

At an early stage of these changes, suspended matter including sulphate-reducing "seed" would begin to settle to the bottom of the rising main. This may well result also in a reduction of the apparent rate of sulphide build-up, due to some loss of sulphide by settling of the insoluble fraction.

The overall effect of these changes should be more or less as shown diagrammatically in Fig. 3.6. The rate of build-up equals the sum of the progressive rates for sewage and wall slimes under quiescent conditions. Finally, if the time is extended, sulphates are exhausted and sulphide generation ceases.

In actual experience it has been well established, that when pumping stops, the instantaneous rate of sulphide generation in the sewage diminishes rapidly

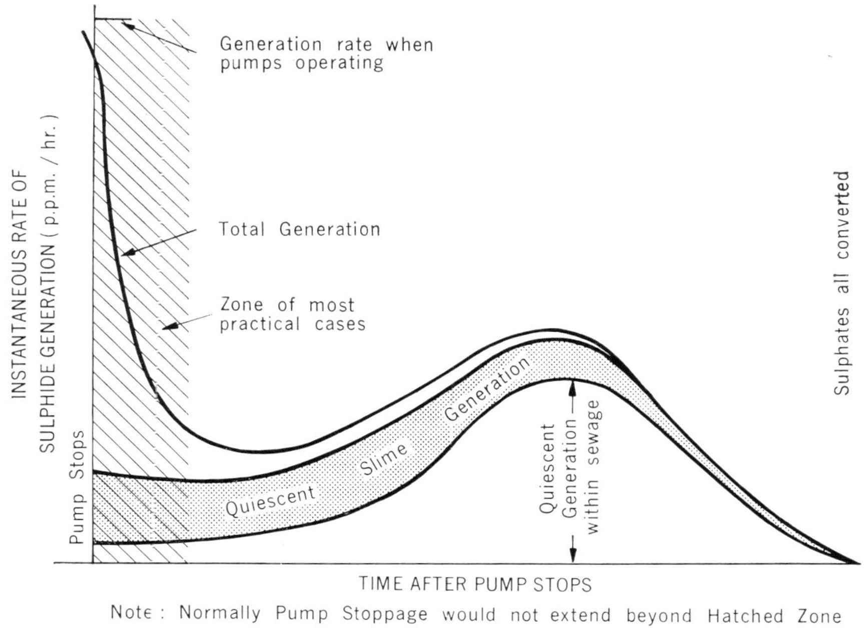

FIG. 3.6 Diagram showing patterns of sewage sulphide build-up in rising mains after pumping stops.

at first and then slowly. A typical case might show a generation rate of 0.7 p.p.m./hr. during pumping but after some hours of quiescence a rate of only about 1/5, say 0.15 p.p.m./hr. or less, as shown in Fig. 3.7.

Sulphide generation along a length of gravity sewer

Apart from average generation rates per unit area, there are two factors which influence the sulphide generated in slimes along a given length of gravity sewer. These factors arise in the diurnal rise and fall of the sewage level and the consequential geometric and hydraulic effects; and in the variations in slime character and thickness caused thereby and by the concurrent effects of sewer grades and geometry, mainly relating to wall boundary (wall shear) conditions.

The first question arising is whether the slimes between minimum and maximum flow depths (i.e. the intermittently submerged slimes) generate sulphide at the same, or at different rates compared with the continuously submerged slimes. This is important because at maximum flows the intermittently submerged slimes may constitute up to 50% of the total slime area under some circumstances, more or less for different sewers and sewer conditions.

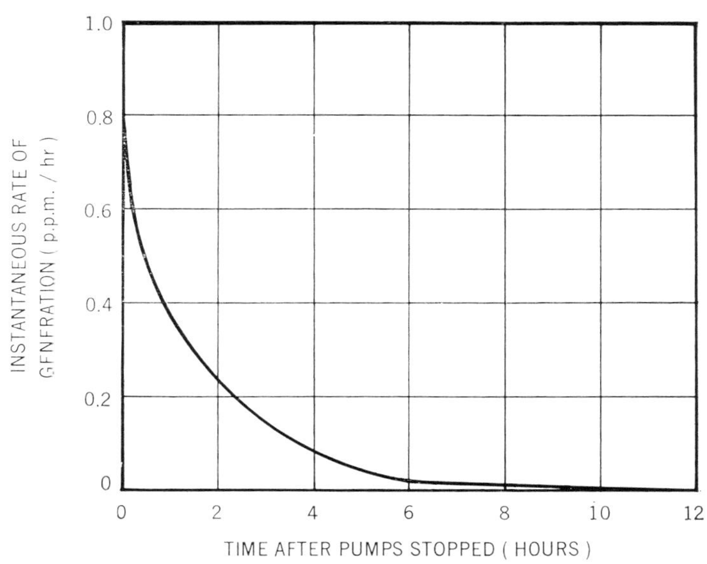

FIG. 3.7 Instantaneous rates of generation (in rising main after cutting-out of pumps).

In Fig. 3.8 the continuously submerged slime area *per foot run of sewer* is denoted by Pc, the intermittently submerged slime area by P_i and the total by $P_t = Pc + P_i$

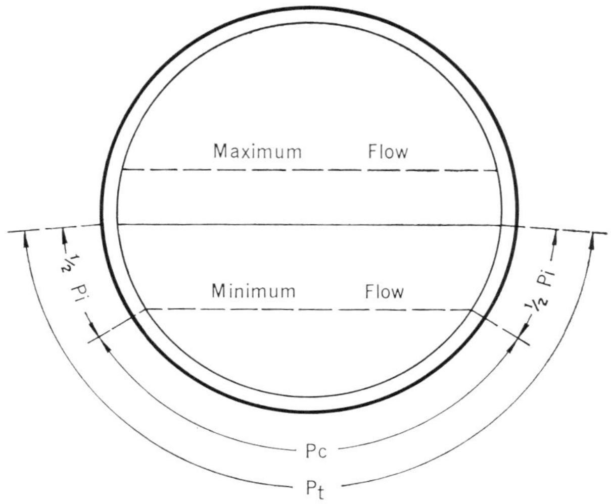

FIG. 3.8 Intermittently and continuously submerged slime areas of sewer wall.

The effective sulphide generation rate in slimes, $(G_s)_{eff}$, reckoned in terms of lb. per hour per foot run of sewer will be

$$(G_s)_{eff} = \frac{(Pc + R_s \times P_i)}{1000} = G_{av}\ \text{lb. } S^{=}/\text{hr.}$$

(see [2.1] on page 7)
where G_{av} is the effective average generation rate of the *fully active* continuously submerged slime and R_s is the factor which represents the average proportional rate applicable over the intermittently submerged slime area P_i.

By substituting

$$F_s = \frac{(Pc + R_s \times P_i)}{P_t} = \frac{Pc}{P_t} + R_s\left(1 - \frac{Pc}{P_t}\right)$$

This may be rewritten in the equivalent form

$$(G_s)_{eff} = F_s \times \frac{P_t G_{av}}{1000}\ \text{lb./hr./ft.}$$

Values of F_s depend upon values of R_s and $\frac{Pc}{P_t}$.

There is inadequate experimental knowledge as to how the value of R_s varies with the depth of flow above the minimum, but the following analysis provides a reasonable basis for assessing its value in any particular case. The wall slime at the level of maximum dry weather flow is exposed to the atmosphere for most of the time and consequently its sulphide producing capacity probably is negligible (i.e. $R_s \to 0$). Near low water level the slime is submerged almost continuously and then its generating capacity must be practically the same as for continuously submerged slimes and value of $R_s \to 1$. Available flow data and geometry indicate that as flow rates increase from low to high flow the *period* of slime submergence, at any level between low and high water, is approximately proportional to the difference between flow at this level and that at minimum flow. It is not unreasonable therefore to assume that *with increasing flow rates* the value of R_s for slimes at the different levels varies uniformly from 0 at surface (maximum flow) to 1 at minimum flow level, the average being close to 0.5. *For flows decreasing* from high flow the values of R_s would be somewhat higher since the hydrograph of the tailing flow is such as to cause submergence for a relatively longer period.

With respect to the influence of the second factor $\frac{Pc}{P_t}$ on the practical effective average value of F_s, an investigation based on the hydraulic considerations involved, has indicated that for normal ranges of proportional depths of flow in dry weather the value of $\frac{P}{P_t}$ is influenced mainly by the actual ratio of maximum to minimum flow. This ratio is in turn influenced by the sizes of the catchments draining into the sewer above the section under consideration. From experience the value of this ratio varies from about 4 for large sewers (serving a population of say 100,000 or more) to 8 for smaller sewers (say serving about 5,000 to 10,000 persons).

When the average dry weather flow requires a proportional depth in the sewer of 0.4 of the diameter or more, the values of $\frac{Pc}{P_t}$ begin to decline rapidly. However, these conditions should not be used for practical design since they indicate overloading of the sewers and could cause sewers to operate as pressure mains at times of the day when maximum dry weather flow occurs. Such conditions should be avoided as they are likely to cause problems.

The figures of Table 3.1 give an indication of values of F_s to be used in equation [3.10].

TABLE 3.1

PROPORTIONAL RATES FOR SULPHIDE GENERATION IN SEWER SLIMES

Ratio Max. Flow / Min. Flow	Value of F_s		
	At. Max. Dry Weather Flow	At Mean Dry Weather Flow	At Min. Dry Weather Flow
4	0.83	0.93	1.00
6	0.77	0.93	1.00
8	0.74	0.93	1.00

The effects of boundary shear conditions will be examined by considering two cases respectively for lower and for higher values of wall shear.

(i) *For lower values of wall shear*, where the value of wall shear is sufficiently low to be conducive to the growth of sulphide generating slimes around the whole wetted perimeter.

As a slug of sewage one foot in length travels along the sewer at velocity V_s ft./sec. the generation of sulphide per foot run of sewer is $(G_s)_{eff}$ where

$$(G_s)_{eff} = F_s \frac{P_t G_{av}}{1000}\ \text{lb./hr./ft.} \qquad [3.10]$$

$$= F_s \frac{P_t \times G_{av}}{1000 \times 3600}\ \text{lb./sec./ft.}$$

Therefore, the total weight of sulphide sulphur entering the sewage slug from the slime as it travels over the distance L ft. is

$$\frac{L}{V_s} \times F_s \times \frac{P_t G_{av}}{36 \times 10^5}\ \text{lb./ft.}$$

This amount of sulphide sulphur enters a slug of sewage weighing $62.24 \times A_s$ lb. where A_s is the cross-sectional area of the sewage flow in sq. ft.

Therefore the build-up, B_s, of sulphide concentration in mg/l as sewage travels through the distance L is given by the expression

$$B_s = \frac{L}{V_s} \times F_s \times \frac{P_t}{A_s} \frac{G_{av} \times 10^6}{62.24 \times 36 \times 10^5}\ \text{mg/l} \qquad [3.11]$$

introducing the value of $G_{av} = G_s$ given by equation [3.8] (p. 15)

we have

$$B_s = 0.143 \times 10^{-6} \times L \times F_s \times P_t \times [BOD_5]^{0.8} \times [SO_4]^{0.4} \times 1.139^{(t°C. - 20)}\ \text{mg/l.} \qquad [3.12]$$

$$\text{and } (B_s)_t = 0.143 \times 10^{-6} \times V_s \times F_s \times \frac{P_t}{A_s} \times [BOD_5]^{0.8} \times [SO_4]^{0.4} \times 1.139^{(t°C. - 20)}\ \text{mg/l sec.} \qquad [3.13]$$

where $(B_s)_t$ is the time rate of sulphide build-up.

As the sewage travels along the sewer the sulphate concentration decreases in proportion as the sulphate sulphur is converted to sulphide. There is, therefore, a limit to further sulphide generation when $[SO_4]$ is exhausted at which stage $B_s \not> 1/3\ [SO_4]$. By setting up the differential equation for this situation and performing the necessary integration it can be shown that the concentration of sulphate declines in accordance with the following equation:

$$\Delta L = \frac{3.86 \times 10^6 \times A_s \{[(SO_4)_0]^{0.6} - [(SO_4)_L]^{0.6}\}}{F_s \times P_t\,[BOD_5]^{0.8} \times (1.139)^{(t°C.\,-20)}} \quad [3.14]$$

where $[(SO_4)_0]$ is the initial sulphate concentration in sewage at the upstream point, $L = 0$, m.g. SO_4 per l.

and $[(SO_4)_L]$ is the sulphate concentration in the same sewage after travelling a distance of "L" feet.

The build-up of sulphide will follow exactly the sulphate sulphur decline. On the basis of this equation the build-up of sulphide sulphur has been computed for two cases and the results are shown on Fig. 3.9.

(ii) *For higher values of wall shear* where the value of wall shear is sufficiently high over some portions of the submerged periphery of the sewer to prevent the development thereon of sulphide generating slimes.

In this case the general equation for $(B_s)_t$ [3.13] becomes

$$(B_s)_t = 0.143 \times 10^{-6} \times F_{ps} \times V_s \times F_s \times \frac{P_t}{A_s} [BOD_5]^{0.8} \times [SO_4]^{0.4}\, 1.139^{(t°C.\,-20)} \text{ m.g./l/sec.} \quad [3.15]$$

Where F_{ps} = proportion of submerged surface on which slimes develop (Dimensionless).

Approximate values of F_{ps} can be obtained from Design Chart No. 10 (following p. 124).

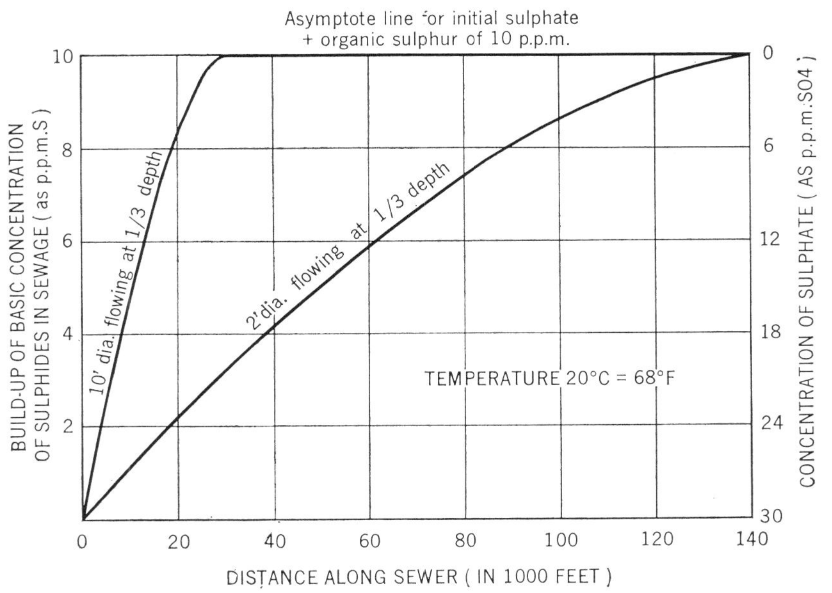

FIG. 3.9 Progressive sulphide build-up along sewers (computed values) comparing 10′ diam. and 2′ diam. sewers each flowing at ⅓ depth.

Generation of sulphide in wet wells, etc.

Whereas structures such as pressure mains and siphons are normally without air contact, in wet walls, storage and sedimentation tanks etc. as in gravity sewers, sewage is able to absorb some oxygen from the air. The ratios of wall surfaces (i.e. potential slime areas), air interface, and relative sewage depth (and volumes), however, may be altogether different. Also, the overall oxygen uptake may be substantially affected by turbulence due to drops, or by air entrainment, etc. The slime generation equations [3.8] and [3.9] provide a basis for preliminary assessments of the likelihood of any sulphide build-up, and should sulphide problems appear to be a possibility, the discussions of pp. 107 et seq. and the consequent formulations might be suitable as a model for more detailed investigations. Experience suggests that for most sewerage systems, under normal circumstances, the sewage reaching pumping stations is unlikely to be septic, or even stale, but most likely will contain at least 1 to 2 mg/l of dissolved oxygen when entering storage. Mostly this will be reduced by about 1 mg/l per hour of storage in wet wells, tanks, etc.

Rates of generation in the body of flowing sewage

It has been found impracticable to measure this rate closely enough, and hence impracticable likewise to formulate the rate. General observation indicates that for most cases the relative amounts of sulphide generated within the flowing sewage do not exceed the amounts corresponding to the errors inherent in estimating the rates of generation in slimes.

The pattern of Fig. 3.4 (p. 13) shows the progress of generation within the body of quiescent *settled* sewage. It has been suggested(*4*) that when raw sewage is flowing in a sewer a "seeding" action takes place. This "seeding" action involves the transfer of sulphide-producing organisms and fragments from the wall slime coating the sewer to the flowing sewage, and also could involve seeding from silt deposited during periods of low velocity flow and so on. If this be so it would be expected that the rate of sulphide generation in the flowing raw sewage would be appreciably greater than that found in quiescent bottle tests with settled sewage only. This notion has not been investigated so far as to indicate whether or not, and in what circumstances, such generation may be significant enough to be taken into account. No doubt, however, generation in the body of the flowing sewage could become significant, especially where the flowing sewage is highly septic, and re-aeration rates unusually low.

It should be noted that even substantial generation of sulphides in the sewage body, e.g. by the "seeding" action referred to in the foregoing, could well accord with the formulations of equation [3.8], since the factors operating are of the same general character.

Chapter 4

RE-AERATION AND THE DISSOLVED OXYGEN BALANCE IN SEWAGE

Introductory

The presence or absence of dissolved oxygen is of significance in relation to the sulphide problem primarily because the presence of dissolved oxygen is inimical to the build-up or persistence of sulphides. The mains water used for flushing or for transport of other domestic or industrial wastes normally carries some dissolved oxygen, and if this be sustained by re-aeration, no sulphide build-up will occur.

A positive or negative balance of the dissolved oxygen depends on the relative magnitudes of (1) the rate of uptake of oxygen from the air above the sewage, *Re-aeration in Sewers,* and (2) the rate of consumption of oxygen within the sewage in chemical and biochemical reactions, *Oxygen Consumption in Sewage.*

Re-aeration in sewers

The absorption of gaseous oxygen at the surfaces of lakes or streams (oxygen up-take or re-aeration) has been the subject of many theoretical and practical investigations since Adeney's early work for the British Royal Commission on Sewage Disposal.

The rates of oxygen up-take have been related to a number of factors, including the following:

(a) The degree of turbulence in the flowing liquid,

(b) The composition of the flowing liquid in respect to:

(i) total salinity,

(ii) the concentration of surface-active substances (detergents) present, and

(iii) the rate of reaction between substances within the liquid and the incoming oxygen (including biochemical reactions).

(c) The temperature of the liquid.

(d) The partial pressure of the oxygen in the atmosphere.

It is convenient to discuss the influence of such factors for sewerage systems by considering first the processes of surface aeration with clean water and then superimposing the influence of sewer structures, sewage composition, etc.

Oxygen absorption at the surface of flowing clean water

The investigations which have been made into the absorption of oxygen from the atmosphere into flowing water have been reviewed by Davy(*15*). Two main types of flow are encountered in sewer systems. The first of these is turbulent flow as met with in channels and the second is free falling flow as met with in vertical drops. Occasionally a type of flow intermediate between these two may occur where grades are sufficiently high to cause air entrainment, but such a situation is not taken into account in the following discussion.

Adeney first established experimentally that the rate of surface absorption of oxygen was directly proportional to the oxygen saturation deficit. The oxygen saturation deficit is the difference between the equilibrium solubility of oxygen at a given temperature and the existing dissolved oxygen content. A chart giving the capacity of water to dissolve oxygen (i.e. the equilibrium oxygen solubility) at various temperatures and pressures is produced as Design Chart No. 4 (p. 119). Table 4.1 is reproduced from Imhoff, Müller, and Thistlethwayte(*16*).

Adeney(*17*) demonstrated that the effective rate of oxygen absorption under quiescent conditions depended also on the humidity and salinity. Low air humidities increase the rate of surface evaporation of water which in turn produces a cooling of the water surface. This cooling therefore increases the oxygen saturation deficit at the water surface and consequently the rate of oxygen absorption. At the same time the cooling and increased salinity of the surface layers due to evaporation cause density increases with density currents and hence further mixing as a result.

In sewers we are practically always concerned with sewage flowing in a turbulent state, that is with eddies as opposed to laminar flow continuously parallel to the channel axis. Under turbulent flow conditions, due to the mixing power of the eddies, oxygen is quickly conveyed from just below the surface to all parts of the body of flowing sewage. The oxygen concentration throughout the main body of the flowing sewage is therefore uniform or very nearly so.

It has long been known that there is virtually negligible resistance to the passage of oxygen from the air into the upper layer, a few molecules in thickness, on the top of clean water. Yet it is also an established fact that the absorption of oxygen by flowing clean water is a comparatively slow process.

In 1923 and 1924 Lewis and Whitman(*18*) suggested the "two-film" theory in their analysis of the interchanges of substances across the interfaces of adjoining fluids. For the case of liquid-gas interfaces, this theory states, in essence, that for turbulent flow conditions the main resistance to transfer of a substance between liquid and gas phases is in the very thin stagnant films which form in each phase adjoining the actual interface (see Figs. 4.1 and 4.2).

Whereas transfer in the main body of flow is relatively rapid due to the operation of eddy diffusion, transfer through the relatively stagnant film near the interface, where molecular diffusion prevails, is much slower.

TABLE 4.1 OXYGEN SOLUBILITY DATA

(a) APPROXIMATE DISSOLVED OXYGEN CONTENTS OF AIR-SATURATED FRESH WATER (TOTAL SALINITY —300 P.P.M.) based on air of normal composition saturated with vapour at the given water temperature.

Elevation above mean sea-level		Total normal air pressure mercury		oxygen dissolved at equilibrium (mg/1) at						
				0°C. (32°F.)	5°C. (41°F.)	10°C. (50°F.)	15°C. (59°F.)	20°C. (68°F.)	25°C. (77°F.)	30°C. (86°F.)
ft.	m.	in.	mm.							
0	0	29.9	760	14.6	12.8	11.3	10.2	9.2	8.4	7.6
1,000	305	28.9	734	14.1	12.3	10.9	9.8	8.9	8.1	7.3
2,000	610	27.8	707	13.6	11.9	10.5	9.5	8.6	7.8	7.1
4,000	1,220	25.8	656	12.6	11.1	9.8	8.8	8.0	7.3	6.6
6,000	1,830	23.9	609	11.7	10.3	9.1	8.2	7.4	6.7	6.1
8,000	2,440	22.2	565	10.8	9.5	8.4	7.6	6.8	6.2	5.6
10,000	3,050	20.5	522	10.0	8.8	7.8	7.0	6.3	5.8	5.2
				40°C. (104°F.)	50°C. (122°F.)	60°C. (140°F.)	70°C. (158°F.)	80°C. (176°F.)	90°C. (194°F.)	100°C. (212°F.)
At sea level		29.9	760	6.5	5.6	4.8	3.9	2.9	1.7	0

(b) APPROXIMATE DISSOLVED OXYGEN CONTENTS OF AIR-SATURATED WATERS OF VARYING SALINITY basis of reference, diluted sea-water, taking saturated normal air at sea level, total pressure 29.9 in. (760 mm).

Chlorinity, lb/gal × 10⁻⁵ (mg/1) of chloride chlorine in diluted sea-water	Oxygen dissolved at equilibrium, (mg/1) of diluted sea-water at						
	0°C. (32°F.)	5°C. (41°F.)	10°C. (50°F.)	15°C. (59°F.)	20°C. (68°F.)	25°C. (77°F.)	30°C. (86°F.)
0	14.6	12.8	11.3	10.2	9.2	8.4	7.6
5,000	13.7	12.1	10.7	9.6	8.7	8.0	7.2
10,000	12.8	11.3	10.0	9.0	8.2	7.5	6.8
15,000	11.9	10.5	9.4	8.4	7.7	7.0	6.4
20,000	11.0	9.8	8.8	7.9	7.2	6.5	5.9

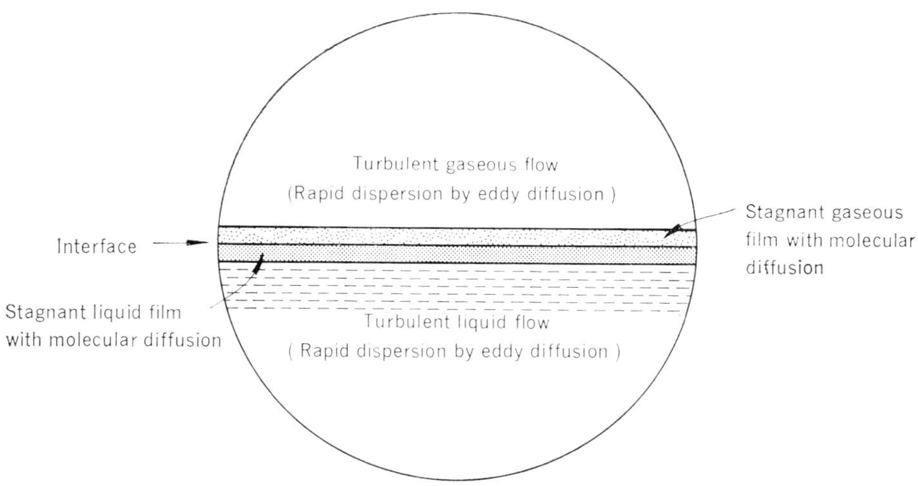

FIG. 4.1 Representation of the "two-film" gas to liquid transfer theory.

The "driving force" causing transfer of a solute substance across a liquid-gas interface arises from the difference of the concentrations of the solute substance in the bodies of the two fluids on either side of the interface. To relate this difference in concentrations it is necessary to first express the concentrations in the two phases in equivalent terms. A suitable relation thus can be found by considering the concentrations of the solute in the liquid and in the gas phase when in equilibrium (i.e. when the number of molecules of gas passing each way across the air interface is the same as the number of molecules passing into and out of solution across the liquid interfacial film). If, for this state of equilibrium, "p_A" denotes the partial pressure of the solute gas in the atmosphere and "C_A" the concentration of the solute in the liquid, then the relationship between "p_A" and "C_A" is given by the following equation,

$$p_A = H_A \times C_A \quad \text{(Henry's Law)} \qquad [4.1]$$

in which "H_A" is the appropriate Henry's Law constant of the gas. For oxygen dissolved in water the value of H_A varies with temperature and at 20°C. has the value of 40,100 if p be expressed in mm. and C_A as molar fraction.

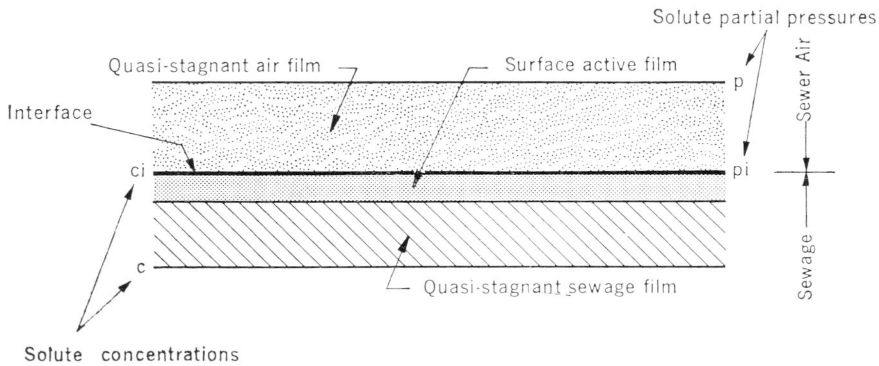

FIG. 4.2 Molar concentrations of solute gas at an air-water interface.

If the mass transfer coefficients for the individual gas and liquid films are denoted by "k_G" and "k_L" respectively and the total amount of the solute transferred across the interface per unit of area and time is denoted by "N_A" then "k_G" and "k_L" can be defined by the following equation:

$$N_A = k_G (p_{Ai} - p_A) = k_L (c_{Ai} - c_A) \qquad [4.2]$$

If K_L and K_G denote the *overall* mass transfer coefficient expressed in liquid and gas phase units respectively, that is to say

$$N_A = K_L (p_A - p_{Ae}) = K_G (c_{Ae} - c_A) \qquad [4.3]$$

where "p_{Ae}" is the partial pressure of solute when in equilibrium with the concentration "c_A" in the main liquid stream and "c_{Ae}" is the corresponding concentration of solute in equilibrium with the partial pressure p_A in the main gas stream;

$$p_{Ae} = H_A \times c_A \text{ and } c_{Ae} = \frac{p_A}{H_A}.$$

The reciprocals of the mass transfer coefficients are referred to as resistances and an essential part of the theory is that the overall resistance to mass transfer between phases is equal to the sum of the resistances of the two individual films. This can be expressed in the following equations:

$$\frac{1}{K_G} = \frac{1}{k_G} + \frac{H}{k_L} \qquad [4.4]$$

$$\frac{1}{K_L} = \frac{1}{k_L} + \frac{1}{Hk_G} \qquad [4.5]$$

This concept of the additivity of the resistances is found to fit some cases, while in other cases unexplained discrepancies have been observed. From equations [4.4] and [4.5] it will be seen that, if the value of H is relatively high, the liquid film resistance becomes dominant and it would be anticipated that the rate of oxygen absorption would be controlled by the liquid film resistance. Most investigations have shown that factors relating to the liquid phase are of major importance for oxygen transfer.

Although workers in the fluid mechanics field now doubt or reject the possibility of a truly stagnant film existing at the surface of water flowing under turbulent conditions the simple two-film concept is still used widely for practical studies. The two films are now seen as being dynamic or statistical in character rather than having a stable structure as pictured by Lewis and Whitman, small elements of water in the film constantly being exchanged with other elements from the main body of the stream underneath. The rate of exchange is a function of stream turbulence.

Attempts have been made also to establish a mathematical relationship between the "time of mixing" and the overall rate of oxygen absorption because the depth of flow is an important common factor. Dobbins refers to a specific relationship between the degree of turbulence and the rate of renewal of the film(*19-21*). Investigations of the absorption of oxygen by liquids stirred in laboratory vessels indicate that the rate of absorption is related to the power input of the stirring mechanism per unit volume of water.

In 1962 Krenkel and Orlob(*22*) made a review of the basic concepts of the mechanism of surface absorption of oxygen in turbulent flow so far as they were known. They suggested that the rate of absorption was influenced also by the longitudinal mixing coefficient.

They derived the following relationship experimentally for a water temperature of 20°C.:

$$K_{O_2} = 2.627 \times 10^{-4} \times \frac{E^{0 \cdot 408}}{h^{0 \cdot 66}} \qquad [4.6]$$

where K_{O_2} is the rate of oxygen absorption, p.p.m./min./p.p.m of oxygen saturation deficit. Note: The terms K_2 and k_2 are commonly used as re-aeration coefficients for oxygen, usually as time units per day, where $K_2 = 2.303\ k_2$

E is the rate energy is dissipated per unit volume of water.

= approximately V × S × g where V is the mean velocity of flow, ft./min.; S is the slope and g the gravitational constant, ft./min.

h is the depth of flow, ft.

In the units used in this Manual for velocity (i.e. ft./sec.) and incorporating the value 116,000 for g, equation [4.6] may be rewritten.

$$K_{O_2} = 0.163 \times \frac{(V \times S)^{0.408}}{h^{0.66}} \quad [4.7]$$

Design Chart No. 6 is a computing chart for calculating the values of K_{O_2} when values of V, S and h are known, Fig. 4.4, p. 26.

Design Chart No. 5 is a computing chart for calculating the value of K_{O_2} when slope S, diameter D and proportional depth are known.

It is sometimes desirable to know the amount of oxygen absorbed per unit area of water surface per p.p.m. of oxygen saturation deficit (K_{O_2D}). This can be obtained from values of K_{O_2} and h by use of the following equation:

$K_{O_2D} = 30.48\ K_{O_2} \times h$ gm/sq. cm./min./p.p.m. oxygen saturation deficit. [4.8]

$= 1830 \times K_{O_2} \times h$ gm/sq. cm./hr./p.p.m. oxygen saturation deficit. [4.9]

The above equations have been found to give values for K_{O_2} agreeing reasonably well with observed values obtained in channels used by other observers, in which the depths of flow differed from those used by Krenkel and Orlob. Churchill, Ellsmore and Buckingham, Dobbins, Owens and Edwards and others (*19-23*) have derived similar relationships.

Thus, Owens, Edwards and Gibbs, taking account of their field observations in United Kingdom as well as data from the United States of America, derived the empirical relationship

$$k_2 20^\circ C = 9.41 V^{0.67} \times h^{-1.85}$$

Where k_2 is the re-aeration coefficient, *per day*, and V, h are respectively mean stream velocity and depth, (ft.—.sec.units).

These equations apply, however, to streams for which the widths mostly are great compared with the depth. It seems that no great error would be incurred in applying these equations for the estimating of the rate of oxygen absorption at the surface of clean water flowing in channels of the shape normally used for sewers provided the depth h is taken as the mean depth defined by $h = \frac{A}{B}$ where A is the cross-sectional area of flow and B is the surface width of flow.

Absorption by water in free-fall drops

Investigations have been made by the Water Pollution Research Laboratory in the United Kingdom with respect to oxygen absorption in the second type of flow, namely free-fall vertical drops. Gameson(*24*) in 1957 reported the results of observations of free-fall weirs in natural streams. Though such weirs differ in design from those encountered in sewer systems, the results are a useful parallel for sewer drops. It was found that little of the oxygen absorption takes place during the free-fall but most occurs as a result of splashing and air entrainment at the bottom of the fall. Gameson puts forward a formula giving roughly the amount of absorption to be anticipated, the formula being adjusted to the units of this Manual as follows:—

$$R_d = 1 + 0.15 \times F_{da} \times F_{db} \times h_d \quad [4.10]$$

where R_d = ratio of the oxygen saturation deficit just above the drop to that just below.

F_{da} = coefficient to allow for degree of pollution of water (surface active substances) having values 1.25 in slightly polluted water, 1.00 in moderately polluted water, and 0.85 in effluents from sewage treatment plants.

F_{db} = coefficient to allow for the type of weir having values 1.00 for a free fall weir and 1.30 for a stepped weir.

h_d = height of fall in feet.

Until more data are available relating directly to raw sewages and to vertical drop structures actually used in sewerage systems it is suggested that the value of F_{da} be taken as 0.5 and value of F_{db} as 0.8, giving

$$R_d = 1 + 0.06\ h_d \quad [4.10a]$$

Gameson's work also indicated that where the head loss at a weir was partly recovered in a steeply sloping channel the oxygen taken up into solution was only about 20% of that absorbed by free-fall over a normal weir.

Effect of salinity on oxygen absorption

Salinity reduces the rate of oxygen absorption because the solubility of oxygen is less in saline than in fresh water. For example the solubility of oxygen in sea-water at 760 mm. partial pressure and at 20°C. temperature is only about 80% of that in distilled water at the same temperature and partial pressure. Ogden, Gibbs and Gameson(*25*) made experimental determinations of the oxygen absorption rates for various mixtures of fresh and ocean water flowing in a small channel. They concluded that the percentage *reduction* in absorption rates could be expressed as follows:—

$$R_{O_2} = 6\sqrt{Sal.} \quad [4.11]$$

where R_{O_2} = percentage reduction in absorption rate

Sal. = Salinity expressed as grams of total dissolved solids content in one kg. of saline water when all the carbonates have been converted into oxides, the Br and I replaced by Cl and all organic matter completely oxidized.

However, these tests indicated a much greater effect of salinity on uptake rates than would have been expected due simply to the changed solubility of oxygen. Using undiluted sea-water, the reduction in the rate of re-aeration was about 34% compared with reduction of only 18.5% based on relative oxygen solubility.

Effect of surface-active substances

In some relatively quiescent solutions certain types of solute or dispersoid tend to become concentrated at the surface. This concentration often proceeds so far that most of those particular types of solute form a film at the surface of the solution which may be only one molecule thick, but may in other cases consist of a complex of many layers. This depends on the nature and variety of these surface-active or other such substances present in the solution, and on their relative amounts.

The structure of a monomolecular film, in its simplest form, is sometimes shown in a diagrammatic form as in Fig. 4.3. This figure indicates that the water-attractive group in the surface-active molecule tends to remain in the water surface, and the water-repellent group tends to shun the water phase.

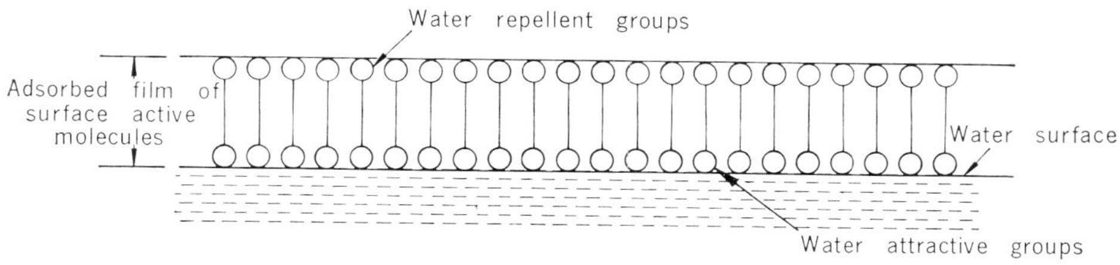

FIG. 4.3 Structure of surface-active films (diagrammatic).

It has been known for some time that absorbed films of surface-active substances retard the transfer of a gaseous solute (such as oxygen) across the liquid-gas interface. However, under conditions of turbulent flow in a channel, segregation at the surface is opposed to some extent by the remixing power of eddies within the flowing liquid, so that the potential retarding effects of surface-active substances on oxygen uptake rates are diminished by increasing turbulence within the liquid.

The effect of surface-active substances in sewage on the oxygen absorption rate at the surface of water flowing in channels has been investigated quantitatively by the Water Pollution Research Laboratory in the United Kingdom (*23–25*). From these experiments, the relative rates of surface absorption by clean and polluted waters were expressed in terms of an "exchange coefficient" f_0, such that the exchange coefficient K_{O_2D} as used for equations [4·8] and [4.9] (p. 24) is synonymous.

Surface-active substances in sewage can be classified into two groups:

(1) Indigenous surface film-forming substances. Sewage contains many substances, characteristic of domestic and industrial wastes, which tend to form surface films, including proteins and some amino-acids. For the purpose of this discussion they are referred to as indigenous surface-active substances. Probably they are relatively more important for raw sewage than for settled sewage. Of course fats and fatty acids also are reckoned as "indigenous".

(2) Synthetic detergents. In recent years a new type of substance has appeared in both domestic and industrial wastes in concentrations now forming significant proportions of the total film-forming potential of sewage. These are the modern synthetic detergents, which are substances specifically manufactured to be surface active (film-forming). While such detergents if considered alone could seriously impede the surface absorption of oxygen, their effect when added to that of the natural or *indigenous* surface-active substances of the sewage apparently causes little change in the overall oxygen uptake patterns.

It is to be expected that the BOD of sewage would be correlated well enough with the concentrations and effects of all surface-active substances, and for that reason the following relationship is proposed as a basis of comparison of oxygen uptake or re-aeration rates for sewage with those appropriate for clean waters.

In connexion with evaporation phenomena, Langmuir (*26, 27*) used the principle of the addition of resistances referred to earlier. Applying this to an analysis of data from Water Pollution Research Laboratory Investigations, and assuming both that the resistance of the air-phase film is negligible, and that the film of surface active material does not materially alter the resistance of the supporting water interface as such, yields the following approximate relationships:—

$$\frac{1}{k_{fi}} = \frac{0.424 \times [BOD_5]^{0\cdot5}}{(K_{O_2D})^{1\cdot37}} \qquad [4.12]$$

where $\frac{1}{k_{fi}}$ is resistance to oxygen entry caused by the adsorbed film of indigenous surface-active substances, using units of mg/1 with K_{O_2D} in gm $0_2 cm^{-2} hr^{-1}$ per unit p.p.m. 0_2 deficit (eqn. [4.9] p. 24).

and

$$\frac{1}{k_{fd}} = \frac{3.3 \times [BOD_5]^{0\cdot33}}{(K_{O_2D})^2} \qquad [4.13]$$

where $\frac{1}{k_{fd}}$ is resistance to oxygen entry caused by the adsorbed film of synthetic detergents, units as for [4.12].

The overall rate of surface re-aeration can be expressed in general terms as follows:—

$$K_{O_2S} = F_{dt} \times K_{O_2} \qquad [4.14]$$

where K_{O_2S} is surface absorption rate of oxygen for sewage in p.p.m./min./p.p.m. oxygen saturation deficit.

K_{O_2} is surface absorption rate of oxygen for clean water in p.p.m./min./p.p.m. oxygen saturation deficit (see eqns, [4.6], [4.7] pp. 23, 24).

F_{dt} is a factor of value less than unity, the actual value correlated with values of BOD_5 and K_{O_2D}.

Fig. 4.4 (Design Chart No. 6, p. 121) enables the value of the factor F_{dt} of equation [4.14] to be read off provided both sewage BOD_5 and the appropriate K_{O_2D} values are known (see equations [4.7] to [4.9] p. 24).

It appears that oxygen absorption rates for sewages are only about 20% to 30% of those applying to clean water. This is highly significant for the sulphide problems of sewerage systems.

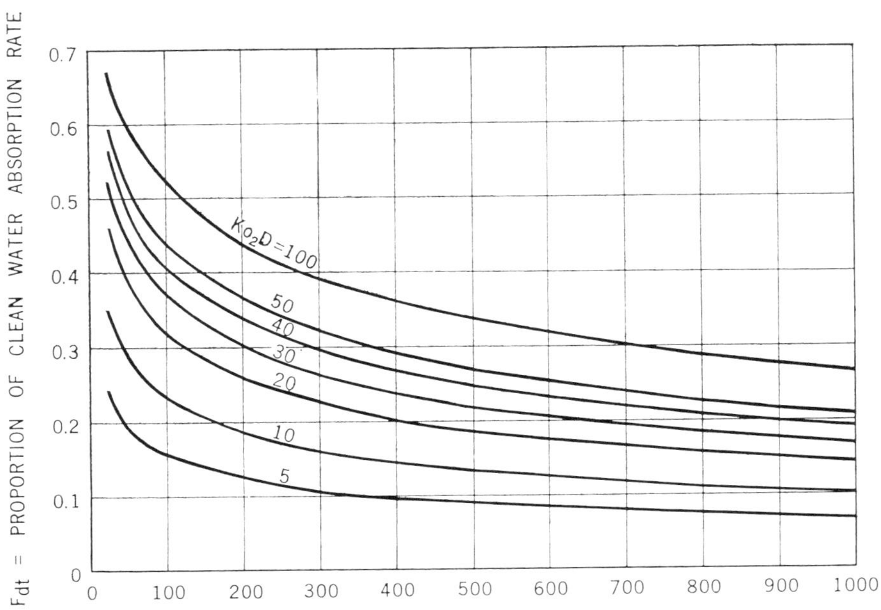

FIG. 4.4 Plot of factor F_{dt} (equation 4.14) which corrects the oxygen absorption rate constant for the effects of surface-active substances in sewage (as correlated with BOD values).

Effects at the surface of (bio-) chemical reactions with oxygen

Some of the oxygen which enters the surface film may be expected to be taken up in chemical or biochemical reactions even within the relatively stagnant film. Any such reactions would thereby reduce the effective oxygen concentration gradients in the film, and in turn increase the overall rate of re-aeration. An analysis of these circumstances indicates that the integrated effects on re-aeration are likely to be small and normally therefore may be neglected.

Effect of temperature

The total rate of re-aeration K_{O_2T} can be expressed by the following equation:

$$K_{O_2T} = O_D \times K_{O_2S} \text{ p.p.m./min.} \qquad [4.15]$$

where O_D = oxygen saturation deficit in p.p.m. (mg/1)

K_{O_2S} = rate of re-aeration of sewage in p.p.m./min./p.p.m. of oxygen saturation deficit, equations [4.12] and [4.13].

The effect of temperature in the re-aeration process can therefore be considered in two parts, the effect on the oxygen saturation deficit O_D and the effect on the re-aeration coefficient K_{O_2S}.

The solubility of oxygen in water decreases with rise in temperature. Thus in the summer the rate of oxygen absorption is decreased and the sulphide problem, insofar as it is affected by the rate of oxygen uptake, is increased.

Oxygen solubility in mg/l at any temperature and barometric pressure is given in Table 4.1, p. 22 and is also given by Design Chart No. 4, which is based on a nomograph prepared by Davis(*28*).

The effect of temperature on the re-aeration rate K_{O_2} for clean water cannot be given properly until the detail of the mechanism of oxygen uptake at the surface of flowing water is known better. However, for most practical purposes the effect of temperature can be obtained from Fig. 4.5. The re-aeration rate K_{O_2} at

20°C. (68°F.) multiplied by F_T gives the corresponding rate appropriate for other temperatures.

There is reason to believe that the re-aeration rate for sewage K_{O_2S} is affected by temperature to very nearly the same extent as the re-aeration rate K_{O_2} for clean water.

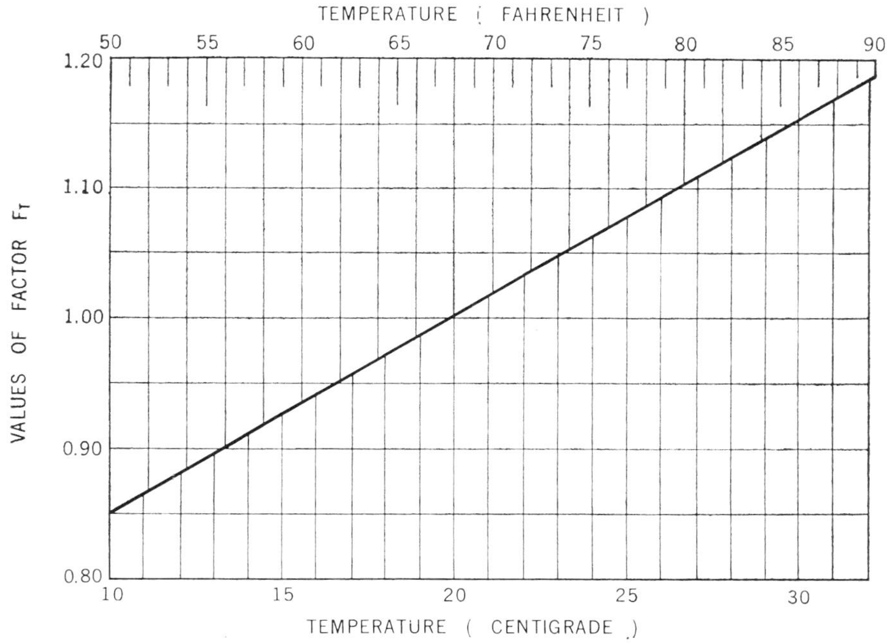

FIG. 4.5 Factor F_T correcting the 20°C. (68°F.) re-aeration rate constant for other temperatures.

Oxygen consumption in sewage

The potential of different sewages to use up dissolved oxygen is commonly measured on a laboratory-scale by tests in which small sample portions of sewage are diluted by relatively large volumes of fully oxygenated water, often using aerobic bacterial cultures (in still smaller proportions) for seed. The sewage is commonly tested at only about 1%, sometimes for weak sewage up to 5% dilution, so that such tests do not represent circumstances applying directly to undiluted sewage.

The Sierp-Fränsemeier (*29*) techniques allow direct short-term or long-term measurements of oxygen uptake rates, using undiluted raw sewage, but again under circumstances different to that found in sewerage systems where sulphide problems occur. Their techniques allow use of relatively large sample portions, nevertheless some respirometer systems used for sewage testing also are restricted to small sample portions. Data from such laboratory scale tests may not be valid for simple direct application to ordinary sewage in sewerage systems, but no other satisfactory basis is as yet available, and the following theoretical discussion is based accordingly on the standard laboratory Biochemical Oxygen Demand test. Test values are expressed in terms of the dissolved oxygen consumed during a specified time period, measured in mg of oxygen per litre of the sample tested. The amount consumed in a period of five days is thus abbreviated as BOD_{0-5}, written simply BOD_5.

From studies with polluted Ohio River water, Theriault and Streeter (*31*) formulated the sequence of oxygen uptake in laboratory tests as a first order reaction, as follows

$$[BOD_t] = L_o(1 - e^{-K_1 t}) = L_o(1 - 10^{-k_1 t}) \quad [4.16]$$

where BOD_t is the Biochemical Oxygen Demand overall from time t_o to time t days, mg. O_2/1.
where L_o is total first stage (ultimate) BOD, mg. O_2/l
t is time in days
K_1 is the BOD rate constant, per day
k_1 is $0.4343K_1$, i.e. $K_1 \times \log_{10}e$, per day

This formulation was derived only for the biochemical oxidation of carbonaceous organic matter (called first stage BOD), but not where nitrification proceeded at the same time. Nitrification is quite unlikely in flowing sewage.

Other formulations provide equally good, if not better, mathematical models for oxygen uptake rates—for example better fits for practical BOD testing may be obtained by assuming second order, not first order, reaction kinetics. Actually such simplified mathematical models can represent only the average of a number of parallel reactions each with a different time pattern. However, because sewages from one area or another are roughly similar in make-up, Theriault's or other such models have proved to be very useful. Theriault's original concepts are mathematically simple, and having now been used generally for sewage testing for about 50 years are considered suitable for use in the subsequent mathematical development.

Immediate (chemical) oxygen demand

Particularly where certain trade wastes may be involved, the above equations for oxygen uptake should not be applied simply, but allowance should be made also for any "immediate" uptake of oxygen due to "chemical" reactions such as oxidation of ferrous to ferric ion, oxidation of sulphides, and so on.

Lag period in BOD tests

Often sewages do not show an immediate uptake of oxygen immediately it becomes available, but rather show an appreciable lag, in some cases of up to many hours, sometimes even after careful seeding. This may be due to toxicity factors or to unusual biochemical conditions and indicates the need for special judgments of each situation, beyond the scope of this particular discussion.

The instantaneous rate of oxygen demand ot sewage. Taking the original equation [4.16], the instantaneous rate of biochemical oxygen demand (rate of oxygen consumption) at time t is given by:

$$\frac{d}{dt}[BOD]_t = 2.303 \times k_1 \times L_o \times 10^{-k_1 t} \text{ see eqn. [4.16]}$$

Putting $\frac{d}{dt}[BOD_0]$ for $\frac{d}{dt}[BOD]_t$, and $t = t_0 = 0$, then

$$\frac{d}{dt}[BOD_0] = 2.303 \times k_1 \times L_o$$

But

$$L_o = \frac{[BOD_t]}{1 - 10^{-k_1 t}}, \quad \text{from [4.16]}$$

$$= \frac{[BOD_5]}{1 - 10^{-5 \times k_1}}$$

Hence

$$\frac{d}{dt}[BOD_0] = \frac{2.303 \times k_1}{\left(1 - \frac{1}{10^5 \times k_1}\right)} \times [BOD_5] \text{ mg } O_2$$

per day per litre [4.17]

Values for $\frac{d}{dt}[BOD_0]$ in terms of analytical data for $[BOD_5]$ may be taken directly from Table 4.2 or read from Figure 4.6 for any selected appropriate value of k, between $k_1 = 0.1$ and $k_1 = 0.24$, the unit here being mg O_2 per second per litre of sewage.

TABLE 4.2 VALUES FOR INSTANTANEOUS RATE OF BOD IN P.P.M./SEC. FOR DIFFERENT VALUES OF K_1

Value of k_1	$D[BOD_0]$ (=Rate of $[BOD_t]$ in p.p.m./sec.)
0.10	$3.9 \times 10^{-6} \times [BOD_5]$
0.12	$4.3 \times 10^{-6} \times [BOD_5]$
0.14	$4.7 \times 10^{-6} \times [BOD_5]$
0.16	$5.1 \times 10^{-6} \times [BOD_5]$
0.18	$5.5 \times 10^{-6} \times [BOD_5]$
0.20	$5.9 \times 10^{-6} \times [BOD_5]$
0.25	$7.1 \times 10^{-6} \times [BOD_5]$

NOTE $D[BOD] = \frac{d[BOD_t]}{dt}$ where $t = 0$, see equation [4.17]

Where oxygen consumption is to be calculated during travel of the sewage over any particular length of sewer $L = L_2 - L_1$ during the time interval $t = t_2 - t_1$, there is a progressive change in the instantaneous rate of demand, requiring a correction factor, which varies as shown in the following according to the length of time t and the relevant value of the BOD rate constant k_1.

Thus we have

$$\left(\frac{\triangle[BOD]}{\triangle t}\right)_{t_1}^{t_2} = \frac{[BOD_{t_2}] - [BOD_{t_1}]}{t_2 - t_1}$$

$$= L_o \times (10^{-k_1 t_1} - 10^{-k_2 t_2})$$

$$= \frac{[BOD_5]_{t_1} \times \left(1 - \frac{1}{10^{k_1 t}}\right)}{t \times \left(1 - \frac{1}{10^{5 \times t}}\right)}, \quad \text{by substitution from [4.17]}$$

The required correction factor, putting $t_1 = 0$, then is given by reckoning as follows:

$$\text{Correction Factor} = \frac{\left(\frac{\triangle[BOD]}{\triangle t}\right)_{t_1}^{t_2}}{\frac{d}{dt}[BOD_0]_{t_1}}$$

$$= \frac{[BOD_5]_{t_1} \times \left(1 - \frac{1}{10^{k_1 t}}\right)}{t \times \left(1 - \frac{1}{10^{5 \times t}}\right)} \times \frac{1 - \frac{1}{10^{5 \times t}}}{2.303 \times k_1 \times [BOD_5]_{t_1}}$$

$$= \frac{\left(1 - \frac{1}{10^{k_1 t}}\right)}{2.303 \times k_1 \times [BOD_5]_{t_1}}$$

For practical problems associated with sewer flows the values of the correction factor, expressed as percentage reductions to initial rates of BOD, are shown in Table 4.3 and Fig. 4.7.

TABLE 4.3 VALUES OF CORRECTION FACTOR TO BE APPLIED TO RATES OF BOD OF TABLE 4.2 (AT TIME T = 0) FOR LONGER PERIODS OF TRAVEL.

Value of $k_1 t$ ($\frac{1}{\text{days}}$ × minutes)	Factor	Corresponding reduction per cent
10	0.994	0.6
20	0.985	1.5
30	0.978	2.2
60	0.957	4.3
120	0.91	8.9
240	0.83	16.9
480	0.70	30.0

If k_1 be 0.17 and t be 60 minutes, then $k_1 t = 10$ nearly, and the correction is only about 1% or less, which may be neglected.

Values of k_1

Determination of k_1. From equation [4.16] the ratio $\frac{[BOD_{t_1}]}{[BOD_{t_2}]}$ is as follows:

$$\frac{[BOD_{t_1}]}{[BOD_{t_2}]} = \frac{1 - 10^{-k_1 t_1}}{1 - 10^{-k_1 t_2}}$$

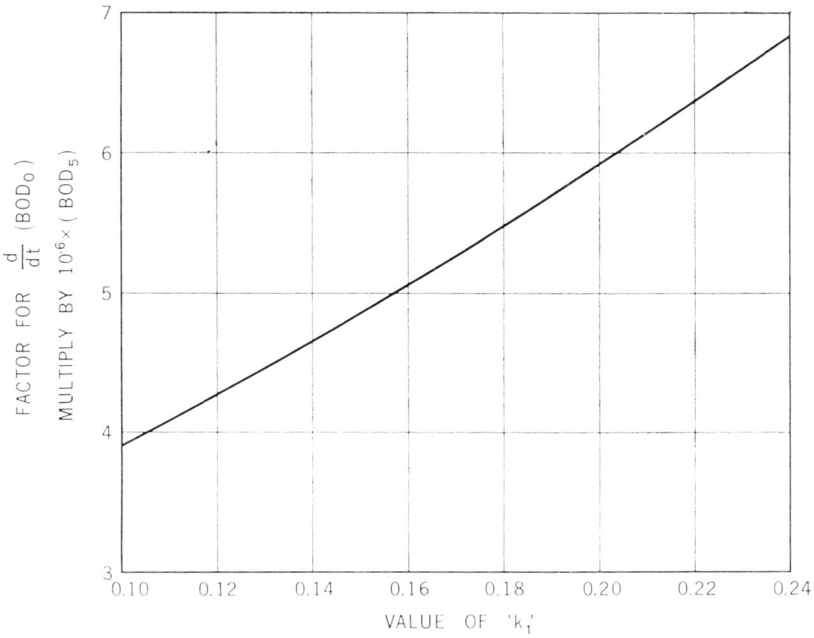

FIG. 4.6 The effect of k_1 on $\frac{d}{dt}[BOD_0]$, the instantaneous rate of BOD.

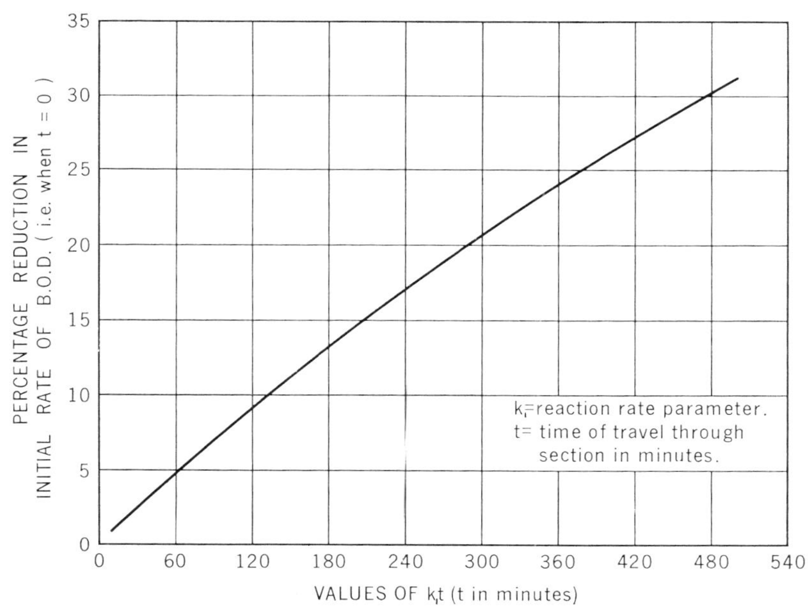

FIG. 4.7 Reduction of $\frac{d[BOD_0]}{dt}$ with time (see Tables 4.2 and 4.3, Fig. 4.6)

Hence, to determine the value of k_1 applicable for a particular sewerage system, sewage BOD tests should be made over two or more time periods, say 1-day and 5-days. Then the value of the ratio is known and k can be reckoned by substitution in equation [4.18]:—

$$\frac{[BOD_1]}{[BOD_5]} = \frac{1 - 10^{-k_1}}{1 - 10^{-5k_1}} \qquad [4.18]$$

Fig. 4.8 plots vlues of k_1 corresponding to particular values of the ratio of 1-day to 5-day BOD.

It should be appreciated however that use of this method for determining k_1 is invalid where the sewage is septic, or for any other reason likewise shows a significant immediate chemical oxygen demand. A direct determination of shorter term oxygen uptake e.g. by using Sierp and Fränsemeier's techniques may be preferable, and obviates lag problems.

There appears to be little data available respecting values of k_1 for raw sewage flowing in sewers. Experience with BOD tests overseas and in Australia indicates values of k_1 ranging from 0.10 to 0.20. However there has been some study of particular factors which affect the k_1 value and this is discussed in the following.

Effect of sewage constitution

Gotaas (*32*), quoting from an unpublished thesis by Thomas Jnr. (1938), states that values of k_1 for different components of sewage at 20°C. (68°F.) are about as follows:—

Colloids	$k_1 = 0.17$
Other Suspended Solids	$k_1 = 0.064$
Dissolved Solids	$k_1 = 0.19$

A typical break-down of BOD of a domestic sewage as it entered a treatment works was found to be roughly as follows:—

BOD of dissolved substances	30% of total BOD
BOD of colloidal matter	30% of total BOD
BOD of non-collodial solids	40% of total BOD

If these be applied to the values of k_1 given in the preceding table for the various component fractions, the computed overall normal value of k_1 for sewage would be about 0.13 Gotaas' values (*32*) for 20°C. however are higher, actually about 0.19.

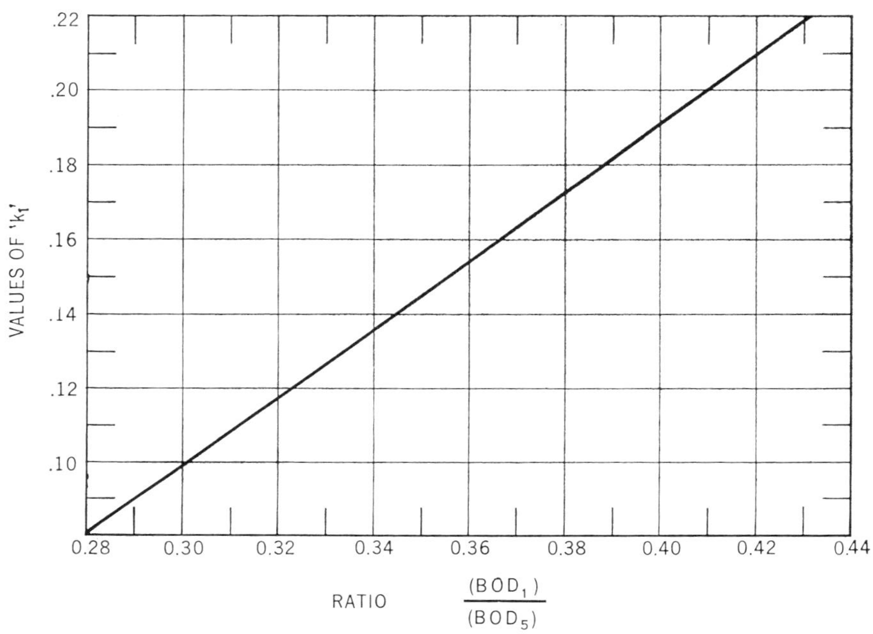

FIG. 4.8 Plot giving k_1 values from BOD test data.

Temperature. Theriault (*31*) related the temperature correction factor to 20°C and, basing it on experiments at 7°, 20° and 37°C, showed a dependence of k_1 on temperature according to an empirical expression $k_1^{(T°C)} = k_1^{(20°C)} \times 1.047^{(T-20)}$. He also reported that the total (ultimate) BOD was the greater, the higher the temperature.

Gotaas however found k_1 increased with rising temperatures to a maximum at about 30°C. Above this, k_1 tended to decrease. His findings are expressed mathematically as

$$k_1^{(T°C)} = 0.0115T^{0.932}$$

The values of the correction factor for temperature effect given in Table 4.4 are based on Gotaas' formulation.

Factors derived by Theriault do not differ greatly from Gotaas' data between about 12°C. and 30°C., except that Gotaas differed from Theriault in finding, as indeed would be expected, that the *total* BOD was not temperature dependent.

Table 4.4 Temperature Correction Factors for k_1 (Gotaas (ref.))

Temperatures		Correction Factor
°C	°T	$\frac{k_1T}{k_120}$
5	41	0.274
10	50	0.525
15	59	0.764
20	68	1.000
25	77	1.231
30	86	1.520
35	95	1.350
40	104	1.107

Circumstances of flow in sewers

The standard laboratory test for BOD does not represent the circumstances in which oxygen demand is exerted by the sewage flowing in a sewer, even in those cases where aerobic conditions prevail. Apart from the obvious differences between the sewer and the laboratory test conditions, other particular differences are noted here.

The necessarily small *sample* of sewage normally used in the laboratory test is unlikely to include the average proportions of the larger solids such as unbroken faeces, paper, rags, fats, etc. characteristic of the normal sewage flow. On the average the tendency probably would be for the *sample* to include less than the due proportion of such larger solids, and the BOD exerted in the sewers, where such solids are gradually broken up by the tumbling action of the sewage, probably would be different likewise.

Flowing sewage in turbulent motion differs from quiescent laboratory bottle tests also because of the frequency of collisions between organisms and their nutrients. Thus even insoluble substances such as fats and soap-curds are much more likely to exert oxygen demand in the sewers.

Thus, in a general way it would be expected that the rate of BOD exerted in flowing sewage would be greater than that determined from the standard BOD test in the manner set out in the foregoing. In fact some exploratory investigations of the short-term (five-minute) oxygen demand of sewages respectively under quiescent and agitated conditions indicated that a significant increase in the short-term rate of BOD did occur when the sewage was agitated. When the degree of agitation was similar to that occurring in a sewer with slope and velocity sufficient to prevent build-up of sulphide, the increase averaged roughly 25%, and this effect increased somewhat more when the concentration of solids was greater.

Value of $\frac{d}{dt}$ [BOD$_0$]

Until such time as more comprehensive and reliable data are available it is recommended that values up to $\frac{d}{dt}[BOD_0] = 6.0 \times 10^{-6} \times BOD_5$ p.p.m./sec. be used for sewer design. This is equivalent to an effective value of k_1 of 0.20. A value of $\frac{d}{dt}[BOD_0] = 5 \times 10^{-6} \times [BOD_5]$ p.p.m./sec. corresponds to an effective value of k_1 of about 0.15, which may be applicable where sluggish velocities occur.

The rates of early stage BOD at other temperatures may be obtained by multiplying the 20°C rate by a factor F_{BOD} computed from the following equations based on work by Gotaas(*32*).

From 5° to 30°C, $F_{BOD} = \left(\frac{T°C}{20}\right)^{0.93} (1 + 0.00066\,[T° - 20])$

From 30° to 40°C, $F_{BOD} = 1.459 \left(\frac{30}{T°C}\right)^{1.06} \times (1 + 0.0013\,[T° - 20])$

The corresponding values of this multiplying factor F_{BOD} may be read directly from Fig. 4.9.

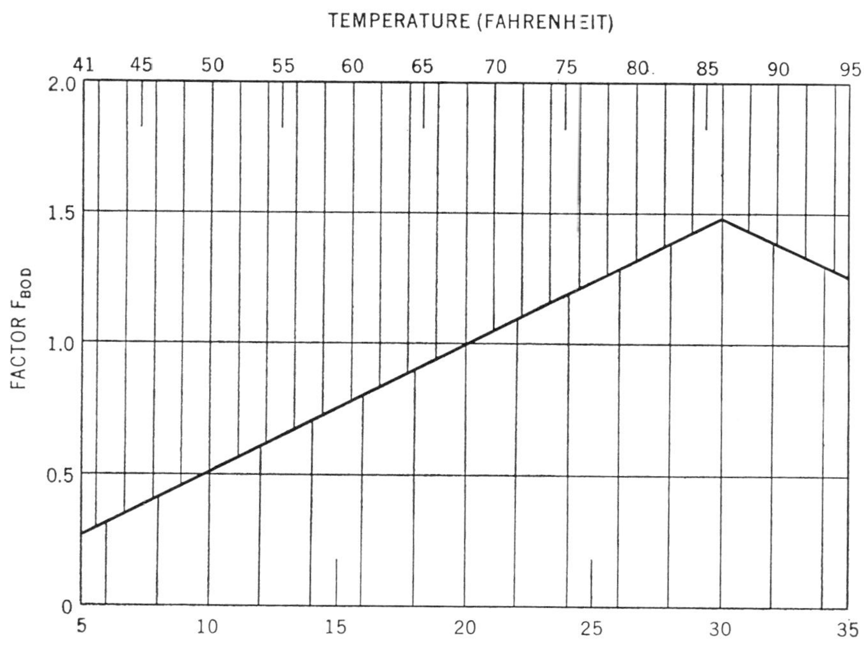

FIG. 4.9 Plot of F_{BOD}, temperature correction factor for k_1 (hence for $\frac{d}{dt}[BOD_0]$)

When the sewage is flowing under anaerobic conditions the effective rate at which the BOD is satisfied may be considerably below that applying under the aerobic conditions of the standard BOD test. There is some evidence (6) that in pressure mains the value may fall to as little as one-third of the corresponding BOD test rate. In Australia a fall of similar magnitude was found in a long gravity outfall carrying septic sewage. Perhaps with septic sewage, because there is not sufficient oxygen present to satisfy the total oxygen demand, the apparent rate of the demand depends rather on the rate of *supply* of oxygen, as well as on other competing demands for oxygen.

However, the use of the BOD_5 concept for computing oxygen uptake in sewers appears to be the best basis at present available both because no other relevant characteristic is normally measured, and because the mathematics are simple to use and to adjust.

The oxygen balance in sewers

Normally dissolved oxygen is present in sewage at the upper ends of the sewerage system because of the original dissolved oxygen in mains-water used for flushing, augmented by splashing, etc. This represents a significant contribution to the oxygen balance for most sewers, the potential saturation level being normally around 8 to 10 mg of dissolved oxygen per litre of sewage. This quantity is sufficient to satisfy the whole oxygen demand of ordinary sewage for about two hours without allowing for any re-aeration. The possibilities of full aeration within the house drainage system should be a first consideration in designing for sulphide control.

Any sulphide entering the sewage with trade wastes or from sites of generation in the system may be wholly or partly destroyed also by oxidation, thus adding to the oxygen demand. However, for this oxidation to occur to any extent, there must be sufficient dissolved oxygen present, and conditions must be favourable. Whereas oxygen may be present in sewage either as freely dissolved oxygen (gaseous), or as chemically bound oxygen (e.g. in sulphates, nitrate, even water itself), under aerobic conditions only free dissolved oxygen is available either for sulphide oxidation or for biochemical oxygen demand. The dissolved oxygen used up in these ways in gravity sewers may be partly or even wholly replaced by the uptake of oxygen from the sewer air, especially at vertical drops where the free-fall of sewage causes air entrainment, but also in the flowing sewers or in pumping station wet wells, etc., wherever there is direct contact between the sewage and atmospheric oxygen.

The design of sewerage systems for sulphide control therefore should be based on considerations of the oxygen balance from place to place along the sewage flow network. Oxygen balances require simply that oxygen uptake be computed, using equations [4.8], [4.9], [4.10], [4.11], and [4.15], with Design Chart Nos. 4-7, also Table 4.1, and Figs. 4.4, 4.5, and oxygen consumptions likewise from equations [4.17] and [4.18] with Tables 4.2, 4.3, and 4.4, and Figures 4.6, 4.7, 4.8, and 4.9, on the general basis (dissolved oxygen increment) = (oxygen taken up — oxygen consumed). In the foregoing, the processes of oxygen up-take (re-aeration) and oxygen consumption (including the effective rate of BOD and also any immediate "chemical" oxygen demand) were considered. The two main processes of re-aeration and effective "flowing" sewage BOD were formulated finally into mathematical terms, from which practical designs can follow. The actual design procedure is dealt with in Chapter 8 and 10 of this Manual.

Satisfactory oxygen balance for sewerage systems

Dissolved Oxygen in Mains Water + Dissolved Oxygen taken up	⩾	Biochemical Oxygen Consumption + Other oxygen consumption (sulphide oxidation, etc.)

Oxygen balance computations

Income	*Debit*	*Measurement or Assessment*
Dissolved oxygen entering with mains water.		Chemical analysis.
Oxygen uptake by re-aeration by flowing sewage, etc., at drops, etc.		Eqns. [4.6] to [4.9], [4.12] to [4.15]. Eqns. [4.10], also [4.11] to [4.15].
	Sewage oxygen consumption: biochemical.	[4.17], [4.18], Tables 4.2 to 4.4, Figs. 4.6 to 4.9.
	Sulphide oxidation, etc.	Chemical analyses.

CHAPTER 5

SULPHIDE BALANCES IN SEWERS AND VARIATIONS OF SEWAGE SULPHIDE CONCENTRATIONS

Introductory

The foregoing has indicated that it is basically the concentration of dissolved molecular H_2S in the sewage flowing in any particular pipeline or other sewerage structure which controls the extent of the sulphide corrosion problem. This concentration at any particular place in a system is the result of a balance between sulphides brought into or generated within the system and their removal either by emission from solution or by reaction within the sewage. Experience has shown that this balance varies from time to time both during any one day or from day to day, and also seasonally.

Sulphide build-up

In Chapters 2 and 3 the origins of the sulphides found in sewerage systems were described as being either derived directly from raw waste inflows (particularly certain industrial wastes), or as deriving from generation within the sewerage system.

Any effective increase in sewage sulphide concentration is referred to as sulphide build-up. It is expressed in simple terms of concentration, as so many milligrams of sulphide sulphur per litre (mg/l) or as so many parts per million by weight (p.p.m.), these two values being practically the same.

The rate of sulphide build-up may be expressed also, in terms of concentrations, for example as p.p.m./sec., p.p.m./min. or p.p.m./hr. It is convenient for some purposes however to use other rate units expressed in terms of lb. (mass) of sulphide sulphur, or as grams or kilograms; for example lb./1000ft.^{2}/hour or lb./1000 ft. run of sewer, and so on.

In Chapter 3, important equations were presented, giving the rates of sulphide build-up for gravity sewers —eqn. [3.5] to [3.9], [3.12] and [3.13]. Using the symbols $(B_S)_T$ for trade waste contribution and $(B_S)_G$ for sulphide generation in the system, with (B_S) representing total sulphide build-up, we have the overall equation for any given length of sewer

$$B_S = (B_S)_T + (B_S)_G - [[S]_{ox} + [S_E]] \qquad [5.1]$$

in which the additional term $[S]_{ox}$ represents the sulphide sulphur portion, if any, oxidized by dissolved oxygen, or precipitated, etc.; and the additional term $[S]_E$ represents the sulphide lost by emission from the sewage flow as H_2S gas.

These two potential losses, $[S]_{ox}$ and $[S]_E$, are discussed in the following pages.

Sulphide removal by oxidation within the sewage

Near the upper ends of reticulation systems, sewage normally contains some dissolved oxygen, but as it travels downstream the concentration of dissolved oxygen may be reduced finally even to zero. Commonly the dissolved oxygen concentrations remain high enough to sustain aerobic conditions within the flowing sewage, at least during travel through the smaller sewers of the network.

Oxidation in clean water.—Mitchell(*34*) has studied the oxidation of H_2S in clean water containing dissolved oxygen (i.e. under aerobic conditions). It was found that the rate of oxidation of H_2S was very low, except where traces of heavy metals, such as iron, nickel, cobalt or copper, were present to act as catalysts. The course and rate of reaction differs with pH and the nature and concentration of the particular metals and the form of the metal compounds and ions which occur in the water. Some metals react so that hydrogen peroxide is an intermediate product which in turn oxidizes the sulphide by reactions also catalysed by heavy metals.

One of the findings which could be highly significant was that whereas reaction proceeds rapidly after first introducing the catalyst, it tends then to slow down and almost stop. This suggests a poisoning or inactivation of the catalyst. The oxidation mechanism may be entirely different in sewage, but present knowledge does not provide a firm statement of what actually happens, nor how fast it happens.

Examination of the mechanics of catalysis suggests that it is probably the hydrated metallic oxide systems which are responsible for the acceleration of the reaction.

The study referred to was extended, to a very limited extent, to studying the destruction of H_2S in the absence of dissolved oxygen (i.e. under anaerobic conditions). It was found that some destruction of H_2S can occur under these conditions, also accelerated catalytically by heavy metals. However these studies were not completed.

Oxidation in sewage.—Sewage consists of a complex mixture of a number of inorganic and organic chemical compounds. This mixture forms a mixed redox system in which compounds at higher redox potential tend to oxidise or partly oxidise compounds at lower potential. Sulphide of very low potential and also normally present only in low concentrations, is likely to be at least partly oxidised even though no dissolved oxygen be present. Moreover oxygen absorbed at the surface of the flowing sewage would increase the general redox level because of partial oxidation of some organic or inorganic compounds (e.g. by removal of some carbonaceous substance as CO_2). Any such increase in redox level would tend to increase rates of sulphide oxidation and in this way dissolved oxygen could cause loss of sulphide by oxidation, as it were, indirectly.

Experience in the field with live sewers, pumping systems, etc. and with laboratory experiments likewise, is consistent with these ideas. In zones of transition from aerobic to anaerobic conditions in the sewage, sulphide does not build up in the sewage, but is detected only in the downstream zones where the condition of the sewage passes from being merely "stale", that is, free of dissolved oxygen but not containing significant dissolved sulphide, to the stage or state of being "septic". This is taken to indicate that the intermediate "anaerobic" state of staleness is associated with redox levels high enough to keep pace with the rate at which sulphide is entering solution, e.g. by diffusion from the slimes. In other words, oxygen uptake then balances sulphide input.

Removal of hydrogen sulphide by emission from solution

The emission of sulphide from the sewage into the sewer atmosphere is an essential step in the overall chain of processes associated with sulphide problems. The process of sulphide emission is analagous to that of oxygen uptake discussed and formulated in Chapter 4. Both are processes of interfacial gaseous transfer.

They involve:

(i) a driving force proportional to the relative differences of concentration of the particular gas respectively in the sewage and in the sewer air. Such concentrations and their relative differences are best expressed in terms of molar concentrations, but in practice either liquid or gas phase concentrations are commonly employed; and

(ii) a series of resistances opposing the driving force. These resistances include those due, first, to the quasi-stagnant films of sewage and air just below and adjoining the sewage-air interface; and second, to the concentration of film-forming substances at the sewage surface.

The "stagnant" interfacial films of sewer air located above and adjacent to the sewage surface become relatively more significant as a resistance to gas transfer the more soluble the solute gas in water. In this connection hydrogen sulphide gas, although not very soluble, is nevertheless relatively much more soluble than oxygen gas.

These films are illustrated in Figs. 4.1, 4.2 (pp. 22-3).

The driving force for H_2S emission.—In considering the possible emission of sulphide from the sewage, only the free undissociated dissolved molecular hydrogen sulphide fraction is significant. The concentration of dissolved sulphide present in the sewage is normally only part of the total sulphide present (pp. 84-5). Only part of this dissolved sulphide normally forms undissociated free molecular hydrogen sulphide (see Fig. 5.1). The free dissolved H_2S concentration $[H_2S]$ which determines the effective H_2S emission rate is given by the relationships of equation [5.2].

$$[H_2S] = [S] \times R_{H_2S} \times F_{H_2S} \qquad [5.2]$$

where [S] is total sulphide concentration in sewage.

R_{H_2S} is the fraction of [S] which is soluble. This fraction, R_{H_2S}, normally ranges between about 0.4 and 0.6.

F_{H_2S} depends on the pH and on the value of the ionization constant, so that it is also temperature dependent, as shown in eqn. [5.3].

$$F_{H_2S} = \frac{1}{1 + K'[10^{pH + 0.00811(T_s - 68)}]} \qquad [5.3]$$

where K′ is the apparent ionization constant, having a value, Pomeroy[1], of about 2×10^{-7} in typical sewage at 25°C. The corresponding figure for 20°C would be 1.7×10^{-7}.

T_s is the temperature of the sewage surface in degrees Fahrenheit. Due to the cooling effect of surface evaporation, this may be taken as approximately equal to the wet bulb temperature corresponding to actual sewer humidity. In sewers where relative humidity approaches 100%, the sewage surface temperature will be very nearly equal to that of the body of sewage.

pH is the standard Sorensen scale value for the actual hydrogenion concentration of the sewage.

In the mathematical development which follows, the units of sulphide concentration employed are in terms of parts per million $\frac{v}{v}$ where applied to the gaseous phase, and in units of p.p.m (=mg/l) where used for the sewage fluid phase concentrations. The final formulations thus are expressed as partial pressures (directly proportional to molar fractions), of the gas phase, p.p.m. by volume and mg/1 for sewage constituents.

The concepts of Henry's Law are necessary to relate gas phase and liquid phase concentrations to the "driving forces" which govern rates of mass transfer across the inter-facial films. Henry's Law states that when a gas (or mixture of gases) remains in direct contact with a liquid solute, an equilibrium will be reached ultimately in which the concentration in the liquid phase of any particular gas considered will vary in direct proportion to its total partial pressure in the gas phase, thus

$$p_{H_2S}\ (\text{p.p.m}) = [H_2S](\text{mg/1}) \times H_{H_2S}\ (\text{p.p.m. per mg/1}),$$

where H_{H_2S} is the Henry's Law constant for H_2S, given nearly enough by the empirical expression

$$H_{H_2S} = \frac{T°F - 3.134}{0.2333} \qquad [5.4]$$

Here T°F is the temperature of the whole system.

Combining equations [5.3] and [5.4] and taking the value of K′ at 1.7×10^{-7}, we have

$$\frac{p_{H_2S}}{[S] \times 0.5} = \frac{T_s - 3.134}{0.2333} \times \frac{1}{1 + 1.7 \times 10^{-7}[10^{pH + 0.00811(T_s - 68)}]} \qquad [5.5]$$

where p_{H_2S} = equilibrium partial pressure of H_2S in sewer air expressed as atmospheres $\times 10^{-6}$ (approximately p.p.m. by volume) corresponding to the total sulphide concentration [S](mg/1) for a temperature T_s, the temperature of the surface of the sewage.

Note that the value 0.5 represents the factor R_{H_2S} of [5.2].

Values of the ratio $\frac{p_{H_2S}}{[S] \times 0.5}$ are plotted against pH and temperatures in Fig. 5.1.

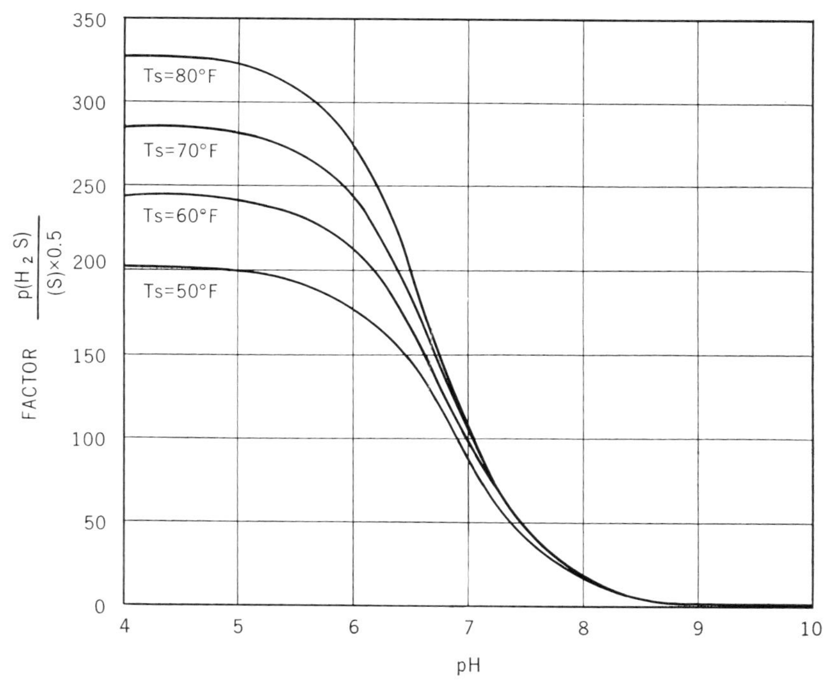

Fig. 5.1 Values of partial pressure of H_2S related to Henry's Law constant, sewage pH and temperature (equation 5.5).

It will be found that actual concentrations of H_2S in sewer air, for which the symbol p_{sa} will be used, are small compared to p_{H_2S} values, as taken from Fig. 5.1, because equilibrium values are seldom, if ever, approached under normal sewer conditions.

The driving force favourable to H_2S emission is then $p_{H_2S} - p_{sa}$, but for most practical cases $p_{H_2S} - p_{sa} \approx p_{H_2S}$ because p_{sa} is relatively small, though it may be more significant in some poorly ventilated sewers, chambers, etc.

Resistance to gas transfer.—The overall resistance to H_2S emission may be considered as being made up of a number of component resistances as was done when considering the process of surface absorption of atmospheric oxygen by flowing sewage (pp. 21-27). In dealing with oxygen absorption, two of the component resistances were considered in detail, first, that due to the quasi-stagnant surface film at the sewage surface and secondly, that due to film-forming, surface active substances also associated with the interface. Owing to the much greater solubility of hydrogen sulphide compared with oxygen, the resistance due to the "stagnant" air film adjoining the sewage/air interface also is brought into account. For this purpose and this particular case the sewage itself may be considered as a solid surface restraining the air-stream. As a reasonable working hypothesis, using the principle of added resistances as before, we may write:—

$$\frac{1}{K_{Sa}} = \frac{1}{k_{lS}} + \frac{1}{k_{Si}} + \frac{1}{k_{Sd}} + \frac{1}{k_{Sa}} \qquad [5.6]$$

where K_{Sa} = overall mass transfer coefficient for H_2S from body of sewage to body of sewer air in lb./(sq.ft.) (hour) (p.p.m. by volume H_2S in sewer air).

k_{lS} = mass transfer coefficient for H_2S through the liquid film below interface in lb./(sq.ft.) (hour) (p.p.m. by volume H_2S in sewer air).

k_{Si} & k_{Sd} = mass transfer coefficients of H_2S through films of surface active substances (indigenous substances and detergents respectively) in lb./(sq.ft.) (hour) (p.p.m. by volume of H_2S in sewer air).

k_{Sa} = mass transfer coefficient of H_2S through air film above interface in lb./(sq.ft.) (hour) (p.p.m. by volume of H_2S in sewer air).

Considering each of these component resistances separately, we have:

$\frac{1}{k_{lS}}$ is analogous to the term $\frac{1}{K_{O_2D}}$ as employed in the development of the oxygen uptake formulation, and with appropriate modifications, K_{O_2D} values may be used to determine k_{lS}. These include, first, an adjustment for the different coefficients of diffusion characterising hydrogen sulphide and oxygen, being respectively 0.000063 and 0.000081. The K_{O_2D} value therefore should be multiplied by 0.78; secondly, corrections for the different units employed, which for K_{O_2D} were gm. $\times$ 10^{-6}/(sq. cm.) (hour) (p.p.m. oxygen saturation deficiency). [The units used for sulphide emission in the present development are lb./(sq. ft.) (hour) (p.p.m. by volume in sewer air). The pound-foot units require a correction factor of 0.685×10^{-8} to be applied to the appropriate K_{O_2D} value]; and finally, due allowance for ionisation of H_2S in aqueous solutions. Transfer of sulphide through the quasi-stagnant liquid film by diffusion includes both molecular and ionised fractions, and since the overall equation [5.6] provides for molecular sulphide transfer only, correction becomes necessary by multiplying K_{O_2D} by the factor, $\frac{1}{F_{H_2S}}$ see [5.2].

Accordingly, we calculate

$$\frac{1}{k_{lS}} = \frac{F_{H_2S}}{0.78 \times 0.685 \times 10^{-8} \times K_{O_2D}} \qquad [5.7]$$

$$= \frac{1}{K_{O_2D}} \times F_{H_2S} \times 1.87 \times 10^{8}$$

The value of K_{O_2D} can be obtained from the procedures shown in the Calculation Chart, Fig. 10.16 (b), and that of F_{H_2S} from equation [5.3] (p. 34).

We may take $\frac{1}{k_{Sf}} = \frac{1}{k_{Si}} + \frac{1}{k_{Sd}}$, this resistance being regarded as made up of two parts, in the same way as for oxygen uptake (pp. 23 et seq.).

For oxygen transfer the respective relationships found were:—

$$\frac{1}{k_{Si}} = \frac{0.424\,[BOD_5]^{0.5}}{(K_{O_2D})^{1.37}}$$

and

$$\frac{1}{k_{Sd}} = \frac{3.3[BOD_5]^{0.33}}{(K_{O_2D})^2}$$

It is assumed that for hydrogen sulphide the same general forms of relationship would apply, but that transfer rates for sulphide through the films would be some multiple of the rates applying to oxygen. If this be so we could write, after adjusting for changes in units as above.

$$\frac{1}{k_{Sf}} = \frac{1}{0.685 \times 10^{-8} \times R_{os}} \left[\frac{0.424[BOD_5]^{0.5}}{(K_{O_2D})^{1.37}} + \frac{3.3[BOD_5]^{0.33}}{(K_{O_2D})^2}\right] \qquad [5.8]$$

where R_{os} = ratio of transfer rate (through surface active substance film) of H_2S to that of O_2. While no value for this ratio has yet been determined accurately, tests of overall emission rates of H_2S in particular sewers suggests that the ratio is approximately 2.5.

Taking the much denser sewage to act as an hydraulically smooth solid boundary so far as the air stream is concerned, the effective thickness of the gas-transfer resisting laminar film of air might be estimated, for example, using a modified form of the Blasius relationship, which states that for Reynold's numbers up to 10,000 the friction factor, $f = \frac{0.316}{(N_{Re})^{1/4}}$.

In a sewer, however, the hydraulic conditions relating to the turbulence pattern are complicated further, first, in that the air stream is bounded partly by stationary surfaces (walls) and partly by a moving surface (sewage surface), and secondly, that there are evaporative processes contributing to convective currents in the air stream.

Taking all this into account, it appears that the empirical equation derived by Powell and Griffiths (*35*) from experiments covering evaporation rates from flat surfaces in a wind tunnel, may be more appropriate, even though these experiments do not allow for the one moving boundary.

$$\frac{1}{K_{Sa}} = \frac{F_{H_2S}}{0.535 \times 10^{-8} \times K_{O_2D}} + \frac{1}{0.685 \times 10^{-8} \times R_{os}} \times \left(\frac{0.424 \times [BOD_5]^{0.5}}{(K_{O_2D})^{1.37}} + \frac{3.3 \times [BOD_5]^{0.33}}{(K_{O_2D})^2}\right) + \frac{m_A{}^{0.125}}{75 \times 10^{-8} \times (1 + 2.209 \times V_R{}^{0.85}}$$

Substituting R_{os} = 2.5 (see above) and simplifying:

$$\frac{1}{K_{Sa}} = \frac{1}{0.535 \times 10^{-8}} \times \left[\frac{F_{H_2S}}{K_{O_2D}} + \frac{0.132}{1} \times \left\{\frac{[BOD_5]^{0.5}}{(K_{O_2D})^{1.37}} + \frac{7.8[BOD_5]^{0.33}}{(K_{O_2D})^2}\right\} + \frac{m_A{}^{0.125}}{140\,(1 + 2.209 \times V_R{}^{0.85})}\right]$$

Incorporating these ideas

$$\frac{1}{k_{Sa}} = \frac{m_A{}^{0.125}}{7.5 \times 10^{-7}\,(1 + 2.209\,V_R{}^{0.85})} \qquad [5.9]$$

where V_R = velocity of air stream relative to sewage (or more correctly relative to sewage surface) which may be taken as travelling at about 1.1 times the mean velocity of the sewage, ft. sec.$^{-1}$.

m_A = hydraulic radius of air stream, ft.

Taking the separate resistances of equation [5.6] as given in the above equations [5.7] to [5.9] we have the following equation [5.10] for estimates of $\frac{1}{K_{Sa}}$

$$\frac{1}{K_{Sa}} = \frac{F_{H_2S}}{0.535 \times 10^{-8} \times K_{O_2D}} + \frac{1}{0.685 \times 10^{-8} \times R_{os}} \times \left(\frac{0.424 \times [BOD_5]^{0.5}}{(K_{O_2D})^{1.37}} + \frac{3{\cdot}3 \times [BOD_5]^{0.33}}{(K_{O_2D})^2}\right). + \frac{m_A{}^{0.125}}{75 \times 10^{-8} \times (1 + 2.209 \times V_R{}^{0.85}}$$

Substituting R_{os} = 2.5 (see above) and simplifying:

$$\frac{1}{K_{Sa}} = \frac{1}{0.535 \times 10^{-8}} \times \left[\frac{F_{H_2S}}{K_{O_2D}} + \frac{0.132}{1} \times \left\{\frac{[BOD_5]^{0.5}}{(K_{O_2D})^{1.37}} + \frac{7.8\,[BOD_5]^{0.33}}{(K_{O_2D})^2}\right\} + \frac{m_A{}^{0.125}}{140\,(1 + 2.209 \times V_R{}^{0.85})}\right] \qquad [5.10]$$

Using ϕ to represent all terms within the square brackets of [5.10]

$$\frac{1}{K_{Sa}} = \frac{1}{0.535 \times 10^{-8}} \times \phi \qquad [5.10a]$$

The reciprocal value K_{Sa} gives the overall transfer rates as follows:

$$K_{Sa} = \frac{0.535 \times 10^{-8}}{\phi} \text{ lb./(sq. ft.) (hour) (p.p.m. by vol. } H_2S \text{ in sewer air)} \qquad [5.11]$$

or

$$K_{lS} = \frac{0.535 \times 10^{-8} \times H_{H_2S} \times F_{H_2S}}{\phi}$$

lb./(sq. ft.) (hour) (1 p.p.m. dissolved H_2S in sewage) [5.12]

where values of H_{H_2S} and F_{H_2S} are given above (pp. 34,35).

The equivalent emission rates for hydrogen sulphide are:—

$$E_{H_2S} = \frac{0.535 \times 10^{-8} \times p_{H_2S}}{\phi} \text{ lb./sq. ft./hr.} \qquad [5.13]$$

$$= \frac{0.535 \times 10^{-8} \times H_{H_2S} \times [S]}{\phi} \text{ lb./sq. ft./hr.} \qquad [5.14]$$

where [S] = concentration of total sulphides in sewage in p.p.m.

The overall sulphide balance in sewage has been discussed in the foregoing. The overall balance has been formulated as shown in Table 5.1.

TABLE 5.1

Formulation of Sulphide Balance in Sewage

Item of Balance Sheet	Procedure for Reckoning
1. Sulphides entering with raw wastes.	No formulation feasible. Requires detailed consideration by chemical experts (p. 84).
2. Sulphide build-up due to generation within systems.	Equations [3.5] to [3.13], Chapter 3, pp. 15 et seq.
3. Sulphide oxidised in the sewage.	Not formulated. May be taken that build-up inhibited if EH remains high, say $-(>50)$mV.
4. Hydrogen sulphide emission rate.	Equations [5.13] and [5.14], p. 36.

If the overall balance operates to keep sulphide concentrations very low, then sulphide problems either will not occur, or will be almost negligible. The amount of sulphide corrosion damage must depend primarily on the amounts of sulphuric acid and sulphate formed on the unsubmerged sewer structures, which must in turn have some dependence on the emission rate of Item 4, Table 5.1. However another aspect of this balance is significant for sewerage systems planning, that of design. Acceptable levels of sewage sulphide concentrations are discussed in Chapter 10 (p. 111).

The sulphide balance in sewer air

The picture of the sulphide balance requires, for a satisfactory overall understanding, that the factors controlling the rate of formation of sulphuric acid or sulphate be taken into account, and, if feasible, formulated, as for the others. The overall balance in this connexion may be expressed thus:

Rate of build-up in sewer air = Rate of emission — Rate of uptake on sewer wall.

It is the rate of uptake on the sewer walls which controls the overall rate of corrosion.

Rate of build-up of sulphide in sewer air along sewer

Case 1: This is the simple model for which it is assumed that no absorption of sulphide occurs on the sewer walls above the sewage flow-line, and also that concentration of dissolved sulphides in the flowing sewage remains constant. Probably such circumstances never arise in practice, but their consideration simplifies the treatment of more complex cases (see Case 2, p. 39).

Consider a "slug" of sewer air of unit length (1 ft.) moving along a sewer, under the influence of ventilating forces, at a uniform velocity of V_A (ft./sec.) as shown in Fig. 5.2.

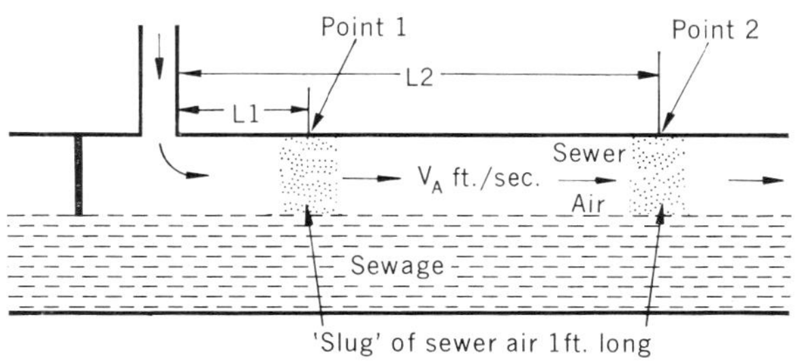

Fig. 5.2

If we consider what happens in a very short interval of time we may write, for the quantity Q_{H_2S} transferred from the sewage into the moving slug of air:—

$$Q_{H_2S} = \frac{A_{Sa} \times K_{Sa} \times (p_{H_2S} - p_{Sa})}{3600} \times dt \text{ lbs.}$$

or $$Q_{H_2S} = \frac{A_{Sa} \times K_{Sa} \times (p_{H_2S} - p_{Sa})}{3600 \times V_A} \times dL \text{ lbs.}$$

where A_{Sa} is the area of interfacial surface between slug of air and sewage, sq. ft. per foot run of sewer = b_{Sa} where b_{Sa} is the width of the sewage surface

K_{Sa} as previously, is the overall mass transfer coefficient in (lb.)/(sq. ft.) (hour) (per p.p.m. by volume of H_2S in sewer air) (eqn. [5.10]).

P_{H_2S} is the partial pressure of H_2S (p.p.m. by volume) in sewer air in equilibrium with the sulphide concentration in sewage [S] of equations [5.2] to [5.6].

P_{Sa} is the actual concentration of H_2S gas in sewer air p.p.m. by volume.

dt, dL are small increments of time or distance of travel of the slug of air related by $V_A = \frac{dL}{dt}$, secs, ft.

The transfer of the quantity Q_{H_2S} of sulphide to the sewer air will cause a change in p_{Sa} which can be expressed as follows:—

Q_{H_2S} is the volume of air over one foot run of sewer, cu. ft. × (Change in concentration of H_2S in air of slug in p.p.m. by volume) × 10^{-6} × (weight of one cubic foot of air in lb.)

$$= A_A \times \Delta p_{Sa} \times 10^{-6} \times 0.0895 \text{ lb.} \quad [5.15]$$

where A_A = cross-sectional area of air stream in sq. ft.

Equating the values of Q_{H_2S} as given by the two preceding equations.

$$\frac{A_{Sa} \times K_{Sa} \times (p_{H_2S} - p_{Sa}) \times dL}{3600 \times V_A} = 0.0895 \times 10^{-6} \times A_A \times \triangle p_{Sa}$$

$$\text{or } dL = \frac{3.22 \times 10^{-4} \times A_A \times V_A}{A_{Sa} \times K_{Sa} \times (p_{H_2S} - p_{Sa})} \times dp_{Sa} \qquad [5.16]$$

which represents the differential equation for the situation.

As the slug of air moves from position (1) to position (2) on Fig. 5.4, the total change in concentration to H_2S in sewer air can be obtained by integration of the preceding differential equation. This integration yields the following relationships:—

$$\triangle L = L_2 - L_1 = \frac{3.22 \times 10^{-4} \times A_A \times V_A}{A_{Sa} \times K_{Sa}} \times \log_e \frac{(p_{H_2S} - p_{Sa_1})}{(p_{H_2S} - p_{Sa_2})}$$

$$= \frac{7.24 \times 10^{-4} \times A_A \times V_A}{A_{Sa} \times K_{Sa}} \times \log_{10} \frac{(p_{H_2S} - p_{Sa_1})}{(p_{H_2S} - p_{Sa_2})} \qquad [5.17]$$

where p_{Sa_1} and p_{Sa_2} = concentrations of H_2S in sewer air at points (1) and (2) respectively.

Note that it is assumed that $p_{H_2S}/[S]$ is unchanged, which is usually close enough over sewer lengths of several thousand feet.

To illustrate the concentrations of sulphide that would arise along an unventilated sewer in the absence of any absorption of sulphides by sewer walls bounding the air-way, the following particular example has been computed using equations [5.10], [5.17] etc.

Sewer details are as follows:

Sewer diameter, 2.5-ft.; Total sulphides [S], 6 p.p.m.;

Dissolved Sulphides = [S] × R_{H_2S} = say 50% of total = 3 p.p.m.;

pH = 7; Factor $\frac{p_{H_2S}}{[S] \times 0.5}$, eqn. [5.5], Fig. 5.1

$= 110$

$\therefore P_{H_2S} = 330$ p.p.m. $\frac{v}{v}$

Proportional depth, 0.25, then $\frac{A_A}{A_{Sa}} = 2.11$

Velocity of sewage 2.0 ft./sec. (from Tables)

$K_{Sa} = 5.4 \times 10^{-8}$ lb./sq. ft./hr./p.p.m. equilibrium H_2S concentration in air, from equation [5.11]. [5.18]

To proceed with calculations for "unventilated" sewers it is necessary to consider the value of V_A, that is the absolute rate of travel of air along the sewer, $\frac{dL}{dt}$.

Two extreme cases arise, the first where there is practically no air movement. In such a case the P_{Sa} must eventually equal p_{H_2S} along the whole length of sewer, and equation [5.17] does not apply. For the case of a 4-ft. deep boat shaped sewer running about 1/3 full, for example, with sewage conditions similar to the above, the saturation value $P_{Sa} = p_{H_2S}$ would be practically attained after about one-half day.

Assuming a mean air flow of 1.7 f.p.s. for the above 2.5 ft. circular pipe operating as indicated above, equations [5.17] and [5.18] yield the data of Table 5.2.

TABLE 5.2

Case (1) Rate of build-up of H_2S in sewer air

Distance L_2	Downstream P_{Sa_1}	Time of Sewage Travel (hrs)	Downstream P_{Sa_2}
10,000	0	1.4	126 p.p.m.
25,000	0	3.5	231 p.p.m.
50,000	0	6.9	300 p.p.m.

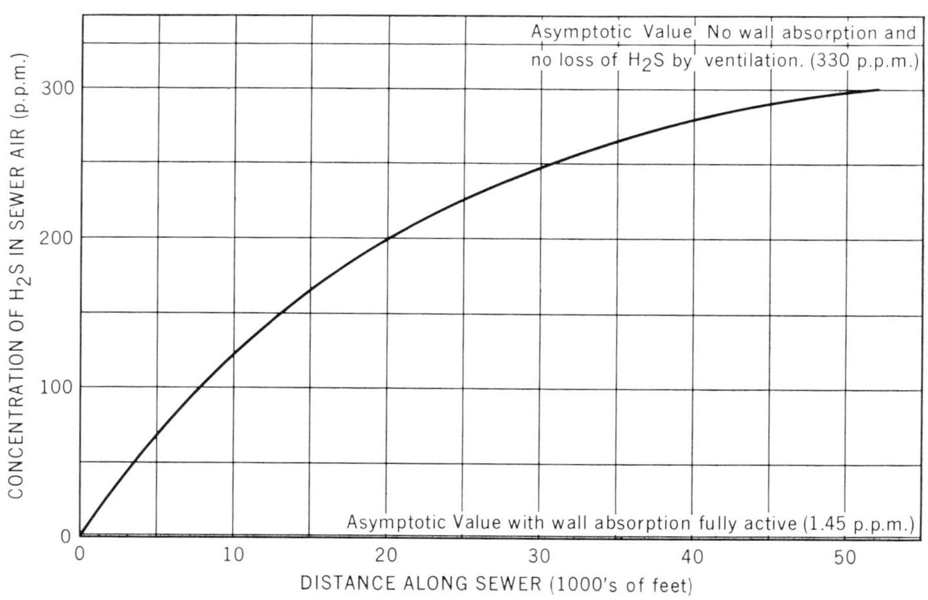

Fig. 5.3 Build-up of H_2S concentrations in sewer air.

Fig. 5.3 has been prepared on the same basis, viz, fresh air entering the sewer at the upstream end free of H_2S, while the sewage containing 6.0 mg/l of total sulphide, and 3.0 mg/l of dissolved sulphide, with pH value 7.0 and temp. 70°F.

Case 2: The second case considered here is the common situation where part of the H_2S emitted is simultaneously taken up and oxidised on the sewer walls, thus initiating and maintaining corrosion activity. The concentration of H_2S in the sewer air from time to time will attain a balancing concentration greater than zero, usually much less than would apply to Case 1. It is important to estimate what proportion of the H_2S emitted from the sewage is likely to be absorbed by the wall moisture, since this determines the rate of attack. However, in those cases where the walls are always sufficiently moist, and the supply of all nutrients always adequate, it seems reasonable to expect that the activity of acid-producing organisms on the walls will be self-adjusted to match the supply of H_2S available. In such cases, then, the controlling influences will be the aerodynamic conditions and the concentrations of H_2S in the sewer air.

Taking the concentration of H_2S in the sewer air ultimately reached at balance under the conditions stated, as equal to the total rate of H_2S emission from the sewage, for one foot run of sewer the reckoning is as follows:—

$$N_{H_2S} \times A_{aW} = E_{H_2S} \times A_{Sa} = K_{Sa} \times p_{H_2S} \text{ lb./hr.} \quad [5.19]$$

where N_{H_2S} = rate of absorption of H_2S by wall moisture, lb./sq. ft./hr.

A_{aW} = area of sewer air—sewer walls interface per ft., ft.^{2} = length of sewer perimeter above sewage, ft.

Furthermore, where

D_{H_2S} = H_2S gaseous diffusivity, assumed constant 0.62 ft.^{2}/hr.

d_{fW} = thickness of wall moisture film, ft.

p_{Sa} = local sewer air H_2S concentration, p.p.m.

H_{H_2S} = Henry's Law Constant for H_2S (p. 34).

$[H_2S]_W$ = concentration of H_2S in wall moisture interface, mg/1.

then also,
$$N_S \times A_{aW} \propto \frac{D_{H_2S} \times (p_{Sa} - H_{H_2S}[H_2S]_W)}{d_{fW}} \text{ lb./unit time}$$

$$d_{fW} = \frac{32.8\nu\left(2 \log \dfrac{r_s}{k_r} + 1.74\right)}{V_A} \quad [5.20]$$

$\nu = 1.63 \times 10^{-4}$ ft.^{2} sec^{-1} (kinematic viscosity)

k_r = surface roughness (say 0.001 ft. for new concrete)

$\dfrac{k_r}{2r_s}$ = relative (hydraulic) roughness

Writing $[H_2S]_f = H_{H_2S} \times [H_2S]_W$

$$N_{H_2S} = \frac{0.0905 \times 10^{-6} \times D_{H_2S}(p_{Sa} - [H_2S]_f) \times A_A}{32.8 \times \nu/V_A \times \left(2 \log \dfrac{d}{2k_r} - 1.74\right)} \text{ lb./hr.} \quad [5.21]$$

also $N_{H_2S} = K_{Sa} \times p_{H_2S} \times A_{Sa}$ as above.

Under all or most circumstances of active sulphide corrosion, it is reasonable to assume that the term $(p_{Sa} - [H_2S]_f)$ may be simplified to p_{Sa} only. The relative roughness $\dfrac{d}{2k_r}$ may be substituted in terms of the hydraulic radius m of the airway, $\dfrac{d}{2k_r} = \dfrac{2m}{k_r}$.

Furthermore, by analogy with heat transfer relationships, V_A probably is better expressed as $V_A^{0.83}$.

Equation [5.21] leads finally to equation [5.22] (note T = 68°F),

$$(p_{Sa})_{bal.} = K_{Sa} \times p_{H_2S} \times \left(2 \log \frac{2m}{k_r} - 1.74\right) \times \frac{A_{Sa}}{A_A \times V_A^{0.83}} \quad [5.22]$$

where $(p_{Sa})_{bal.}$ is the H_2S level in the local sewer air at which the rate of emission from the sewage is exactly balanced by the rate at which it is taken up by the wall moisture film. Note that p_{H_2S} is the equilibrium value for sewer air H_2S concentration corresponding to the effective proportion of molecular H_2S in the flowing sewage.

Applying the circumstances set out above for Case (1), which assumed NO uptake by the wall moisture film (p. 57), equation [5.22] gives the other extreme of Case (2), where emission is balanced exactly by the rate of uptake by the wall moisture film. It should be noted that neither extreme would normally occur in operating sewerage systems.

Taking the air stream velocity as averaging 1.7 f.p.s., for Case (2) the H_2S concentration at balance, computed from [5.22] is as shown in Fig. 5.3 (p. 38), viz. 1.45 p.p.m. at 50,000 ft.

Transfer of hydrogen sulphide from the sewer atmosphere to the surface of the sewer structure above sewage level

In the preceding discussion of the emission of sulphide through the surface of the flowing sewage into the sewer atmosphere, the next step in the corrosion process was also referred to hydrogen sulphide transfer from the sewer atmosphere to the structural surfaces above sewage level. Such surfaces are usually damp, because of condensate from the sewer atmosphere, infiltration of ground-water, or incidental wetting by the sewage, including capillary water drawn up the walls. The hydrogen sulphide transferred from the sewer atmosphere brings the necessary sulphur into the moisture film where it can be utilized by the acid-forming organisms to produce sulphuric acid. The degree of dampness must be sufficient, otherwise corrosion cannot proceed.

When a fluid is in a state of turbulent flow in a conduit, such as sewer air moving through a sewer, there will be a thin film of the fluid air adjoining the conduit wall or wall moisture in a state of laminar flow, so that the filaments or layers of flowing air move parallel to the wall without mixing eddies (see Fig. 5.4).

The hydrogen sulphide in the turbulent body of the air stream passes through this laminar layer to the moisture film on the wall by molecular diffusion, as opposed to eddy diffusion in the main air stream.

The thickness of the laminar layer for a fluid flowing in a conduit of circular cross-section is given approximately by equation [5.20] above.

k_r for new concrete is of the order of 0.001 to 0.002 ft. Assuming a value of 0.001 and an air velocity of 1.50 ft./sec., for a new concrete sewer of 2 ft. diameter, the thickness of the laminar film would then be about 0.33 inches, and for a new 10 ft.

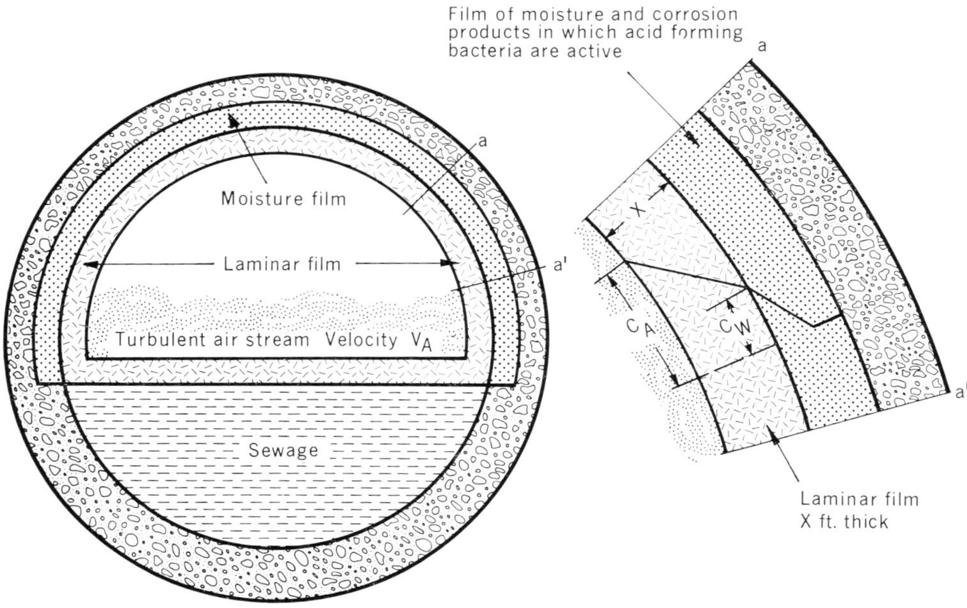

Fig. 5.4 Transfer of H_2S from sewer air to sewer walls (diagrammatic only).

diameter sewer, about 0.40 inches. As the sewer wall surface of the airway becomes rougher due to collection of sewage debris on the walls above water level during periods of extra high flow or due to corrosion having commenced, the laminar air film increases somewhat in thickness. Finally, when the dimensions of the roughness so caused become sufficiently great the larger projections will penetrate the laminar layer and be in direct contact with the turbulent body of the air stream.

Fig. 5.4 above shows diagrammatically the sulphide concentration gradients in the airway adjoining the wall after corrosion of the concrete has commenced but has not proceeded to the point where surface irregularities penetrate the laminar layer.

From equation [5.21] above we take

$$N_{H_2S} = 0.0905 \times 10^{-6}\, D_S \frac{(p_{S_a} - H_{H_2S} \times [H_2S]_W)}{d_{fW}} \qquad [5.23]$$

Where acid formation by bacteria in the moisture film is proceeding actively, it is practically impossible to estimate $[H_2S]_W$ values on a reasonable basis. In terms of liquid phase units of mg/1 the value must be very small because under equilibrium conditions 1 mg/1 of molecular hydrogen sulphide corresponds to about 300 p.p.m. in air. Hence with a concentration say 10 p.p.m. by volume of H_2S in the sewer air, for there to be any net transfer of sulphide to the wall film, however little, the concentration in the moisture must be less than 0.03 mg/1. Also from available knowledge of corrosion rates the concentration of molecular H_2S in the wall moisture must be less than 0.1 mg/1. A suitable assumption might be that $[H_2S]_W \times H_{H_2S}$ is related either directly to p_{S_a} or to $(p_{S_a} - H_{H_2S} \times [H_2S]_W)$.

Since the thickness d_{fW} of the laminar air film is inversely proportional to the air velocity it would be anticipated that the driving force, either p_{S_a} or $p_{S_a} - H_{H_2S} \times [H_2S]_W$ would cause a transfer of sulphide from the body of the air stream to the wall moisture film at a rate also directly proportional to the air velocity. However, in the case of heat transfer from the walls of a tube to the body of the fluid flowing in it, it has been found experimentally that the rate of heat transfer is proportional to $V^{0 \cdot 83}$ rather than V^1.

Taking $[H_2S]_W \approx 0$, or $[H_2S]_W \propto p_{S_a}$,

Then $$N_{H_2S} = k_{SW} \times p_{S_a} \times V_A^{0 \cdot 83} \qquad [5.24]$$

where k_{SW} is a constant of proportionality.

If the velocity of the air stream is increased to $V_A{}^1$ without significant alteration to the sewage geometry and velocity and therefore without alteration to the concentration of sulphide in the sewage, then the value of p_{S_a} is decreased, due to the extra air dilution, to $\frac{p_{S_a} \times V_A}{V_A{}^1}$ (However p_{S_a} will be decreased somewhat less than this ratio indicates because the extra velocity of the air relative to the sewage probably results in a slight increase in the rate of sulphide emission); p_{S_a} becoming $p_{S_a} \times \frac{V_A{}^1}{V_A}$ accordingly, whence the overall effect is negligible,

$$p_{S_a} \times \frac{V_A}{V_A{}^1} \times \frac{V_A{}^1}{V_A} = p_{S_a}$$

A change in the air velocity has practically no overall affect on the rate of transfer of sulphide to the wall moisture film because any consequential drop in sulphide concentration tends to be compensated by a corresponding reduction in thickness of the laminar film.

In line with the foregoing discussions it may be assumed for most practical circumstances that if the sewer is unventilated and the rate of bacterial conversion of H_2S to sulphuric acid in the sewer wall is not hindered by lack of moisture or nutrients (other than H_2S), or by lack of sufficient alkaline substances in the wall materials to maintain the reaction with wall acid (i.e. sewer corrosion), then the rate of H_2S

transfer to the walls above the sewage tends to balance closely the rate of its emission from the sewage. This rate of emission is proportional to the concentration of molecular sulphide in the sewage. It appears therefore that the rate of transfer of H_2S to wall moisture per unit area of wall depends only on:

(i) the concentration of molecular H_2S in the sewage,

(ii) the factors which directly influence its emission rate,

(iii) the ratio (width of sewage surface) : (peripheral length of sewer wall exposed to sewer air).

In terms of equation [5.19] the rate of absorption of H_2S by wall moisture is given by the following:

$$N_{H_2S} = K_{Sa} \times p_{H_2S} \times \frac{A_{Sa}}{A_{aW}} \text{ lb. Sulphur/ft.}^2\text{/hr.} \quad [5.25]$$

As roughness increases (pp. 39-40) and a significant proportion of the wall surface penetrates through the laminar air film, the transfer of sulphur also increases. This results in a lowering of air-stream sulphide concentration, and eventually a new balance is established with the net rate of transfer to H_2S to the walls likewise somewhat increased.

It might be noted here also that as acid production increases and the wall moisture pH is lowered, essentially all H_2S therein occurs in undissociated molecular form. It appears, however, that some elemental sulphur, and perhaps other oxidized forms of sulphur, may be formed and are to be found also.

Temperature effects.—The thickness d_{fW} of the laminar film is shown to be directly proportional to the kinematic viscosity ν of the air and equation [5.23] shows the rate of transfer of sulphide from the air stream to the wall to be inversely proportional to d_{fW}. The rate of sulphide transfer is also directly proportional to the coefficient of diffusion of H_2S in air D_{H_2S}. Both D_{H_2S} and ν are affected by temperature and to much the same extent so that over the range of temperatures normally encountered in sewers the effect of temperature on transfer rate may be neglected.

Significance of the sulphide balance in sewer air

The foregoing shows that at least for most cases the rate at which sulphuric acid formation proceeds on the structural surfaces of sewerage systems—practically synonymous with the rate of sulphide corrosion—is related directly to the rate of emission of H_2S from the flowing sewage.

This rate of emission depends, in turn, mainly on the overall sulphide balance within the flowing sewage. A factor of major importance for the sulphide balance is the overall oxygen balance discussed in Chapter 4. The design of sewerage systems therefore must take account particularly of these two items, along with, of course, the whole geography of the area to be sewered. The same two items are paramount factors in the control of sulphide problems in existing systems.

Variations in sewage sulphide concentration

Sewage sulphide concentrations vary with time, and usually vary from place to place in the sewerage system, and usually also according to changes in the weather and from season to season.

Variations with time

Whereas the presence of soluble sulphide is a necessary prerequisite for the occurrence of any sulphide problem, it is important to realize that sewage composition and character are not continuously uniform at any given place and therefore sulphide concentrations may vary considerably from time to time. This is illustrated in Figure 5.5 which shows variations in sulphide concentrations in a pumping main as determined from samples taken at five-minute intervals during mid-afternoon.

Also because of variations in the various circumstances which influence the concentration of sulphides in sewage, it has been found that systematic patterns of variation of sulphide concentrations occur in the sewage within the time of day, day of week, and season of the year. Some knowledge of the patterns of such variations is necessary for a proper overall appreciation of the sulphide problem in any particular sewer, and is essential as background for the proper and economical planning of sulphide surveys, and for the evaluation of any measurements of sulphide concentrations, whether as a series or in isolation.

The following examples illustrate what is to be expected in systems which have been designed to standards hitherto generally considered as representing good practice, but which did not incorporate any measures directed towards sulphide concentration control.

Effect of time of day

Consideration under this sub-heading is restricted to observations of sewage from areas used mainly for residential purposes, and the following two cases have been selected as fairly typical examples. Hour-to-hour concentrations of sulphide as measured are plotted in Figs. 5.6 and 5.7.

It will be noted that peak concentrations of sulphides occur near times of maximum flow, when BOD is usually also at its maximum. Peak concentrations in the sewage however do not necessarily result in peak concentrations of H_2S in the sewer air.

Effect of day of the week, including some industrial wastes influence

In most larger cities and towns there is much less industrial and commercial activity at the weekend

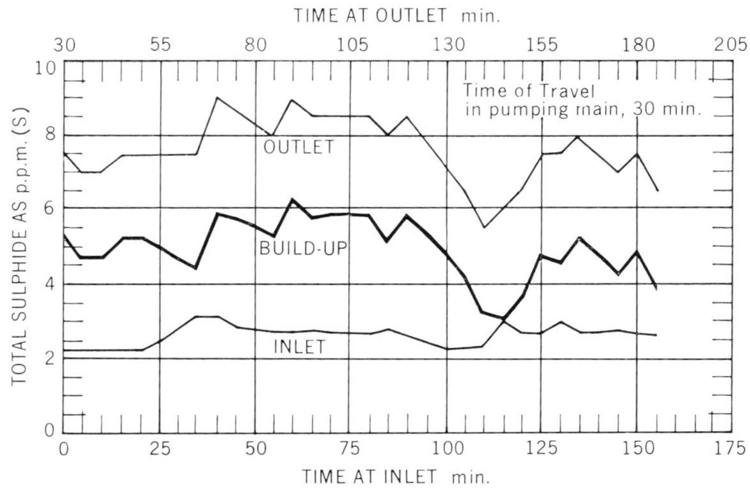

Fig. 5.5 Variation of sulphide concentrations in samples taken from a pumping system at intervals of 5 minutes during mid-afternoon.

(Saturday and Sunday) than on weekdays, and if industrial wastes contribute a significant amount to sulphide concentrations in the sewage, it would be anticipated that this would be indicated by lower average sulphide concentrations, e.g. on Sundays. However, the industries which contribute most sulphides may be among those which operate continuously, and in such cases the sulphide concentrations conceivably could be higher on Sundays, not lower.

Figures available from the records of one Australian city having a population of approximately 1,000,000 at the time of the survey are given in Table 5.3. These figures apply to the sewage at the downstream end of the city's main outfall sewer.

TABLE 5.3 SULPHIDE CONCENTRATIONS FOR DIFFERENT DAYS OF THE WEEK

(Averaged over 12 months)

Day of week	Sulphide Concentration (as mg/l H_2S)	Hundreds of pounds of Sulphide per day
Monday	6.8	37
Tuesday	7.6	41
Wednesday	7.9	42
Thursday	8.0	41
Friday	8.1	43
Saturday	6.9	34
Sunday	4.9	21
Average for weekdays	7.7	41

The difference in average sulphide concentration on weekdays and on Sundays is seen to be 2.8 mg/l or 36% of the weekday average. For this particular city at the time of the survey industrial wastes were therefore responsible for adding 57% to the total average concentration of sulphide compared with the basic residential sewage. Allowing for the increase in volume also, the total additional amount of sulphide due to industrial wastes is about 92%. This does not mean that industries discharged the equivalent of these percentages directly as sulphides but that also in some other way, such as by increasing the BOD, temperature, or concentration of sulphates or surface active substances, they have been responsible for increased sulphide content. This example emphasizes the possible special significance of industrial wastes for sulphide problems, particularly when it is realized that the figures given apply only to the overall effect on a large system. For sewers receiving larger relative volumes having an adverse influence on the sulphide problem, the effects are likely to be much more significant. Many industrial wastes are not harmful in this way, but may be even beneficial as diluents.

On the other hand, industrial wastes, or trade wastes as they are frequently termed, are very diverse in nature, also in times and rates of discharge, both as between and within individual establishments. Therefore no generalized pattern of the effect of trade wastes can be postulated. Changes in the patterns of such discharges and their composition may be very significant, however, and some assessment of the possibilities should always be attempted and allowed for.

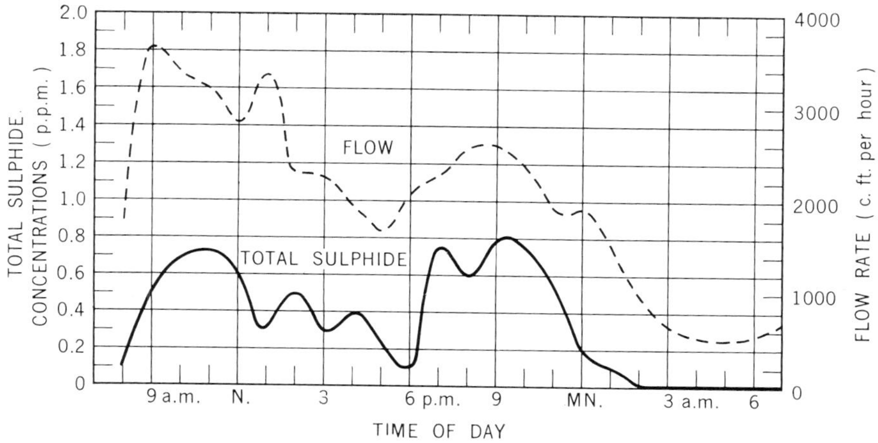

Example No.1: Sewer 27" Diameter, Serving Population of approximately 13000 persons and 1400 Acres. Temperature of Sewage 65.3°F

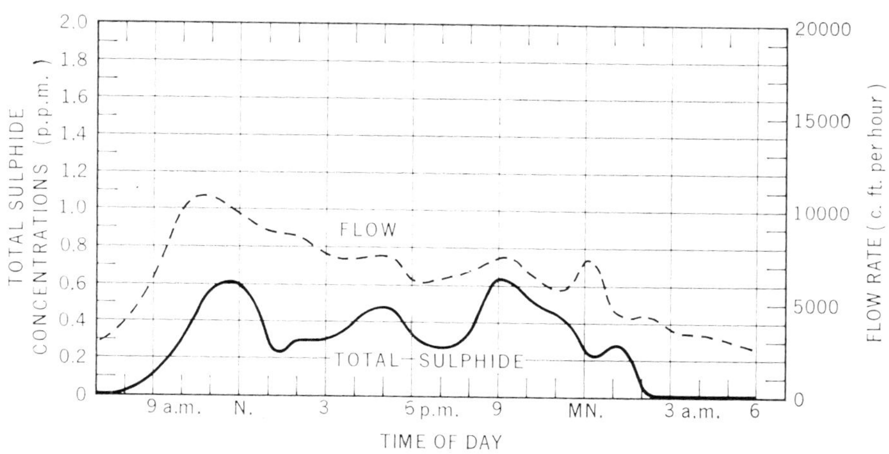

Example No.2: Sewer 2'4" × 3'6" Ovoid, Serving Population of approximately 42000 persons and 3700 Acres. Temperature of Sewage 68.7°F

Figs. 5.6 and 5.7 Variations of sulphide concentrations with flow during 24-hour periods, examples 1 and 2.

Industrial wastes may have detrimental or beneficial effects on sulphide conditions in local or downstream parts of the sewer system. Some wastes produce toxic conditions in sewers dangerous to maintenance personnel.

Effect of time of the year or seasonal effects

For the same observation point as referred to above, values of sulphide concentrations are available for one day in each month over the period 1956-61. The average value for each month, together with temperature, is shown in Fig. 5.8, which clearly indicates a seasonal effect.

Effect of rain periods

Even in modern sewerage systems designed on the basis of separate drainage systems for stormwater, a considerable proportion of stormwater commonly reaches the sewers during and following precipitation. The resulting increased flow first produces a flushing effect. As a result septic detritus is scoured from sewers and causes first increased sulphide concentrations then later simple dilution, during which concentrations decrease. These effects are illustrated in Fig. 5.9. Experience indicates that reduced sulphide concentrations commonly are prolonged even after flow and BOD return to their normal dry-weather values. This reduction in sulphides is related primarily to reduced slime generation activity, and also to oxygen balance factors being more favourable, the effects lasting for some days at least, or even several weeks.

Rainfall is an intermittent phenomenon and though the proportional duration of rainfall may determine to what extent the overall average sulphide concentration is affected, during longer periods without rain extreme sulphide concentrations can develop. It is during such periods that the rate of attack is highest.

Experience has shown that combined sewer systems, that is, systems designed for stormwater as well as polluted drainage, commonly are free of corrosion problems, especially in areas of more uniform rainfall incidence. In hot dry climates, this may not be so.

Wherever sewerage sulphide problems are under study it is important to understand the effects of rainfall on the particular sewerage system in question. This is important for the planning and conduct of investigations, sulphide surveys, etc., and for the interpretation of all relevant data.

Influence of degree of building development

It is common practice to design sewers to provide for a much greater degree of development in their catchments than exists at the time of construction. Consequently in their early life the dry weather flows and velocities normally are significantly less compared to those which will occur in later stages of development. It has been found that sulphide concentrations in sewage often are considerably greater during the early life of a sewer than they are when development has increased and the sewers are operating with increased flows and velocities, presumably closer to those used for design purposes. It is therefore essential when designing for sulphide control to give due consideration to the probable rate of catchment development and the corresponding increases of flow and velocity.

The increase in sewage flows and decrease in total sulphides concentrations experienced in a system during sixteen years after commencement of operation is shown in Fig. 5.10. The main sewers of this system were designed with fairly flat grades and some deposition of silt was experienced during the early years.

The decrease in sulphides during the first eight years is particularly noticeable.

Variations with location in the system

The following statement is consistent with general Australian and overseas experience.

The dissolved oxygen in upper reaches of the system persists generally in reticulation sewers and sometimes in small branch mains though at diminishing concentrations. The distance downstream for which some dissolved oxygen persists depends upon the temperature, sewer grades, and the height and frequency of vertical drops, etc., which in turn depend generally on the topography of the area. Concrete pipes have been in use in Australia for construction of reticulation sewers for about fifty years, but there has been little evidence of sulphide corrosion in such pipes except where they are laid on flat grades or where they carry discharges from certain industrial wastes, or septic sewage from pumping systems, etc. It would appear, therefore, that the zones of dissolved oxygen, or more correctly the zones wherein significant sulphides are absent, normally extend to the downstream limits of the reticulation sewers provided these are designed to normal standards (excepting of course where special circumstances may prevail).

As reticulation sewers up to nine inches in diameter comprise at least about 70 per cent. of the total length of sewers even in larger sewerage schemes (and roughly 60 per cent. of the total cost), this conclusion is of considerable importance.

Downstream of the dissolved oxygen zone there is little or no build-up of sulphide concentrations when the grade of the sewer mains is sufficiently steep, but where the grades are flat and velocities low, serious build-up of sulphides may occur.

Pumping systems (or other pressure mains such as long siphons) can be a very important source of sulphides. Experience has shown that many serious sulphide problems originate in pumping systems.

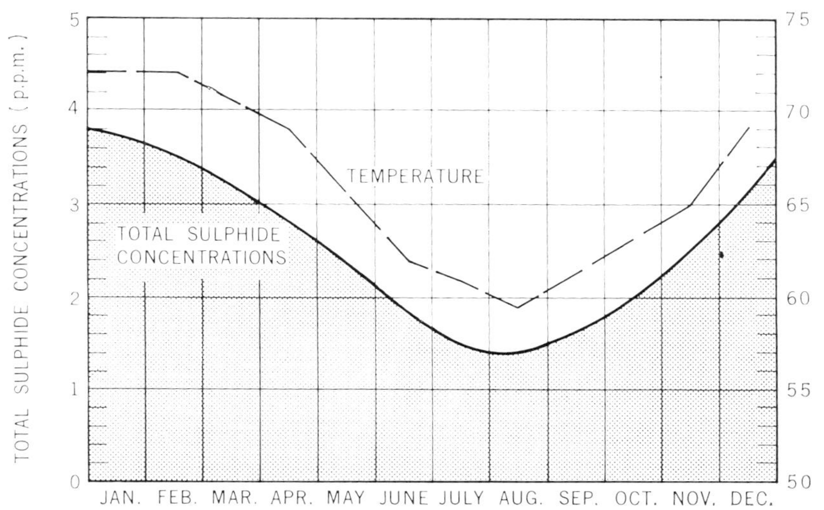

Fig. 5.8 Seasonal variations in sulphide concentrations.

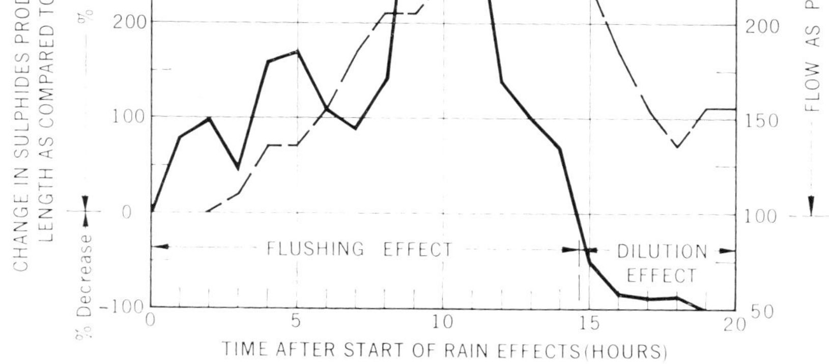

Fig. 5.9 The effects of rainstorms on sulphide concentrations.

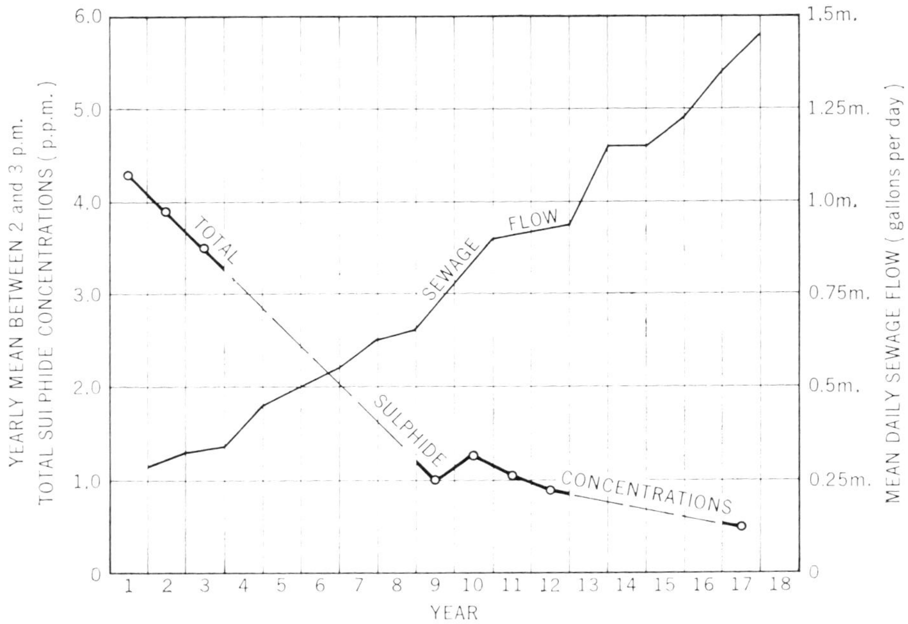

Fig. 5.10 Effects of increased sewer flows on sulphide concentrations, being a result of building development.

The problems caused by high sulphide build-up in one part of the system may manifest themselves in parts of the system downstream from the actual site of the sulphide generation.

Detention tanks may increase sulphide concentrations significantly also, unless operated and maintained so as to obviate substantial sulphide generation.

Vertical drops and other features causing splashing or increased turbulence may be beneficial or otherwise according to whether they occur in the dissolved oxygen or dissolved sulphide zones. This emphasizes the importance of determining, for each particular area, just where the boundary between the zones occurs.

The overall sulphide balance in sewers

It is clear from the foregoing discussions of this and earlier chapters that the presence or absence of sulphides and other sulphur compounds from place to place in any particular sewerage system, and any consequential corrosion problems, are the result of an overall balance, again usually different from place to place, of relatively complex situations; which generally may be summarized as follows:

Residual Sulphide = (Sulphide taken in + Sulphide generated) — (Sulphide lost).

OVERALL INPUTS

(i) Sulphide brought in by wastes, de-oxygenated groundwater, etc.; this sulphide tends to be evolved (in small part) from the sewage to the sewer air.

(ii) Sulphide generated in slimes and diffusing into the flowing sewage; this also is partitioned between sewage and sewer air.

OVERALL LOSSES

Sewage

(i) Sulphide precipitated by heavy metals in sewage or trade wastes.

(ii) Sulphide oxidised by sewage dissolved oxygen, especially in presence of metal oxides.

(iii) H_2S lost by emission (evolved from solution) to sewer air.

Sewer Air

(i) H_2S carried out of system by ventilating air.

(ii) H_2S dissolved and held on sewer walls.

(iii) H_2S oxidised on sewer walls and held as corrosion products.

CHAPTER 6

VENTILATION IN RELATION TO THE SULPHIDE PROBLEM

Introductory

The volumes of sewage discharged into the sewers vary so much from time to time throughout each 24-hour day that it is necessary to make provision for the movement of air into and out of sewers, tanks etc. to ensure free movement of sewage. Provision also must be made for the access of personnel as may be required from time to time. Three aspects arise, all certainly inter-related, but nevertheless different, and demanding separate consideration because each generates specific parameters respecting both quantitative requirements and design details. These three aspects are considered in the following pages under headings (i) physical aspects, (ii) chemical and biological aspects and safe working, and (iii) ventilation in relation to sulphide corrosion control and control of odour nuisance.

Physical aspects of sewerage ventilation

Each arrangement of sewers and pumping stations calls for corresponding ventilation arrangements. Among the many determining factors are environmental conditions of temperature and pressure, pipeline and channel gradients and sizes and the nature of the waste flows and their temporal variations. Safety and odour problems are often paramount considerations for which the physical aspects play a very direct part.

Broadly, the ventilation of sewers or other sewerage structures may be natural or artificial.

Natural ventilation is that where movement of air into and along sewers etc. occurs as a consequence of natural barometric or gravitational forces, or of the "drag" exerted by the flow of the sewage, all without any application of mechanical power.

Artificial, mechanical (or forced) ventilation is that where air is forced through or drawn along sewers etc. by the application of mechanical power-driven equipment (usually an electrically driven air-moving plant operating at relatively low head).

At any given time, the air-way of any sewer may be taken as being enclosed by incompressible peripheral boundaries, including the surface of the sewage. However, allowance must be made for the movement of the sewage surface boundary relative to the sewer walls. This situation is summarized in the following diagram of air-flow in a sewer(*36*).

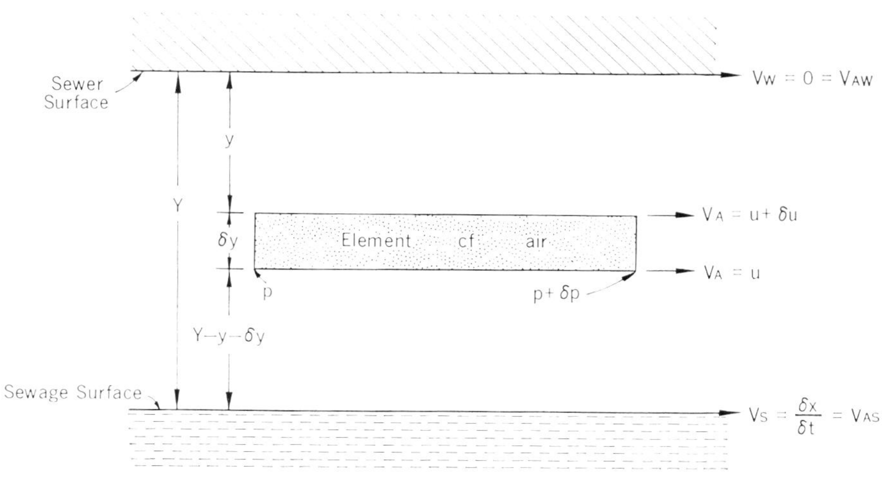

DIAGRAM REPRESENTING CONDITIONS GOVERNING AIR-FLOW IN SEWERS.

$$u = \frac{\delta p}{\delta x} \times \frac{y^2}{2\mu} - \frac{y}{\mu} \times \frac{\delta p}{\delta x} \times \frac{Y}{2} - \frac{\mu V}{Y}$$
$$= \frac{\delta p}{\delta x} \times \frac{(y^2 - y \times Y)}{2\mu} + \frac{V_S \times y}{Y} \quad [6.1]$$

where p is piezometric pressure, taken to be constant across the flow; μ is Viscosity of sewer air; v_S, V_{AS} and V_A are respectively the velocities of sewage surface and sewer air; V_{AS} being the air velocity at the surface of the sewage.

This relationship includes the combined effect of barometric gradient and sewage drag. In the absence of any barometric gradient along the sewer, a circumstance which is often approached closely enough, then

$$\frac{\delta p}{\delta x} = 0 \quad \text{and} \quad u = V_S \times \frac{y}{Y} \qquad [6.2]$$

This last relationship indicates that the effective air stream velocities due to "drag" vary linearly, between a value equal to the sewage surface velocity (about 1.1 × mean sewage velocity) adjacent to the surface of the flowing sewage, and zero velocity adjacent to the non-submerged sewer wall surfaces, unless assisted in some way by piezometric pressure gradients induced naturally or artificially along the sewer airway.

Natural ventilation of sewerage systems

The natural forces which may produce movement of fresh air into and along a sewer at any given time include the following:

Change in barometric pressure along sewer

A glance at any daily weather map will show that barometric pressure varies from place to place. Barometric pressure is transmitted to the sewer atmosphere and if the equivalent pressure gradient is significant, movement of the sewer air will be affected accordingly. One millibar potentially equals about 300 f.p.m. of air flow, but usually this is greatly reduced by friction losses at entries and exits, manholes, etc. (see also p. 143). Equation [6.1] applies.

Wind velocities past induct and educt vents

Wind velocities may be taken as proportional to the logarithm of the height above ground level, although the pattern may be modified by the presence of buildings or other local topographical features (such as cliffs, hills etc.). Based on this general law of increased wind velocity with greater height, the production of air pressure gradients along sewers by the alternation of high and low vents has been common practice (especially in older designs). Nevertheless, air-flow data for such systems have not been documented.

Form of vents

At least partly related to wind velocities and vent locations are the factors concerned in the design of vent-stacks and ducting, including special shapes for intakes and outlets. Various designs for cowls have been marketed, the makers claiming increased suction or pressure effects for their products. Some types have swinging cowls of the shape used for ship ventilators, the air opening always facing the wind for inducts and trailing it for educts, the direction relative to the wind being controlled by air vanes.

Other designs use no moving parts but rely on fixed vanes to deflect passing air to give maximum suction upwards or pressures downwards at vent stacks. The more reputable makers should have aerodynamic test data available respecting their products.

For emergency ventilation, arrangements of canvas sails erected on frames are used to direct wind movement into and out of ordinary manhole openings.

Sewage "drag"

Because of the frictional drag of flowing sewage on the sewer air above there is always a force component tending to move the air in the same direction as the sewage. The magnitude of this effect depends directly on the velocity of the sewage and since this varies throughout the day for most sewers there will be corresponding variations in the effect of the drag. Equation 6.2 applies in the absence of any significant barometric gradient along the sewer.

Rise and fall of sewage

At times of high sewage flow the sewage occupies a greater proportion of sewer volume than at times of low flow. As a consequence some of the air in the sewer must be displaced and find its way out of vents. Conversely at change to lower flows, fresh air is drawn into the sewers. Although these volume changes may not amount to much, averaged over the whole 24 hours, their effect is by no means negligible.

Relative density of sewer air and outside air

When the sewer air is of less density than the outside air the effectiveness of high educt vents is increased. The reverse holds when the sewer air

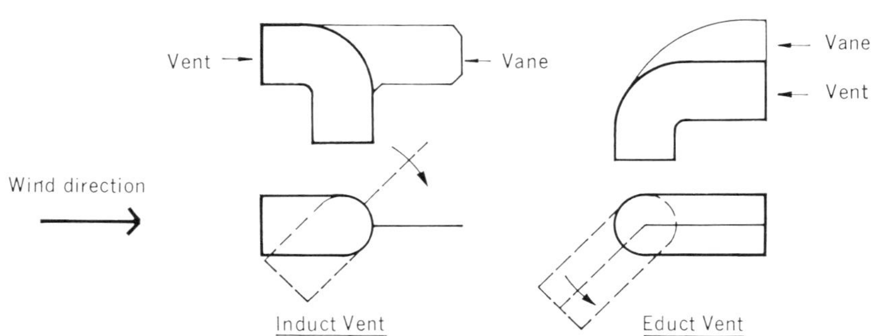

Fig. 6.1 Self-adjusting ventilator systems.

density is greater than that of the outside air.

The density of sewer air is dependent mainly on its temperature and humidity, the density falling with increased temperature or humidity. In winter during the daytime, sewer air temperatures tend generally to be greater than those of the outside air, but in summer the reverse may hold, while on a cold night the sewer air may be less dense than the outside air. As a consequence, natural ventilation may be less effective in summer during the daytime periods when sulphide corrosion problems are likely to be more severe.

Of these various forces tending to produce movement of the sewer air, that which acts continuously, and in the one direction, is the drag with the surface of the flowing sewage. As a consequence except when there are forces of corresponding relative magnitude acting to oppose this drag effect, the direction of air flow will be commonly downstream with the sewage. However, it is not to be expected that all of the above factors will operate favourably in concert, therefore in practice the rate of natural ventilation must vary considerably, certainly in amount, and sometimes in a contrary direction.

Probably the best arrangement is to concentrate natural ventilation by ensuring a multiplicity of inducts and educts opening more or less at a similar height above ground, for example by venting the sewers freely through each house connexion, i.e. eliminating all or most boundary traps. This would assist the air to be moved freely by the sewage flow. However, overall arrangements might need to provide specially for breaking up the system into zones, for example, by the use of brattices, or using specially designed manholes, so that the air is completely renewed in each zone separately. Then it would be feasible to reckon on a minimum continuous downstream movement of air so as to achieve at least a certain "number of changes per hour" of the air contained in the various sewer lengths. This normally should be not less than 1 to 4 changes per hour.

Fig. 6.2 illustrates the type of system once common practice in ventilating main or outfall sewers, based on the erection of special vents at intervals along the sewer; alternate vents being designed to act respectively as inducts and educts by virtue of differing heights, perhaps assisted by special cowl designs. In such a system the condition may arise in which the pull of the educt vents in each alternate section of sewer tends to balance out the effective drag of the flowing sewage. As a consequence, the air in each alternate section would be stagnant most of the time, producing an undesirable set of conditions. Such effects can be prevented by flexible bulkheads, as shown in Fig. 6.3. An advantage of providing for the air to flow always in the same direction as the sewage is not only that the drag of the latter (continuously acting downstream) assists any natural or artificial draught, but also that the average velocity of the air relative to sewage is lower, resulting in a lower emission rate of sulphides and a lower rate of evaporation of moisture from the sewage surface.

In reticulation sewers the practice of omitting boundary traps, now current in most localities, should allow admission of air to the sewers through the house connexion system vents. Some air is also introduced to the system as air entrained in sewage by the splashing which takes place at sewerage fittings.

In main or branch sewers with infrequent connexion of house branch or reticulation sewers, provision of vents is generally desirable.

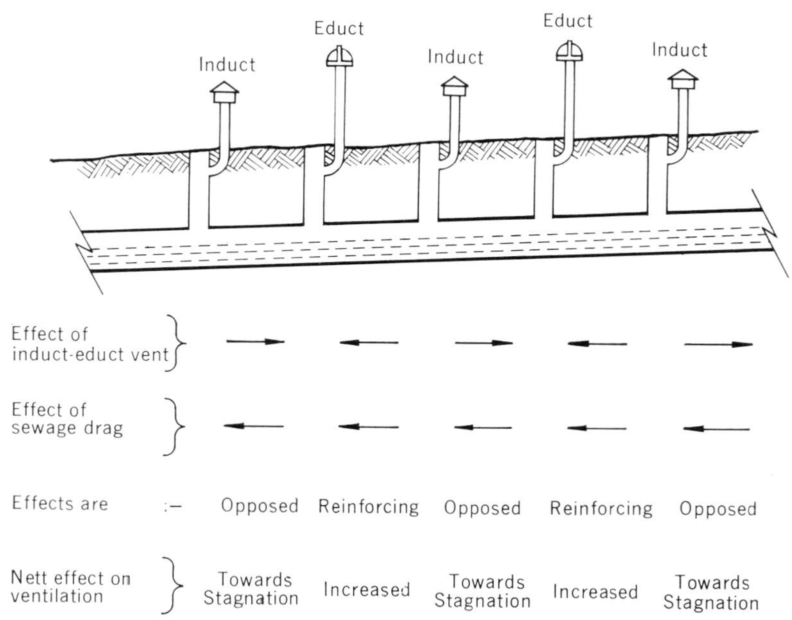

Fig. 6.2 Simple uncontrolled alternating sewer ventilation system.

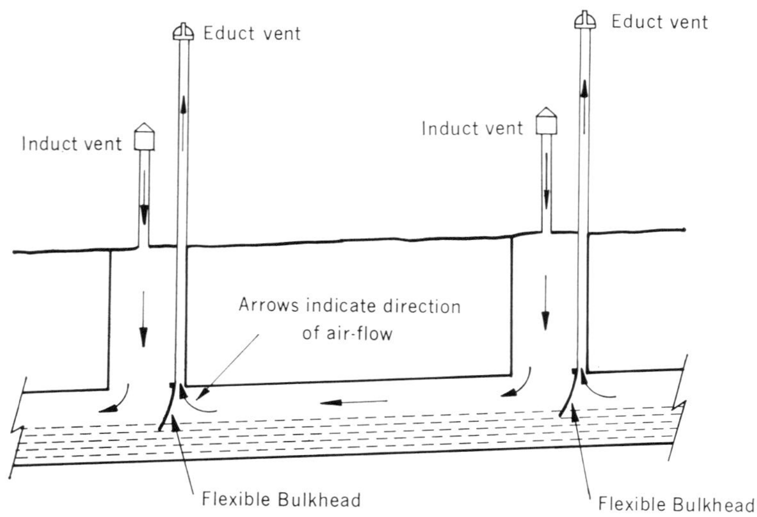

Fig. 6.3 Controlled air-flow sewer ventilation.

Normally oxygen is present in the sewer atmosphere of free flowing sewers, but conditions can occur when the oxygen content of the air may be seriously depleted. Two examples of conditions where oxygen depletion may occur are shown in Fig. 6.4 which refers to the upstream approach of an inverted siphon and in Fig. 6.5 which refers to conditions that can arise when ponding in a gravity sewer results in complete filling of the sewer (p. 52).

Under these conditions, educt vents should be provided, not only to ensure safety of personnel working in the sewer and to obviate local corrosion, but also so that the ventilation of the upstream sections of the sewer may function effectively, providing an outlet for the air moving downstream under the influence of the surface drag of the sewage.

Practical experience with natural ventilation

The London County Council in 1933 reported observations made along a two-mile length of 2-ft. to 9-ft. diameter sewer. Under natural ventilation conditions sewer air flowed out of openings at road level at average velocities between 200 and 400 feet per minute (f.p.m.) consistently. No barometric or temperature data were reported, but one Australian sewerage authority made observations of natural ventilation in several of its larger outfall sewers over a period of several years. The following generalities were derived from these studies, which included observations of barometric and other natural conditions as well as corresponding data for the sewer air.

Natural conditions vary from time to time during any one 24-hour period. Mostly, conditions overnight between about 9 p.m. to 10 p.m. and 6 a.m. to 7 a.m. tend to be relatively steady. The outlflow of sewer air during this period tends to be consistently high, in a relative sense, reaching say 500 f.p.m. or even 700 or 800 f.p.m. when temperature conditions are most favourable. This is due mostly to appreciably lowered atmospheric temperatures which develop following the hours after sundown, compared with sewer air temperatures which change little, if at all, throughout the 24 hours. The morning hours, say from 7 a.m. up to about midday, or even later, would be expected on a similar count to be rather variable by comparison and measured sewer air upcast velocities show that this is characteristic.

Practical data do not show any simple correlations of sewer air flow with barometric conditions, however, and can be related only in a very general way (see pp. 146, 147). Likewise, the relative temperatures of atmospheric and sewer air do not provide simple correlations from which natural sewer air flow could be reckoned. Changing patterns of atmospheric/sewer air temperatures and humidity acted variously to confuse the overall effects on upcast velocities. Much more data would need to be available for analysis if an attempt were to be made to establish useful correlation.

The data of Table 6.1 are taken from records of the Australian investigations referred to above. In the case of Sewer No. 1, the outflowing sewer air was discharged through a high ventstack with outlet level about 260-ft. above the soffit of the sewer itself. Of this height 160-ft. is the actual depth of the sewer below ground level. Sewer No. 2 was located about 180-ft. below ground and the ventstack opening was about 30-ft. above ground.

TABLE 6.1

Typical Natural Ventilation Patterns

(during Autumn in Eastern Australia)

Outfall Sewer	EXTERNAL WIND CONDITIONS			SEWER AIR RELATIONSHIPS		
	Direction	Velocity m.p.h. = f.p.m.	Temperature °F.	Temperature °F.	Velocity f.p.m.	Volume c.f.m.
No. 1	S (across sewer)	10 = 900	62	66	100	4600
	do:	10 = 900	60	66	150	6500
	W (down sewer)	7 = 600	58	65	40	1700
	do:	18 = 1600	56	67	190	8500
No. 2	S-W (45° down sewer)	4½ = 400	59	63½	640	TIME 4 p.m.
	N-W do:	15/20 = 1300/1800	53–56	61 to 62	400–850	6 a.m.–2 p.m.
	N-W do:	10/15 = 900/1300	48–49	62	600–700	2 p.m.–10 p.m.
	W (down sewer)	14 = 1200	62	65	500	3 p.m.
	W do:	2½ = 220	59	64	280	4 p.m.

A concept developed during these investigations was that of the "airosur", an abbreviation for "air-oversurface", the rate of ventilation expressed as c.f.m. of air flow per sq. foot of sewage surface ventilated. It was reckoned from the data collated that sewer conditions probably would be satisfactory if natural ventilation resulted in about 0.05 airosurs. This corresponds, roughly, to from about 8 to as little as ½ changes of sewer air per hour, more or less, for small or large sewers respectively from 9in. up to say 12ft. diameter. The values 0.05 airosurs and 2 changes of air per hour are about equivalent for sewers of about 5ft. to 7ft. diameter.

Mechanical ventilation

Mechanical ventilation may be designed to achieve a substantially constant rate of airflow, which can be predetermined and also is constant in direction. Where it is desirable to prevent polluted air from a main sewer passing through or into reticulation lines or escaping from manholes, mechanical ventilation should be by the exhaust system, with the educt taken preferably from the main sewer. Airflow from branch sewers, or from an adjoining length of the sewer being ventilated, can be prevented from entering the mechanically ventilated section by installing in such branch sewer a hinged or flexible bulkhead which closes the air-way without significantly impeding the sewage flow at any stage, or by installing a water trap. An educt vent should be provided at the upstream side of the bulkhead or trap and an induct at the downstream side. With an exhaust system, the fans and ductwork should be constructed of corrosion-resistant materials, or allowance made for relatively frequent replacement of corroded parts.

Many cases of severe corrosion occurring in the sewage wells at pumping stations have been reported. Although mechanical ventilation alone may not necessarily prevent corrosion, its provision is desirable, both to control the rate of corrosion and as a safety measure for the protection of personnel who need to enter the sewage wells for maintenance work purposes. One authority found that hydrogen sulphide corrosion of the walls and roof of a sewage wet well was effectively controlled by the provision of mechanical ventilation which properly distributed inlets ensuring five changes of air per hour. This is about the upper limit of acceptability for satisfactory working comfort under routine conditions. The control of all relevant problems, which includes problems of toxicity and odour nuisance, usually is not practicable unless mechanical means of ventilation are available powerful enough to ensure that the air moves always in the same direction at a rate sufficient for most eventualities. However, good practice necessitates also that the working conditions engendered be held within suitable limits so that any discomfort caused by air movement in working areas will not be intolerable. Commonly, this limits mechanical ventilation to comparatively short zones of sewers, for any one fan installation.

Rates of ventilation may be calculated using various relationships set out in pp. 142 et seq.

Ventilation for special cases

Special provisions must be made to counter the dangers inherent with ponded sewage. Figs. 6.4 and 6.5 illustrate two such cases. The former is that of a type of structure normally used only where unavoidable, but for which special ventilation provisions would need to be made during design. The latter case is more likely to be fortuitous. If it also should prove unavoidable, special temporary ventilation arrangements should be made, e.g. by introducing an adequate flow of compressed air so as to prevent dead-end effects immediately upstream of the pond, pending some permanent provision.

Other special circumstances arise in connexion with the discharge of ventilating air, particularly in situations where the sewage is septic or where trade waste discharges cause unpleasant odours to enter the sewer air.

Aspects of toxicity and odour nuisance

Whilst there are indications that adequate ventilation contributes to the decrease in the intensity of sulphide problems in sewers, it does, in turn, initiate a nuisance problem with the discharge of odorous sewer air into the atmosphere. This can lead to a spate of complaints from residents in the vicinity of the

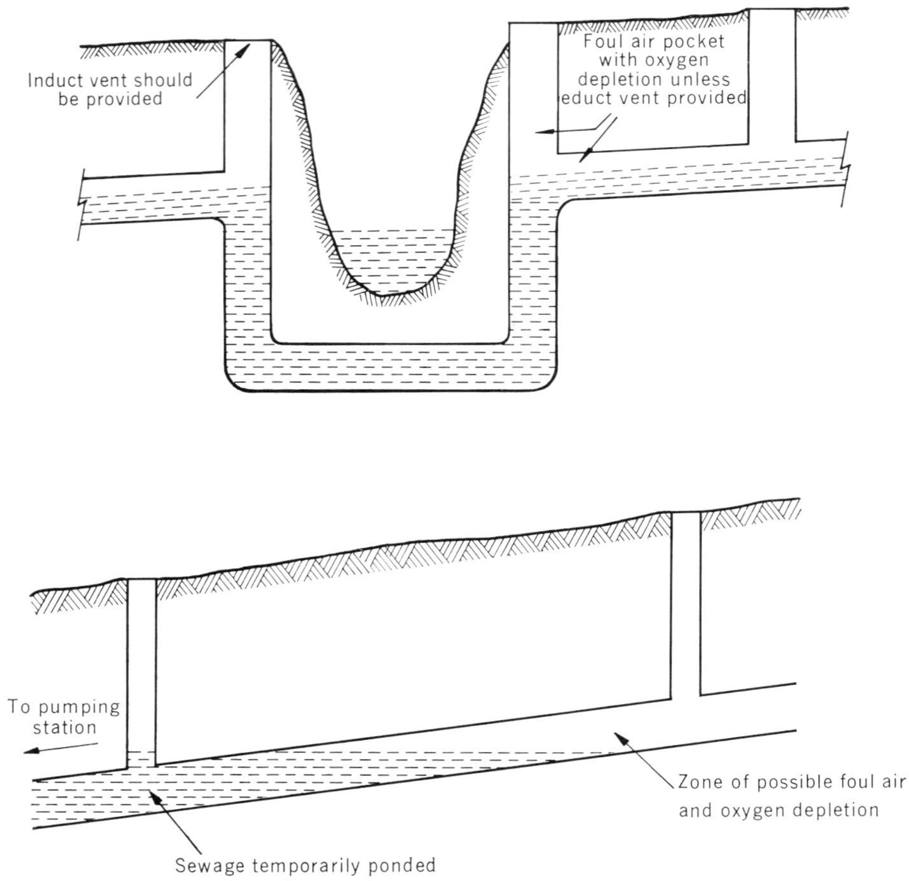

Figs. 6.4 and 6.5 Ventilation problems of filled sections of sewers.

point of discharge. These complaints may relate not only to the odours accompanying the discharge of gaseous sulphides, but also to the tarnishing effects of sulphides on metal articles in homes or in local factories. In some locations adequate ventilation may not be practicable unless such nuisance can be prevented.

The control of sewer air conditions to avoid any likelihood of toxic levels of gas concentrations is a most important consideration.

Odour nuisance at educts

Nuisances arising from the discharge of sewer air to the atmosphere are accentuated under certain atmospheric conditions. The worst meteorological condition is that known as atmospheric thermal (or temperature) inversion. (See Fig. 6.7). Under normal conditions, air temperature gradients of around 1°F for each 300ft. or so of height above ground are maintained, the upper air being cooler accordingly. Certain circumstances, however, produce a barrier layer of *warmer* air some hundreds of feet above ground, called an inversion layer. In some particular localities this occurs relatively frequently. Under these conditions, which are akin to those producing fog, dispersion of discharged gases is slow and they may travel quite long distances without much loss of concentration.

Whether or not H_2S be the principal cause of these nuisance problems, H_2S concentration levels are used normally as a measure of sewer air concentrations. On this basis, the control of sewer air odour problems by dilution techniques usually requires arrangements to ensure that H_2S concentrations anywhere in the potential nuisance area do not exceed 0.03 p.p.m. per volume.

Often the situation is such that concentrations of hydrogen sulphide in the sewer air are many times that permissible in the outside air. To avoid nuisance it is desirable to understand something of the natural processes of dispersion and dilution of the effluent sewer air into the atmosphere outside. This is illustrated diagrammatically in Figs. 6.6 and 6.7, which represent two extremes of the atmospheric conditions which are important in this connexion.

Fig. 6.6 showing conditions where there is sufficiently strong wind to set up turbulent conditions in the atmosphere with normal temperature gradient decreasing with height above ground.

Fig. 6.7 showing near-quiescent conditions in which air flow is essentially laminar. Temperature inversion conditions may also be present, i.e. where temperature

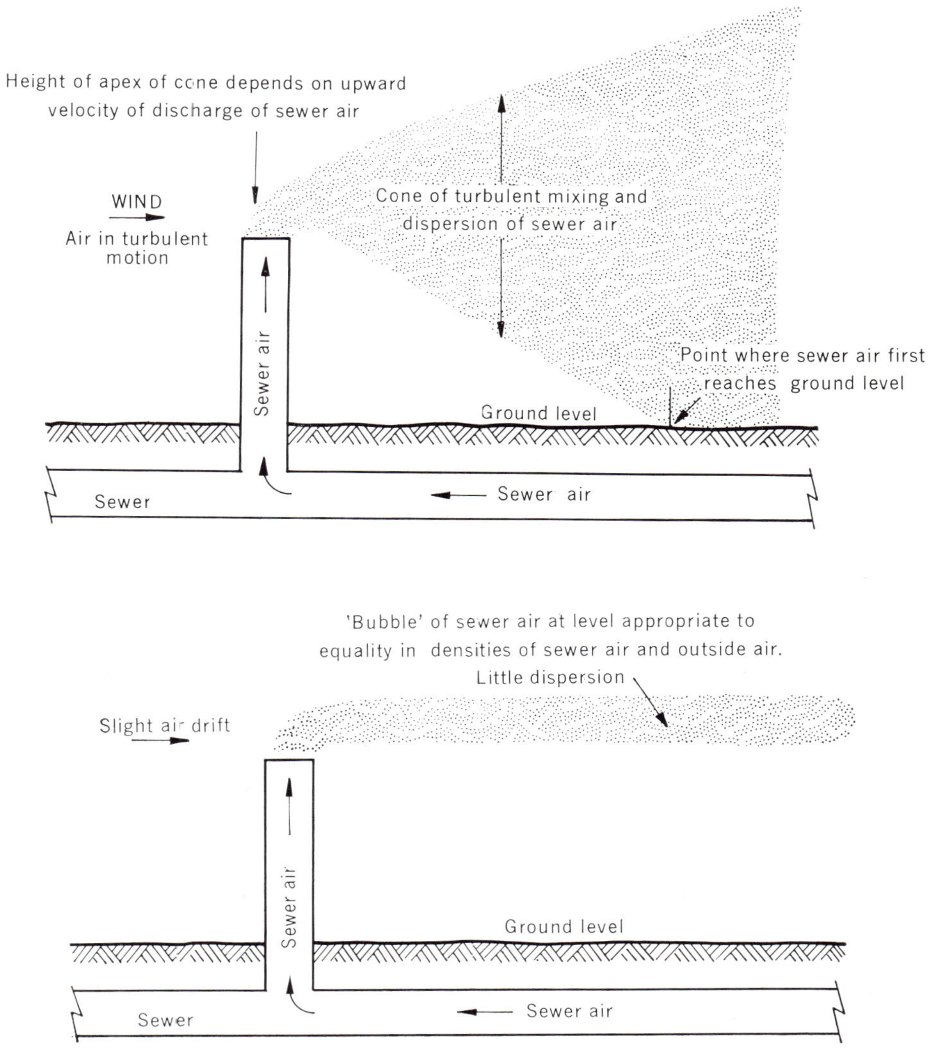

Figs. 6.6 and 6.7 Patterns of gaseous dispersion down-wind of vent stacks.

increases with height above ground until the boundary inversion layer is reached (Figs. 6.6 and 6.7), and patterns of gaseous dispersion down-wind of vent stacks are seen to be entirely different.

Under the conditions depicted in Fig. 6.6 a considerable reduction of sewer gas concentration of sulphide takes place before these gases reach ground level. This reduces the risk of odour complaint. The actual concentrations of sulphide present in the sewer air mixture reaching ground level (p. 146) depend primarily on the vent height, wind velocity, and volume and upward velocity of the sewer air, at the exit from the vent; also on topography, distribution of buildings, trees, etc. Relative densities of sewer air and the outside atmosphere have some effect but normally mixing due to turbulence rapidly eliminates any differences. There is now a large body of data available relating to gaseous pollution from chimney stacks and its dispersal, which can be applied to sewer vents. The literature is reviewed and summarized by Summer(*2*). For this application to sewer problems it is necessary to know the weight of sulphide discharged from the sewer vent per unit of time (say lb./hr.). For proposed sewers where some septicity of the sewage is unavoidable, an estimate can be based on the rate of sulphide emission for each ventilated section. For existing systems measured sulphide concentrations and sewer air flow in the particular vent stack concerned give values for actual operating conditions.

Experience has shown that the conditions depicted in Fig. 6.7 are those which give rise to many if not most sewer air nuisance complaints. They occur during periods of relative calm when temperature inversion also arises. Sewer odours are often particularly noticeable under such conditons in the early evening. The mechanism involved may be explained as follows.

In the periods during which the conditions of Fig. 6.7 prevail, the slow-moving air streams travel under conditions of laminar flow and a mixing cone as in Fig. 6.6 is absent. Instead, the sewer air emitted from the vent stack forms a non-dispersing "bubble" which slowly drifts along, with some spread due to the slight air movements. The bubble rises or falls to the levels at which the emitted gases and the air generally are at the same density. This may or may not be close to ground level, but the lower limit of inversion layer may be only 100 or 200ft. up. In the evening, as the general air temperature falls, the air nearer the ground is slightly warmed by ground radiation and rises, to be replaced by air from above, now colder. This brings the sewer gas bubble down progressively with the upper air until it finally reaches ground level. Under such conditions wherever the sewer air is slightly heavier than the outside air it will drift directly downwards to ground level. The conditions necessary for the above sequences to operate develop most commonly in spring and autumn, and it is also in these seasons that the densities of sewer air and the atmosphere outside are most likely to approach equality. Under other conditions when a non-dispersing bubble is formed as above, it may drift for some miles before the polluted air reaches ground level, not much diluted.

This indicates how odour nuisance and the resulting complaints may come not only from "people in the vicinity of vent-stacks", but also from people further away. Then, less frequently exposed but rather only intermittently affected, these people may be more sensitive to the unusual odours, and the more likely to complain.

Odour nuisance is a problem usually associated with septicity of sewage and the other sulphide problems of sewage works. The problem arises only when the odour of a substance, as conveyed to a person through his or her nasal cavity, is considered by that person to be objectionable or offensive. "Taste" may be part of this reaction.

Two functions are involved in the overall mechanism by which odorous substances in the air are detected, recognized (classified) and appraised by a human being, namely:—

(a) first the odour is converted into a primary perception by the olfactory system which controls the sense of smell, and

(b) secondly the sensory perception is conveyed to the brain, by nerve signals, where the odour is classified, on the basis of previous experience and associations, as pleasant, unpleasant or even repugnant; and perhaps actually attributed to a particular substance or type of substance, or to a particular origin.

With respect to (a), the exact method of operation of the olfactory nerve receptors is not completely understood as yet, but it seems clear that the "strength" of an odour from a particular substance depends directly, in respect both of intensity and of character upon the concentration of such substance in the air entering the nasal cavity. Theories covering the detailed operation of "the sense of smell" are reviewed at some length by Summer(*2*).

With respect to (b), a particular "odour" might be classified by one person as objectionable but by another as agreeable as in the case of garlic, for example. However, sewage odours are considered highly objectionable to most people and consequently, when odours characteristic of sewage occur in a neighbourhood, complaints must be expected, and corrective action often is required.

According to Gage(*37*) the following relationship summarises the main influences controlling public nuisance levels due to odour emanations—

$$M = \frac{D}{t} \times \frac{RH}{t} \times \frac{P}{1000} \times I = \frac{D \times RH \times P \times I}{1000 \times T^2}$$

Where M = nuisance magnitude
t = total number of days of surveillance
D = number of days odour observed
$\frac{RH}{t}$ = time weighted mean humidity
P = population involved
I = average odour intensity level

The nature of the odorous gases in sewers

Scientists engaged in investigations concerning sewage problems early in this century appear to have attributed the variety of odours met with in sewerage systems to only a short list of "gases and vapours", including mainly the sulphide-sulphur group, specifically hydrogen sulphide and ethyl mercaptan, and the amino-nitrogen group, viz. primary, secondary and tertiary methylamines. The presence of indole and its 3-methyl derivative skatole was also referred to commonly, without firm evidence. However, modern chromatographic techniques have revealed that there are upwards of 50 different gaseous substances whose presence in sewer air can be detected (but by no means all identified chemically). More than half of these are distinctly odorous, these odours being attributed not only to compounds such as the abovementioned, but to other compounds which appear to include simple hydrocarbons, paraffinic and aromatic aldehydes and ketones (saturated and unsaturated), carboxylic acids and esters, etc., terpenoid substances, and several substances such as are normally associated with earthy and musty odours (Actinomycetes' odours) and so on.

Thistlethwayte and Goleb have supplied some details (private communication 1971, from a publication in preparation) in amplification of the above, based on experimental studies of the composition of sewer air samples from Sydney and Melbourne. Of the "mercaptan" group, they found methyl mercaptan highest in volume proportion, with methyl sulphide odour also prominent, the total sulphur content occurring in such compounds being equal normally to between about one-fiftieth and one-hundredth only of the H_2S sulphur content, on a volume basis. Trimethylamine was the most prominent of the amino group, which occurs in volume concentrations also of the order of one-hundredth of the H_2S volume concentrations. A third group of substances, again normally related to H_2S content in similar proportions as the above-mentioned, are essentially composed of aldehydes. In some instances these are even more significant. Neither indole nor skatole were detected. Hydrocarbons were found to be present in volume concentrations hundreds of times greater than the H_2S; some of them associated with industrial wastes in particular.

A general conclusion is that the composite "sewer air odour" is due mainly to sulphidic gases together with paraffinic amines, aldehydes and body acids. These appear to be correlated with H_2S levels, confirming the traditional use of H_2S concentration as a useful overall measure of sewer air levels, ignoring the solvent vapours of relatively low odour potential. An important characteristic of H_2S however, is its toxic potential. It is especially dangerous because it rapidly de-sensitizes the odour sensory centres, and this has the effect of reducing its significance relatively as a source of odour nuisance. On this account the other constituents such as mercaptans, organic sulphides, trimethylamine and lower aldehydes probably are the main determinants of sewer odour potential. As noted above, their concentrations in sewer air are generally correlated with H_2S levels.

Of the other constituents, carbon dioxide is very important. Concentrations normally are between about 0.5% and 1%. About 10% of the hydrocarbon-type vapours are made up of the chlorinated solvents, mainly trichlor- and tetrachlor- ethylene.

However, it appears also that in the case of municipal compared with purely domestic sewerage systems, so simple a picture of the odour problem could be grossly misleading, especially where particular types of industry may contribute larger proportions of wastes.

Odour threshold.—For odours to lead to individual or collective complaint of nuisance the air breathed by the person or persons concerned must contain odorous gases or vapours in such concentrations as can be detected by the individual's sense of smell. The concentration in the air at which the odour of a particular substance can just be detected is known as the threshold concentration. This threshold concentration will vary from person to person and also possibly with other factors such as air humidity, so that it is not feasible to determine precisely the threshold concentration for a particular gas. As a rough guide, data for some of the particular gases involved in sewer odour problems, mainly taken from Stern (*37a*) are given in the following Table 6.2.

TABLE 6.2

Gas or Vapour	Odour Threshold Concentration (p.p.m. by volume of air)
Hydrogen Sulphide	About 0.10
Methyl Mercaptan	„ 0.04
Methyl Sulphide	„ 0.004
Iso-valeric Acid	„ 0.001
Butyric Acid	„ 0.001

The above threshold concentrations were in turn derived from laboratory test data using observers specially alert to detect odour. In practice concentrations well above those tabulated in the foregoing table might be tolerated or pass unnoticed by all but persons extremely sensitive to odours. The threshold concentrations given in the literature by different observers often give values for the same gas differing by factors of several hundred. It is suggested that practical permissible concentrations may be taken as between 10 and 30 times the values given in Table 6.2 pending better information becoming available.

Odour levels in sewer air

Just as threshold concentrations have been measured for pure gases, so it is feasible to measure the odour threshold for sewer air samples. If we take the H_2S threshold as suggested, say, at 10 times the value of Table 6.2, we have the generally detectable level as above 1.0 p.p.m. by volume, the value for most people. If the H_2S concentration in a particular sample of sewer air contains, say, 30 p.p.m. v/v of H_2S, the corresponding threshold concentration for H_2S would be reached by diluting the sewer air thirty times, but other odours come into question, and should be allowed for. In the case of one particular outfall where the maximum age of the sewage was about 1½ days, sewer air containing up to about 50 p.p.m. H_2S showed threshold dilution ratios (T.D.R. values) up to about 5000 to 10,000 max. for the sewer air. This and other experience indicates that whereas the T.D.R. values for sewer air may be correlated with H_2S contents, the actual values are largely determined by other constituents having more odour potential, that is to say, lower odour threshold concentration characteristics. Nevertheless it is satisfactory and is common practice to use the concentration of H_2S as a working basis for ventilation designs, whether the particular concern be toxicity, odour, or sulphide control, although moisture control may not be correlated with H_2S. Regarding odour, the effects of different gases are likely to be additive to some extent. A rough working rule for sewer air threshold dilution ratios, which is the proportion of clean air required to dilute sewer air odours below ordinary detectability levels, is to multiply H_2S concentrations in p.p.m. by between 100 and 200 times.

It might be noted here that the toxicity effects of the different gases also are to some extent additive.

Toxicity factors in sewerage ventilation

Toxic effects of sewer air

The possibilities of toxicity hazards for sewerage maintenance personnel arise not only because sewerage structures generally are located below ground and are enclosed, but also because almost any sort of substance used by mankind may be found in sewage at least sometime or somewhere, while many industrial or household poisons may be present in sewage practically all the time. Some of these substances are readily soluble in water, others relatively insoluble; some are volatile; some acidic; others alkaline in character, so that the pattern of their behaviour as a constituent of sewage varies from place to place in the sewerage system. Probably all are affected to some extent likewise by sewage temperature; and so on.

Among the substances commonly present in municipal sewage and which tend to escape as toxic gases or vapours into the sewer air, are hydrogen sulphide, hydrogen cyanide, ammonia, and solvents such as benzole, ethylene dichloride, etc. Dangerous conditions associated with such substances may arise in the sewers not only near the point of entry of the wastes, but further downstream where temperatures or other circumstances may change their characteristics. Thus for example cyanide wastes may be discharged as components of wastes from electroplating works, or sulphides in highly alkaline unhairing liquors or soaking pits from tanneries. The wastes then may become strongly acidified fortuitously on mixing with an intermittent highly acidic discharge in a downstream sewer, whereupon toxic conditions may develop quite suddenly and quite unexpectedly.

Dangerous conditions sometimes develop simply because of the accumulation of carbon dioxide. The atmosphere in most sewers usually contains concentrations and proportions of oxygen close to the normal level, 20.9% by volume of the open air, but CO_2 is usually relatively higher—up to about 10,000 p.p.m. compared to about 300 p.p.m. in the open air. However, if the airflow is restricted, oxygen uptake and CO_2 accumulation together can produce irrespirable or otherwise dangerous conditions. Many such cases have indeed been experienced in operating systems, and even under stagnant conditions without flowing sewage.

The accidental entry of external toxic gases such as town gas also must be allowed for. Many old sewerage systems traverse areas where badly corroded or abandoned gas mains cause serious leakage of toxic (and sometimes explosive) gases into the sewers. Leakage from underground fuel tanks can produce dangerous conditions in sewers while digester gas and similar products of stagnant conditions in sewerage systems might in some cases lead to explosive concentrations of methane. Accidental spillages of volatile solvents or fuels are other hazards.

Dangers such as these are obviated by proper ventilation practice, coupled with control of waste discharges, but accidents remain always a possibility, and require careful vigilance on the part of all members of maintenance teams working below ground, and also their attendants at the surface.

It might be noted here that a future hazard as yet of little account in most sewerage systems, will be that of radio-activity (whether in sewage or in sewer air), but this problem is beyond the scope of this Manual.

The above discussion makes reference to explosion hazards in addition to various hazards due to toxicity factors, H_2S and hydrogen cyanide being referred to as particular cases. H_2S is certainly a common potential hazard. Other possible hazards of toxic or irrespirable sewer atmospheric conditions include oxygen depletion in the sewer air at points where ventilation is inadequate, and the presence of coal gas, petrol fumes, other poisonous gases originating in industrial wastes, and so on. It is beyond the scope of this Manual to deal further with these aspects, which are referred to in some of the references given in the bibliography.

According to Bowlus and Pomeroy(*38*) the toxic potential of hydrogen sulphide in sewage and in sewers is greater than that of hydrocyanic acid because of the relatively greater solubility and low escape rates of hydrocyanide acid gas.

Air containing hydrogen sulphide in sufficient concentration when breathed into the lungs acts as a systematic poison and will be rapidly lethal if the H_2S concentration is above about 1000 p.p.m. Many fatal accidents to sewer workers have been reported as attributable to hydrogen sulphide poisoning, but the levels of concentration at which the gas becomes lethal cannot be stated precisely because it will depend both on the actual circumstances and on the individual. The presence of other sewer air constituents could add to the dangers due to H_2S alone especially at very high H_2S concentrations.

The information available from the literature is to be taken only as a guide to the possibilities of danger and to the levels of concentration likely to produce symptoms as indicated in Table 6.3 below. Where H_2S concentrations exceed about 100 p.p.m. by volume, conditions are to be regarded as altogether too dangerous for normal respiration.

In lower concentrations, hydrogen sulphide exerts an irritant action on the respiratory tract which resembles the effect of chlorine but is less intense, and also on the cornea of the eye. High concentrations in excess of 1000 p.p.m. will tend immediately to paralyse sensory nerves. Since even at lower

TABLE 6.3

Symptoms Produced by Exposure to Air Containing Various Concentrations of Hydrogen Sulphide

Symptoms	Exposure time and conditions	Concentration of Hydrogen Sulphide in Sewer Air (p.p.m. by volume)
Little or none	Prolonged exposure, light work	5-10 (some persons less)
Slight eye and respiratory irritation; headaches	One to two hours light work	10-50 (some persons less)
Temporary blindness	Six hours hard manual labour	about 50
Maximum limit without serious consequence	One hour hard manual labour	about 100

N.B. The values given above include some allowances for conditions in sewers compared with industrial atmospheres polluted by H_2S alone.

concentrations the sense of smell is rapidly inactivated, it must be assumed that no self-protective odour-detecting mechanism is available to anyone entering or working in spaces where hazardous levels of H_2S might arise.

It is strongly recommended that wherever men are required to enter sewers, the normal range of sulphide conditions should be known from regular testing programmes. Where work is to be extended, special temporary forced positive ventilation should be provided, unless any permanent installations have been tested and shown to be adequate. Safety precautions are dealt with in Chapters 12, 13.

Ventilation for control of sulphide corrosion

Sulphide corrosion control aspects

The principal aspects requiring consideration in this connexion are hydrogen sulphide concentrations and the moisture content of sewer air. Normally, the major factors determining the rate of corrosion of sewer walls are the concentration of sulphide in the sewer atmosphere and the degree of wetness of walls. In a given sewer both of these factors are influenced by the amount of ventilation provided. Increased quantities of ventilating air drawn from outside the sewer act as a diluting agent and so reduce sulphide concentrations. Also, as air traverses a length of sewer its humidity increases. Consequently, whilst at the time of entrance to the sewer, and for some time after, its humidity may be such as to produce a drying effect on the walls, this drying capacity will be lost as it progressively picks up evaporated moisture from the sewage. Eventually the sewer air may become saturated with moisture and develop a capacity to wet rather than dry the sewer walls. This tendency will be increased when fresh air admitted to a sewer is at a temperature higher than that of the sewage and therefore gradually cools. Also, as air travels along the sewer, sulphide emitted from the sewage is picked up and sulphide concentration in the sewer air builds up until the rate of absorption of sulphide per foot run of sewer walls equals the rate of emission per foot run of sewage surface.

The question of humidity and the relative degree of wetness of the sewer walls relate to bacterial activity, and to attack on the concrete dealt with in the follow-

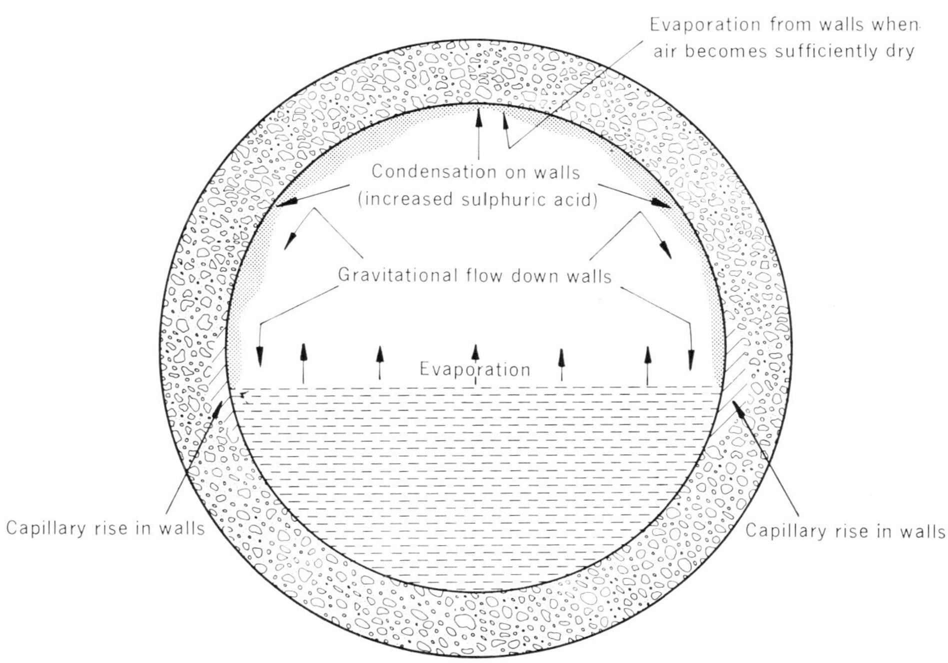

Fig. 6.8 Paths of moisture uptake and condensation in sewers.

ing chapter. The other aspect which arises is that of odour nuisance, discussed later in this chapter.

Moisture uptake by sewer air

It has been mentioned already that the rate of corrosion can be reduced or even stopped entirely by reducing the moisture content of the sewer walls and roof sufficiently. When the temperature and humidity of the outside air are suitable, ventilation will decrease the relative humidity in the sewer atmosphere, and so reduce the amount of moisture present on these structural surfaces. In some cases it may be practicable to keep the walls and roof sufficiently dry to prevent or at least considerably reduce the activity of the bacteria responsible for producing sulphuric acid, and hence control the effective rate of corrosion.

Reference to Fig. 6.8 will show that some difficulty may arise in achieving this result for those portions of the sewer wall just above sewage level because a certain amount of moisture is being continuously lifted above sewage level by capillary action. If the rate of lift by capillary action is beyond the evaporative capacity of the sewer air, then the lower wall area will be always sufficiently moist for some bacterial action to proceed even though perhaps less actively. The difficulty of keeping dry the portion of the wall just above sewage level is increased also by the higher humidity of the sewer atmosphere just above the sewage surface. The adsorbed surface film of surface-active substances reduces evaporation from the sewage surface and is an advantage in this respect.

Should mechanical ventilation be installed for an existing sewer with the aim to stop or reduce corrosion already established on the concrete walls, it must be assumed that when the ventilation system is brought into operation the sewer walls will be saturated. Under these conditions the ventilating air will collect moisture from the surfaces both of the flowing sewage and of the sewer walls above the sewage. If the walls are to be dried out progressively from positions downstream from but adjacent to the fresh air inlet (or inlets), more and more in proportion to the total moisture taken up into the ventilating air will come from the sewage surface. Finally, when the whole of the sewer walls above sewage level are fully dried, almost all of the added moisture would come direct from the sewage surface, a little from capillary wall moisture above the changing sewage level and a little from residual wall moisture as the sewage level falls from its daily maximum to minimum flow levels. Thus greater ventilation capacity is required to dry walls already corroding than is required to keep them dry. During some stages of the drying out, it may be that the sewer walls downstream towards the air exit actually become wetter, but if the ventilation system is of adequate capacity this effect will eventually disappear and the drying will extend progressively downstream along the sewer. It may take many months before complete drying of the full length of walls is accomplished. If ventilation be inadequate the corrosion will be controlled only near the air inlets, and possibly accelerated near air exits because of condensation of moisture from the air.

It has been established(*39*) that for acid production by bacteria on sewer walls to be stopped because of low wall moisture, the relative humidity of the sewer air should be kept below about 85%. The

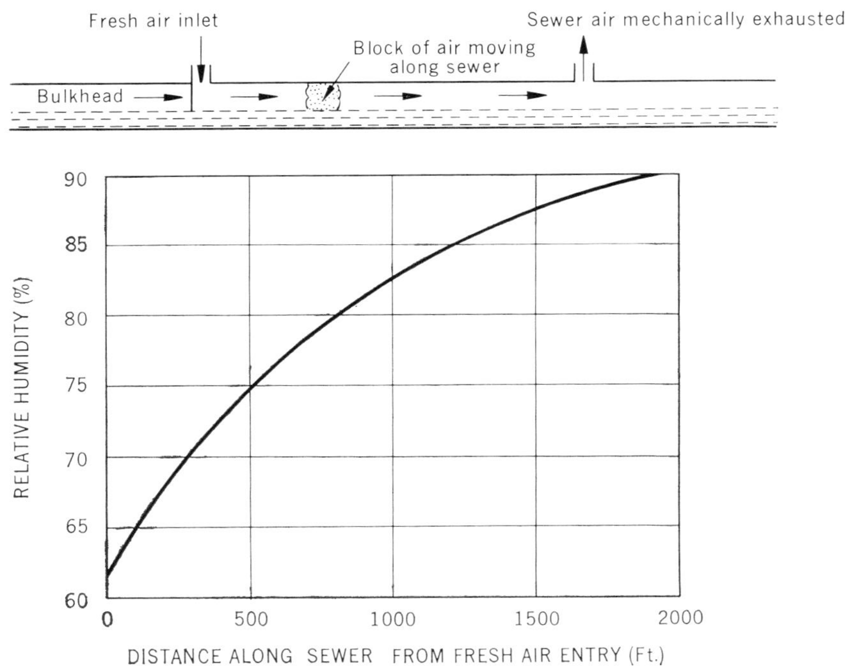

Fig. 6.9 Change in sewer air humidity along a sewer.

design of the ventilation system should therefore be aimed at achieving this. It is obvious that it becomes increasingly difficult to achieve any such humidity control the more closely the inlet fresh air humidity level approaches this value of 85% relative humidity. However, even though atmospheric humidity may be as high, or higher, for part of the time, if it be substantially lower especially throughout most 24-hour periods, day and night, of summer, full or partial corrosion control by mechanical ventilation may nevertheless be achieved as a practicable and economically satisfactory control measure.

Fig. 6.9 shows, for a particular case, how the relative humidity of the sewer air increased progressively as the ventilating air travelled along the sewer. In this case the sewer was circular 11ft. 0in. in diameter with mean sewage velocity 3.7ft./sec. and sewage depth practically constant at 3ft. 6in. Sewer air velocity was 8.3ft./sec. in the same direction as the sewage.

Thus, where local climatic conditions, particularly with respect to humidity, are suitable and if local odour or other nuisances can be avoided, mechanical ventilation provides an effective and economical means for controlling sulphide corrosion in existing problem sewers. It may also be a desirable supplementary measure for corrosion control in new systems. Moisture uptake in such systems is accordingly analysed as follows, and an adequate mathematical model developed.

Rate of moisture uptake.—The system dealt with is not isothermal, i.e. sewage and air temperatures are not necessarily the same and moreover cannot remain constant throughout any given length of sewer if evaporation is to occur. First, the physical phenomenon of evaporation causes a cooling of the sewage surface; consequently the surface of the sewage is cooler than the body of the sewage, however small the difference may be at any time. As a first approximation, the temperature of the sewage surface is assumed equal to that of the wet bulb reading corresponding to the relative humidity of the sewer atmosphere. Secondly, the temperature of fresh air brought into the sewer is seldom the same as that of either sewage or sewer air, and consequently it will change gradually until some sort of equilibrium with sewage and sewer air is reached. In order to predict air humidities from place to place along the sewer, it is necessary first to determine the corresponding air temperatures at these points. Equation [6.4] (see below) enables air temperatures along the sewer to be estimated with reasonable accuracy when the temperatures of the sewage and of external air entering the sewer are known. The temperatures of the air and sewage travelling along the sewer will approach a common temperature asymptotically, which finally will not be much different from that of the upstream sewage.

The primary factors entering the problem are related as follows:

$$L_{1-2} = \frac{A_A \times V_A}{A_{Sa} \times A_{wW}} \times \frac{1}{K_{wa}} \times \frac{1}{V_{Sp} \times (p_t - p_{wSf})} \times \log_e \left(\frac{p_{wSf} - p_{wa_1}}{p_{wSf} - p_{wa_2}}\right) \qquad [6.3]$$

where L_{1-2} is the length measured along sewer between stations 1 and 2.

A_A is the cross-sectional area of air current perpendicular to sewer axis, sq. ft.

V_A is the mean air velocity, ft. per sec.

A_{Sa} is the relative area of sewage surface exposed to air, sq. ft. per ft. length of sewer.

A_{wW} is the relative area of wet wall surface above sewage level, sq. ft. per ft. length of sewer (= 0 when walls are dried out).

K_{wa} is the mass transfer coefficient for water vapour, between sewage and sewer air, lb. per sq. ft. per sec, per in. Hg partial pressure difference between actual and saturated conditions.

$= 4 \times 10^{-6} \times [1 + 2.209 \times (V_{AR})^{0.85}]$ lb.ft.$^{-2}$ sec.$^{-1}$ (in. Hg)$^{-1}$

where V_{AR} is the relative velocity between sewage and sewer air ft. per sec., and values of K_{wa} for different values of V_{AR} are reckoned accordingly.

V_{Sp} is the specific volume of water vapour, at the appropriate temperature and barometric pressure, cub. ft. per lb.

p_t is the barometric pressure in the sewer, in. Hg.

P_{wSf} is the water vapour pressure corresponding to the sewage film temperature which is taken to be T_{wa} °F, the sewer air wet bulb temperature. Values for p_{wSf} may be read from humidity charts accordingly, in. Hg.

The value used in [6.3] is $p_{wSf} = \dfrac{(p_{wSf})_1 - (p_{wSf})_2}{2}$ in. Hg.

p_{wa} is the partial pressure of water vapour in the sewer atmosphere, in. Hg 1 and 2 respectively are stations 1 and 2, in. Hg.

(RH) is the relative humidity of sewer air.

$= \dfrac{p_{wa}}{P_{STa}}$

P_{STa} is the saturation pressure of water vapour corresponding to the temperature of the sewer air, T_a°F, in. Hg.

For isothermal conditions where the sewage surface film and temperatures are the same,

$$(RH) = \frac{p_{wa}}{P_{STa}}$$

The last term of equation [6.1] then becomes

$$\log_e \frac{[1 - (RH)_1]}{[1 - (RH)_2]}$$

Note that the value of K_{wa} for moisture loss from sewage should be less than for clean water because sewage contains considerable concentrations of film-forming substances (see pp. 23, 25, 26). The value might reasonably be adjusted in the same way as for the rate of re-aeration (pp. 25, 26), assuming a similar overall relationship to the BOD of the sewage, using Design Chart No. 6 for a correction as for the re-aeration constant.

The manipulation of the preceding equation [6.3] requires a knowledge of air temperature throughout the ventilated length of sewer. These can be predicted by the use of the following equation

$$L_{1-2} = \frac{A_A \times V_A}{A_{Sa}} \times \frac{1}{U_{Sa}} \times C_{pa} \times \rho_a \times \log_e \frac{T_S - T_{a_1}}{T_S - T_{a_2}} \qquad [6.4]$$

U_{Sa} is the heat transfer coefficient for sewage and sewer air, B.T.U. per sq. ft. per degree Fahrenheit temperature difference between sewage and sewer air.

$= 2 \times 10^{-3}$ B.T.U. (sq.ft.)$^{-1}$ (Sec.)$^{-1}$ (°F)$^{-1}$, when $V_A = 8.8$ fps

C_{pa} is the specific heat of sewer air at constant pressure, B.T.U. per lb. per degree Fahrenheit, and may be taken as 0.24 B.T.U./(lb.)(°F).

ρ_a is the density of moist air at the temperature, barometric pressure and partial vapour pressures of sewer air, lb. cu. ft.$^{-1}$

T_a is the temperature of sewer atmosphere, deg. F.

T_S is the temperature of sewage, deg. F.

The remaining symbols have the same meaning as for equation [6.3].

Air-velocity factor.—If, for a particular case, the sewage depth and temperature, initial inlet air temperature and humidity and the final permissible relative humidity at the outlet all have definite values, then the length of sewer which can be ventilated without the relative humidity of the sewer air rising above the permitted value is determined by the value of the term V_A/K_{wa} for equation [6.3]. For existing sewers the term V_A/K_{wa} is in fact the only factor which can be adjusted practically. The greater the value of V_A/K_{wa} the greater will be the length of sewer in which the relative humidity of the sewer air can be controlled within the limits desired. The quantity of air passing through the sewer influences both V_A and K_{wa}. This can be seen when we replace K_{wa} by the value given under equation [6.3] above. Substituting this value

$$\frac{V_A}{K_{wa}} = \frac{1}{4 \times 10^{-6} \times [1 + 2.209 \times (V_{AR})^{0.85}]}$$

but V_{AR} = relative airstream velocity = $V_A - V_S$

hence $$\frac{V_A}{K_{wa}} = \frac{V_A}{4 \times 10^{-6} \times [1 + 2.209 \times (V_A - V_S)^{0.85}]} \quad [6.5]$$

Fig. 6.10 is a plot of V_A/K_{wa} for different values of sewage and air velocities.

Inspection of Fig. 6.10 indicates quite clearly that the following are optimal criteria:

(a) The relative air velocity V_{AR} should approach zero. This requires that the air velocity should equal sewage surface velocity, roughly 10% above the mean sewage velocity.

(b) The higher the sewer air velocity otherwise the greater the length of sewer which can be dried out.

The problem thus broadly resolves into determining the progressive lengths of sewer which can be dried out sufficiently by a series of ventilation units, which usually will require an air velocity in the sewer equal to or rather greater than the mean sewage surface velocity. These lengths may determine the spacing of the ventilating fans, where more than one fan is required, the normal arrangement being that the ventilating air moves in the same direction as the sewage flow. It will however be more economical to ventilate in both directions to common exhaust points, if by so doing the total required number of fan installations can be reduced. When ventilating air moves in the opposite direction to sewage, the value for K_{wa} is increased and therefore the overall lengths of sewer dried out will be reduced accordingly. Such an arrangement is shown diagrammatically in Fig. 6.11 below.

Where the design provides for the ventilating air to travel always in the downstream direction, but more than one fan installation is required, flexible bulkheads or curtains are necessary as shown in Figs. 6.3 and 6.9, otherwise the system will not operate effectively due to short circuiting. The curtains must be of sufficient weight to prevent lifting out of the sewage due to differential air pressures. Where there may be doubt as to the effectiveness of ventilation because of limited drying potential of the fresh air introduced (e.g. high humidity of fresh air), the installation should allow for all ventilating air to move in the same direction as the sewage.

If a ventilating system be designed to keep walls dry —A_{WW} = 0, eqn. [6.3]—it will usually be capable accordingly of drying out the wet walls of a corroding sewer from the inlet end progressively, if only gradually, along the downstream sewer length. During this

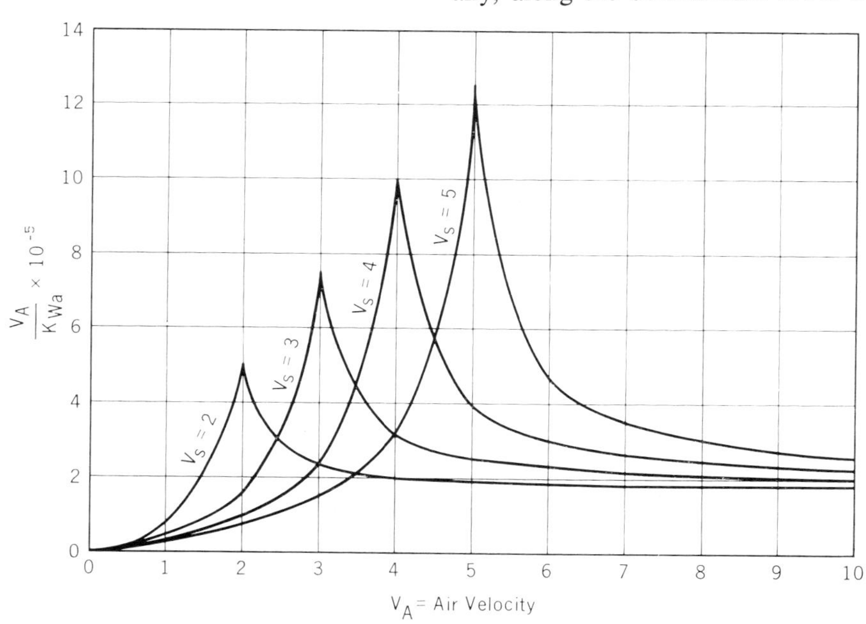

FIG. 6.10 Plots of $\frac{V_A}{K_{wa}}$ vs. V_A for V_S 2 to 5— see equation [6.5]

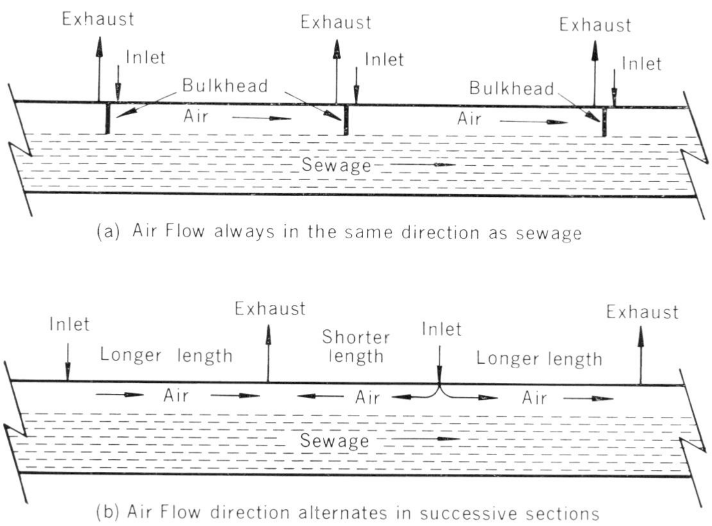

Fig. 6.11 Humidity control by mechanical ventilation. (a) Air flow all downstream with sewage flow. (b) Air flow partly countercurrent.

drying out period the relative humidity of the exhaust air may nevertheless be greater than the 85% necessary to prevent bacterial production of sulphuric acid on the walls, and accordingly the value of A_{WW} may be taken likewise to be zero. The rate of progress of the drying out can be reckoned by using suitable values for A_{WW} progressively from place to place, using equation [6.3] appropriately.

The humidity and temperature of the fresh air introduced to the sewer will vary considerably from time to time and at times may be such as to make the attainment of the desired humidities within the sewer impossible. From the practical point of view it is the longer term weighted averages of humidity and temperature which decide whether a ventilating system will finally become effective in controlling bacterial activity.

The fan installation should include motors of capacity well in excess of that computed for minimum air requirements so that if found desirable, as a result of tests after installation, air quantities may be increased relatively easily.

Measures for control of odours and odour nuisance

Nuisances due to hydrogen sulphide are reduced by the various measures, already described, for retarding its generation in sewers and its emission to the sewer atmosphere. Proceeding on the foregoing suggestion that H_2S concentrations in sewer air provide a sufficient measuring stick for most, if not all, of the odorous constituents, prevention or reduction of odour problems requires the elimination or at least substantial reduction of sulphide concentrations which occur in a free atmosphere at the point where it is likely to make contact with human beings in their homes or elsewhere. To do this it is necessary to ensure that the sewer air is free from odour when vented, or that concentrations are reduced to scarcely recognizable or at least to tolerable levels in living areas.

Gage, see p. 54, suggested that the odour nuisance level, while related mainly to the relative frequency and intensity of odour, is also more noticeable when the atmosphere also has a high relative humidity.

Some control may be achieved by dilution with sufficient fresh air in the sewer ventilation system or subsequently by the atmosphere between vents and nuisance areas; or by treating the sewer air by chemical or biological means before discharge, or by adsorption techniques. These various means of controlling or preventing odours, which include odours of hydrogen sulphide also, are listed hereunder accordingly.

By dilution

A high dilution ratio is often necessary since the odour limit for hydrogen sulphide is very low, probably lower than about 0.03 p.p.m., and even much lower for other gases. Thus, if the educted air contains 10 p.p.m. hydrogen sulphide, each cubic foot might need to be diluted with at least 300 cu. ft. of *clean* air to render it inodorous on account of its H_2S content alone (but see also pp. 54-6). There are several ways in which the degree of dilution may be to some extent predetermined.

Progressive dilution in the atmosphere may be ensured in many cases by discharging the sewer air through stacks of appropriate height.

The use of high vent stacks is based first on the common occurrence of greater wind speeds as the height increases, therefore increased draught results within the sewer on that account as well as on account of the greater draught due purely to density effects. Both effects produce a greater primary dilution of the sewer air. Second, where the top of the stack is a greater distance from the ground the effluent gases follow a longer path, hence, are likely to be better dispersed and more diluted before they reach ground level again. The greater wind speeds at higher levels also increase the effective length of the travel path, because the downwind cone of mixing is extended in consequence. High stacks should be located and designed particularly so that prevailing winds are unlikely to carry the sewer air down to public places or occupied premises without sufficient dilution having occurred. This may not be effective however under the conditions of temperature inversion previously referred to, unless perhaps when the stacks are high enough to penetrate the inversion layer. Fig. 6.12 shows the down-draughts which occur near the top of a cliff. Buildings, etc. may present similar hazards, and for such problems, wind tunnel tests may be worthwhile. The effective height of a stack is increased by using high velocity jets for discharge, at the cost of increased power, and in addition to this increase in effective height, greater dilution is achieved.

Direct dilution immediately before discharge may be obtained by inducing a substantial air-flow into the ventilation system at or near the vent discharge. This commonly provides direct dilution with lower power costs than would result by bringing the whole flow through the sewer; and usually ensures that the final discharge carries only the same total gaseous pollution, although more air-flow through the sewer usually picks up more gases (Table 12.2, p. 147). This pre-discharge intake of diluting air must not include much short-circuiting recirculation flow, otherwise the advantages sought will be circumvented. Such short-circuiting may be practically unavoidable during thermal inversion conditions, even using very special and maybe very costly, although carefully designed, arrangements. This could be indeed so difficult to avoid that when conditions for odour dispersal are least favourable, the system itself might well make matters worse, not better. If the ventilating air discharge were to be heated, however, inversion layer problems could be avoided entirely. Only a few degrees temperature rise might be enough.

Other physical (mechanical) means

That is, control methods which perhaps involve physico-chemical or chemical techniques, but in which no specific chemical reactions other than incidental combinations with atmospheric oxygen are essential features of the processes involved.

(a) **Odour destruction by combustion** requires air temperatures of about 1200°F or more using furnaces and heat regeneration systems, perhaps using a combination of digester gas and other combustibles, to be reliable for removing all odours. Special attention to possible explosion hazards are required also. The temperatures at which most of the odorous constituents are destroyed are lowered if combustion catalysts are used, and a number of installations of this sort are reported for some sewerage systems, mostly in overseas countries. However, so far as is known, none of these is on a large scale.

(b) **Adsorption on activated carbon.** Most odorous substances can be removed from sewer air readily by passing it through beds of granulated activated carbon. Three beds, each three inches thick and arranged in series have been used for this purpose, the air velocity (based on the gross plan area of a bed) being about 1/3ft. per sec., for a total head loss through the three beds (including the wire gauze supporting the carbon) of about 0.6in. water gauge. Some economy in operation is achieved by periodical regeneration of the spent carbon at relatively high temperatures (e.g. using super heated steam) from which a most objectionable odour results, thus requiring secondary deodorisation. High regeneration temperatures are necessitated by the transformation of H_2S to elemental sulphur on the charcoal surfaces, and acid conditions also result. Provision for corrosion prevention adds to the overall costs, but the process may be suitable nevertheless for "polishing" treatment of sewer air from which, particularly H_2S is first removed, and the relative costs for smaller scale installations may well be tolerable for specially difficult cases. Appropriate

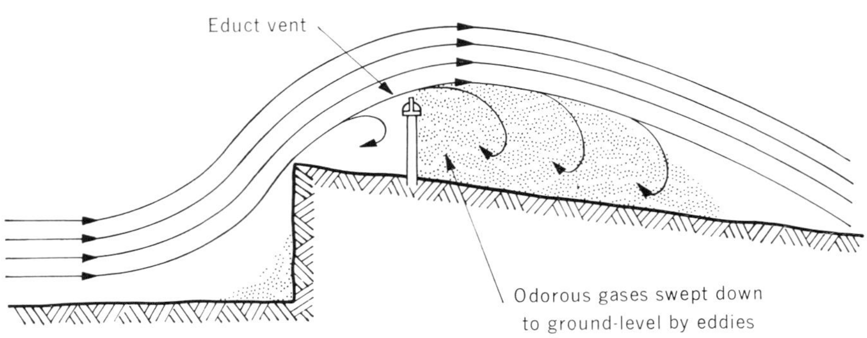

Fig. 6.12 Downdraughts caused by obstructions upwind of ventstack.

grades of carbon have capacities to purify raw sewer air adequately using three beds in series as above, of up to say 200,000 to 500,000 cu. ft. per lb. activated carbon per p.p.m. sewer air H_2S, before regeneration becomes necessary, at least for the leading bed.

Other adsorbents, even simple solvents, etc., also have been used for small scale applications, including some industrial air pollution control installations involving H_2S, mercaptans, etc.; but mostly where the concentrations of these odorous constituents are considerably higher than with sewage. Some of the costs then are offset because of this and may be lower due to recovery of saleable by-products. Simple water scrubbing is quite effective for the more soluble vapours for example alkaline odorous compounds such as alkylamines from fish processing, but something more is necessary for low concentrations of relatively insoluble gases such as H_2S.

More exotic systems such as those based on ultra violet light also have been used, mostly on a relatively small scale, but actual data respecting their efficacy are not yet available.

Odour control by chemical methods

A large number and variety of chemical substances have been proposed and tried, and some used, for control of odorous gas and vapour concentrations in industrial gas streams, including sewer air. The following provides only a much shortened representative type-list.

(i) **Scrubbing with alkalies.** Hydrogen sulphide can be removed by scrubbing the polluted air with milk of lime (an aqueous suspension of slaked lime). This can be distributed through the air stream as fine droplets for example by a rotating brush dipping into a trough of suspension, but since lime also reacts with carbon dioxide, the concentration of the latter in the sewer air may determine the effective cost of the process. Other alkalies have been used, and also are affected by CO_2.

If any of the odours present in significant proportions are themselves alkaline in character, it may be necessary or desirable to provide a supplementary washing with water sprays, which may be acidulated.

(ii) **Injection of reactive vapours into the sewer air stream** is possible for many locations in sewerage systems, and chlorine and sulphur dioxide have been used in this way. Such processes however usually are uncontrollable and are not recommended. It may be that the use of properly controlled amounts of alkaline hypochlorites or simple alkalies in fogs carried along the sewer under the influence of suction fans would be specially suitable for intermittent applications, provided water sprays were used finally to retain the aerosol particles before venting.

(iii) **Odour destruction by chemical oxidation includes** processes such as those based on chlorine, hypochlorite, and sulphur dioxide, also ferric chloride, permanganate and arsenate or other such catalysed air-oxidation systems. Most of these have been developed primarily for either H_2S or thiol removal (or both together), mainly for specific applications in the gas or oil industries. While no specific data appear yet to have been published for such systems used for sewer air, enough appears to be known of several such systems to ensure that they can be made effective. $KMnO_4$ vapours, generated in-situ from the crystalline substance to which concentrated sulphuric acid was added, were used as part of the Reeve's patent ventilation system more than 100 years ago.

On account of the rapid fluctuations which are likely to occur in the concentration of the odorous gases, it is to be expected that some under-dosing or over-dosing of the deodorising chemicals will occur. If over-dosing occurs, for example using chlorine, some chlorine will be discharged to the atmosphere unless special measures are taken, and such gases in addition to being toxic could be even more intolerable than the odorous conditions that prevailed before their use.

Most of the abovementioned oxidants are used as aqueous solutions or suspensions. In addition, industry has developed a number of processes, essentially chemical, based on the use of beds of reactive solids. A well-known example is the use of iron oxide for removing sulphide from town gas before delivery to gas mains. For the coal gas industry, the iron oxide is mixed with sawdust, and is regenerated by exposure to air. When the sulphur content is high enough, it is used as a source of sulphur for acid manufacture. Though effective for removing most of the sulphides present in town gas this method is unlikely to be economical for removing sulphide from sewer air. Such beds were used widely towards the latter part of the 19th Century, but it was concluded finally that they were not effective for sewers because sewer air contained too much moisture.

A great variety of special sulphide-removing solids have been patented, especially for use also in the mineral oil and gas industries, based mostly on iron oxide or on coke, but including metal "catalysts", etc. Sulphur bacteria may play a part adventitiously, but as above no data of practical value for sewer ventilation practice have been yet published.

Odour control by biological methods

The possibility of using soil beds, trickling filters and so on, for deodorization of air from sewerage ventilation systems, has long been realised. A patent was granted in the U.S.A. years ago(*40*) for using soil and has been developed particularly for sewage pumping systems. It has long been known also that odours may be removed by passing foul air through a trickling filter, or into activated sludge tanks, but no sound data are yet available on which engineering designs might be based, but for larger air flows, overall costs may be relatively high, especially using activated sludge. Some studies are now in progress under the sponsorship of one of the participating authorities (Goleb and Thistlethwayte(*41*). See also Carlson & Leiser(*42*)).

The discharge of foul sewer air into buried beds of broken stone has also been used, the air being dispersed from the broken stone into and through the surrounding soil. Information is not yet available for rational designs of such systems.

An odour control filter bed has been constructed in a large American sewage treatment plant in the manner of a tile-drain field, the foul gases being discharged into the tile drain system to filter through a crushed rock bed which underlies a stratified cover of straw and sand. Experimental work has also been carried out on the diffusion of foul gases through activated sludge cultures, but as before, no information is yet available from these various experiments.

The discharge of foul sewer air into beds of broken concrete within which active bacterial corrosion has been encouraged is another possibility, but such a method has not yet been tried in practice.

A great advantage of biological methods of odour control in general is that rapid variations in sewer air composition usually do not require any operational changes; overall efficiency being substantially constant even for large variations in odour levels.

The use of odour-masking chemicals has received considerable publicity for odour control. Such methods are unacceptable for the control of nuisances from the venting of sewer air which may be potentially harmful or dangerous, but may be useful in certain particular cases where unpleasant but not harmful conditions may arise. For example where sludge is drawn intermittently from sedimentation tanks into open drains or wells, an important daily task in many plants; unpleasant conditions may be alleviated locally by introducing, say, an odour of violets to improve the otherwise extremely malodorous environment.

CHAPTER 7

BACTERIAL ACTIVITY AND THE RATE OF CORROSION OF SEWERAGE STRUCTURES

Introductory

The last two items of the sulphide corrosion sequence (pp. 4-5) are respectively the formation of sulphuric acid on the surfaces above normal sewage levels (Item 4) and finally the direct corrosion process, the attack by the acid and its reaction products (Item 5). Significant aspects of the latter are discussed in this Chapter under the separate headings of *Bacterial Activity* and *Construction Materials' Relationships.*

Bacterial activity

The nature of the bacteria involved

The sulphur bacteria which effect the oxidation of the sulphur or sulphur compounds to sulphuric acid are autotrophic, that is they require only elementary substances such as CO_2 as the primary source of carbon, NH_3 as a source of nitrogen and so on. As shown diagrammatically in Fig. 7.1, however, several types of organisms usually are involved.

As early as 1900 it was known that concrete sewers conveying H_2S-bearing sewage were subject to rapid and extensive damage, and the damage was attributed to sulphuric acid attack. It appears to have been recognized that this sulphuric acid was produced by the oxidation of hydrogen sulphide from the sewer air, but the oxidation mechanism was not clear. Originally it was thought the process was essentially chemical, except that some workers recognized the possibility of bacterial action playing an important part.

In 1945 C. D. Parker (*43-5*) published papers describing the isolation from corroded concrete of an acid-forming species of thiobacillus to which the name *Th. concretivorous* was given. His experiments proved that in a suitable chemical environment including hydrogen sulphide, corrosion could be brought about by the proliferation of this organism, whereas under sterile conditions and in the same environment, no corrosion occurred. This species could proliferate under conditions of very low pH (pH about 1 or even less).

There is now strong evidence from investigations carried out in Australia, England and South Africa (*43-7*) that the only organisms significantly active in the advanced highly acidic high-rate stages of concrete corrosion in sewers are *Th. concretivorous,* other species not surviving the low pH conditions (say below pH 3).

The pH of the concrete surface must however be reduced first to a value of about pH 6 before *Th. concretivorous* will proliferate fast enough to account for the high rates of corrosion observed in sewers. Parker

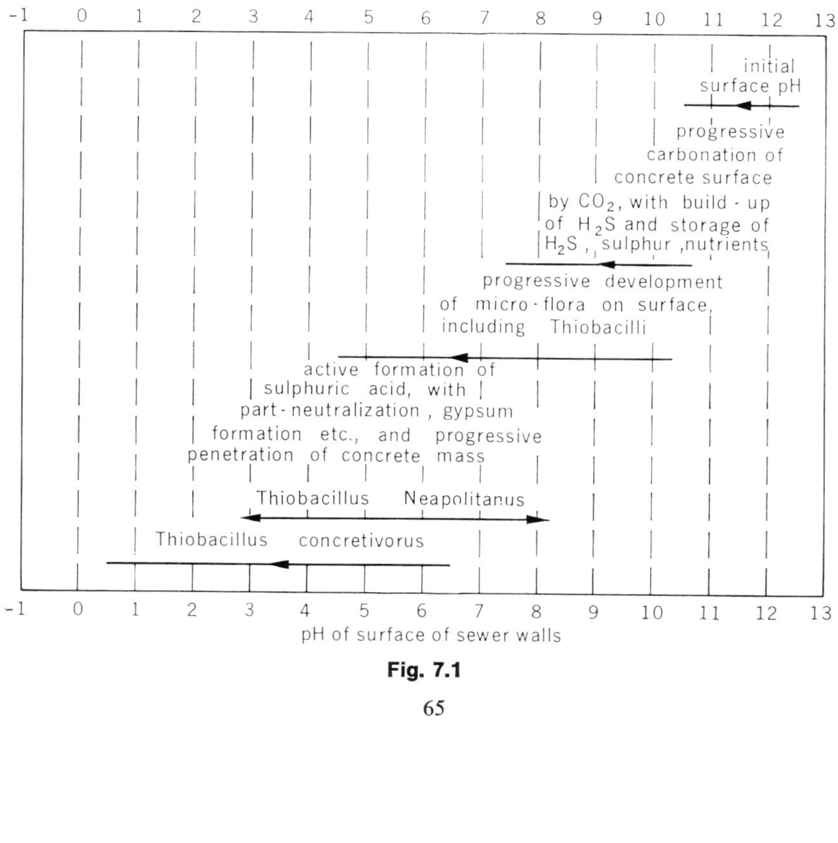

Fig. 7.1

(*45*) later described the chain of bio-chemical and chemical processes which bring about this progressive lowering of pH. This chain of processes is shown in diagrammatic form on Fig. 7.1.

Growth requirements

For the chain of bacteriological process to commence and continue to the final stage involving *Th. concretivorous,* and for this final stage to remain active, certain environmental conditions must obtain. These include the following:—

(i) there must be a primary source of infection of the concrete surface with the necessary bacteria,

(ii) the concrete surface must be always moist enough,

(iii) the necessary nutrients must be available for proliferation of the various bacterial species, including the supply of sulphide,

(iv) the temperature must be suitable likewise.

These are discussed in more detail as follows:

Source of infection.—It has been established as a result of a comprehensive examination of concrete surfaces of sewers and stormwater drains and of the waters in contact with them, mains water and so on, that the organisms active during the various stages of the corrosion sequence probably come originally from the water supply but multiply in sewage and on sewer pipes. The sources of infection of the concrete surfaces are present in the sewerage system, and presumably could be carried onto the concrete surfaces directly in one way or another.

Moisture on concrete surface.—It has been found (p. 57) that the amount of moisture present on a concrete surface materially influences the number of active thiobacilli present and hence controls the rates of corrosion to some extent. Corrosion ceases if the surface is sufficiently dry, as discussed in the foregoing Chapter 6 where it is reckoned that for wall moisture levels to halt corrosion the relative humidity of the sewer air must be less than about 85%. It might be noted that sewer air at a relative humidity of 85% would be in equilibrium with wall moisture containing about 23% of sulphuric acid by weight. Therefore wall moisture would continue to evaporate into sewer air until the concentration of free sulphuric acid in the wall moisture rises to 23% (neglecting the effects of the corrosion reaction products).

The moisture on the sewer walls may be derived largely from the condensation of moisture in the sewer air which in turn may come mainly from evaporation at the surface of the flowing sewage. Capillary action in the concrete also may draw some moisture up the walls from the surface of the flowing sewage and this effect may contribute to the increased corrosion rates often observed in the vicinity of sewage T.W.L. In some cases however, moisture may infiltrate directly by seepage through the sewer walls from ground waters outside the sewer.

Nutrient supply to bacteria.—The nutrient requirements of the autotrophic sulphur oxidizing and nutrifying bacteria, the only groups likely to be involved directly in the corrosion sequence, are generally simple. Among these various requirements are the following.

Sulphur: The primary materials for sulphuric acid and sulphate formation is the H_2S evolved from the sewage into the sewer air.

Nitrogen: The commonest sources of nitrogen usually are ammonia or ammonium salts. For sewers the primary source usually is the gaseous ammonia (NH_3) which also may be emitted from the sewage into the sewer air, especially from certain trade wastes or alkaline sewage.

Carbon: for cell synthesis is obtained in the form of CO_2, normally present in ordinary air but in much larger relative proportions in sewer air, because of evolution likewise from the flowing sewage.

Mineral Salts: Small quantities of mineral salts are essential but the particular elements required vary from organism to organism. Magnesium, iron, potassium and manganese appear to be necesary at concentrations of about 1 mg/l or perhaps less, for some organisms calcium likewise, while phosphate is also necessary for active growth perhaps at somewhat higher concentration levels. Investigation has shown (*43, 45*) that most if not all mineral requirements are available by solution from the concrete. Near the surface of the sewage some additional dissolved mineral compounds presumably are drawn up with the capillary water from the sewage and probably occur in detritus perhaps remaining above sewage T.W.L. from storm flow.

Temperature.—The effect of temperature on the rate of acid formation has been determined by Parker, for one strain of *Th. concretivorous,* as shown in Table 7.1.

TABLE 7.1

Effect of Temperature on Acid Formation

Temperature °C	Relative amount of acid formed in 50 days
4	0
10-12	7.33
20	9.05
30	34.42
37	26.18

The temperature for optimum production is about 30°C which is above that normally applying to sewer walls. The marked increase in activity between 15°C and 25°C, which covers the range of sewage and sewer air temperatures commonly found in Australian sewers, indicates that corrosion rates in summer in Australia are likely to be much higher than for winter, except where factors such as humidity levels operate to reduce the corrosion rates which otherwise would be reached.

Energy and oxygen requirements

The energy requirements for bacterial metabolism are derived from specific oxidations of sulphur or nitrogen compounds, mostly using atmospheric oxygen.

The possibility that the rate of acid production by *Th. concretivorous* might under certain circumstances be controlled by the concentrations of ammoniacal nitrogen or phosphate present also has been investigated by Parker, using laboratory cultures. It was found that the amount of acid formed at a given time increases as the concentration of ammonia nitrogen present increases, up to a limit of about 50 p.p.m. N equivalent. Above this concentration there appeared to be some slight inhibition.

The amount of acid formed in a given time was directly proportional also to the phosphate content up to a concentration of 0.5 p.p.m. equivalent phosphorus. Above this level only a little further increase was observed in the amount of acid produced, other things being equal.

Analyses of corroded concrete where high corrosion rates have been observed showed that ample supplies of ammoniacal nitrogen and of phosphorus have been available to support these high corrosion rates.

A varied bacterial flora similar to that found on concrete also develops in the moisture condensed on "inert" surfaces such as wood and stoneware. However, on these materials the pH drops much more rapidly as there is no lime to neutralize the acid as it is formed. Activity is therefore brought to a stop much more rapidly by the extreme acidity resulting, or by the lack of essential nutrients. This does not perhaps apply to steels, in which other metals such as phosphorus are available, and where neutralization may occur also.

At present there appears to be no information available with respect to rate of acid production on such substances as plastic linings, where nutrient supply also may be limited.

Progressive deterioration effects

In a new sewer, deterioration by sulphide corrosion requires first a lowering of the pH of the sewer walls, as shown in Fig. 7.1. This period under mildly aggressive conditions may be a year or more. During the early stages of attack initiation of corrosion comes about according to two different patterns. Where conditions are mildly aggressive a generalised powdery condition develops which then becomes wetter and intensified near the water line. Where conditions are highly aggressive a typically infective form of attack can be observed. Corrosion commences at foci of infection and then spreads as an increasing larger circular area. These areas merge to form a continuous band of corrosion.

In the later stages of attack the effect is the complete destruction progressively deeper into the surface of the concrete which is converted to an amorphous mass without any effective strength. The outer strata may be crusty in appearance, and mostly are dark grey in colour, due to organic impurities. Underneath the crust, where a crust does form, there is a putty-like paste, usually very acid. Below this soft and obviously deteriorated surface the underlying concrete may appear unchanged, but in order to establish the overall effect on the strength and stability of the remaining structure some investigation of the underlying concrete is required. Chemical analysis will show that there is some penetration of sulphate in to the underlying surface, which commonly shows white in colour, rather than grey. Such penetration has been found to vary considerably, from as little as 1/10in. to ½in. to ¾in. or even more.

Compressive strength tests indicate that some deterioration of strength will also have occurred even at depths below surface where no significant change in appearance can be seen. It has been suggested that this loss of strength is due to formation and crystallisation of calcium-sulpho-aluminate in the body of the concrete ahead of zones of complete destruction of concrete. This salt can only form in an alkaline environment, and the formation of its relatively voluminous crystals is disruptive to otherwise sound concrete. It is possible, therefore, that the inward progress of the attack is spearheaded by the formation of crystals of this salt and associated cracking of concrete ahead of more general penetration. While this cracked zone probably is relatively thin, such an assumption is compatible with the evidence of loss of strength in small samples, say about 2in. squares. With larger samples cut from the walls of a sewer tunnel, 12in. thick, the average loss of strength beyond the zone of obvious deterioration after 40 years exposure to severely aggressive conditions has not been such as would warrant concern.

Analysis of corrosion products indicates that cement constituents are completely broken down to calcium sulphate, basic iron and aluminium sulphates and silica. If this be so it follows that, based on analysis of a modern Portland cement, one pound of sulphuric acid (expressed as sulphur) will destroy about 2¼ pounds of cement.

In circular and ovoid sewers the relative progress of corrosion is commonly much greater immediately above the daily high water level. In sewers where the ventilating conditions have kept most of the wall surface above sewage level reasonably dry, corrosion near the daily high water level has been marked by a clearly defined continuous groove of corroded concrete, with the concrete above still comparatively sound. Coatings which have been applied to the internal surfaces of new concrete pipe before being laid as an experimental pipe line have been observed mostly to fail first near sewage T.W.L. These observations that maximum corrosion tends to occur just above sewage level are particularly important because this position often coincides with a high bending moment, and corrosion in this area may therefore be especially significant for the structural stability of sewers subject to soil loading stresses. It should also be noted that steel reinforcement may be endangered by the progress of corrosion, and the placing of steel, indeed the question of using reinforced concrete for major sewers demands the most careful consideration.

Factors which are known to produce or encourage greater corrosion just above sewage level include:—

(a) **The greater rate of acid formation near the water line,** the environmental conditions there commonly being more favourable than elsewhere, including increased wall moisture due to capillary rise from the sewage and from detritus adhering to the walls, which also may lead to improved nutrient supply locally, favoured by increased sewer-air moisture, H_2S, CO_2, ammoniacal nitrogen, etc., immediately above the sewage. Actual determinations made in circular sewers have indicated that the moisture content of the walls decreases progressively from water line to crown.

(b) **The gravitational flow of condensate downwards on the sewer walls.** Some of the highly acidic moisture may drip from the horizontal or near horizontal surfaces near the top of sewer directly into the sewage, directly reducing the effective amount of attack on the sewer walls generally. Some part of the remaining moisture will travel down the walls, and in circular or ovoid sewers the result will be that the amount of acid available for concrete destruction, per square foot of wall surface, gradually increases from crown to water level. Of course, some of the acid moisture running down the sewer walls is also likely to enter the sewage without reaction with wall concrete, but certainly the reaction will be intensified near sewage T.W.L. In the case of ovoid sewers or other sewers with flat crowns, a greater proportion of acid formed would drip directly into the sewage from the crown than would occur in a circular sewer. This may result in a more uniform corrosion pattern. While there are insufficient data from which firm conclusions may be drawn, respecting the proportion of the acid entering the sewage un-reacted, the indications are that where the sewer walls above sewage are very wet, the proportion may be relatively high, perhaps of the order of 50%.

Rates of corrosion

The environmental conditions discussed in the foregoing not only determine whether the overall sequence proceeds to the final stages or not, but also determine the intensity of corrosion at each stage, including the final rate of production of acid if any and consequently the maximum rate of corrosion.

This maximum rate is reached generally when the pH on the sewer walls approaches the lower limit of the range over which *Th. concretivorous* is active. This will occur especially if the rate of neutralisation of the acid is less than the rate of acid build-up. This indicates, in turn, that the concentrations of H_2S in the sewer air are above some particular value related to the proportions of alkaline substances and the rates of access of and reactions with sulphuric acid; but that excessive concentrations of H_2S are not accompanied by proportionally higher rates of corrosion. The particular values for maximum corrosion rates will depend on the composition of the structural materials in question and also at least to some extent on the residual local corrosion products. The rates of corrosion are usually expressed in inches depth per year, in terms of concrete lost or reduced significantly in strength measured inwards from the original profile. Areas where the rate of corrosion reaches maximum may show several times the average.

The average rate calculated for one large circular sewer, from measurements around the whole periphery of the airway above sewage T.W.L. varied up to a total of about 6 inches in 50 years, or say up to 1/8 inches per year from station to station.

However, it is the maximum rate of corrosion which is more important, and observations in sewers have indicated that under severely aggressive conditions this rate may be as high as ¼ inch per year.

If certain assumptions are made it is feasible to develop a useful mathematical model giving the order of the average rate of corrosion around the periphery of the sewer air-way. The first assumptions are that the sewer is unventilated and all of the walls consequently continuously wet; that no H_2S escapes from the sewer to the outer air, and that under these conditions likewise all of the H_2S emitted from the sewage is converted to sulphuric acid on the sewer walls.

Assuming further that:—

(i) the rate of acid production is not hindered or inhibited by any unusual conditions, and all circumstances are normal;

(ii) all of the acid formed reacts with concrete;

(iii) the aggregate used for the concrete is completely inert and does not react with acid;

(iv) the hydrated cement is the only material reacting with the acid;

(v) the proportion of cement used in the concrete mix is known; and

(vi) the average rate of acid formation around the peripheral wall of the air-way of a sewer could be equated to the rate of transfer of H_2S to the sewer walls, then, from eqn. [5.19] p. 39.

$$N_{H_2S} = E_{H_2S} \times \frac{A_{Sa}}{A_{aW}} = K_{Sa} \times p'_{H_2S} \times \frac{A_{Sa}}{A_{aW}} \qquad [7.1]$$

where N_{H_2S} represents lb.S in sulphates per sq. ft. of wall per hour.

On the further assumption that 0.44 lb. of sulphuric acid (expressed as sulphuric acid sulphur) are required to destroy one pound of cement, the weight of concrete destroyed per square foot of wall surface would be

$$\frac{N_{H_2S}}{0.44 \times M_C} \text{ lb./hr. per sq. ft.}$$

or

$$19400 \times \frac{N_{H_2S}}{M_C} \text{ lb./year per sq. ft.} \qquad [7.2]$$

where M_C = weight of cement in one pound of concrete. In practice the value of M_C varies according to the proportions used for the concrete mix.

The average rate of corrosion then would be given by the following equation:

$$(R_C)_{av.} = 19400 \times \frac{N_{H_2S} \times 12}{M_C \times \rho_C} = 233000 \times \frac{N_{H_2S}}{M_C \times \rho_C} \text{ inches depth per year} \qquad [7.3]$$

where $(R_C)_{av.}$ = average rate of corrosion, inches/year
ρ_C = density of concrete, lb./cu. ft.

Introducing the value of N_{H_2S} from (5.21) and (7.1),

$$(R_C)_{av.} = 2.3 \times 10^5 \times \frac{K_{Sa} \times P_{H_2S} \times A_{Sa}}{M_C \times \rho_C \times A_{aW}} \qquad [7.4]$$

Just above daily high water level the rate of corrosion will be higher than the average rate for the reasons previously explained. This must be regarded as the upper possible limit of the rate of corrosion under conditions of no ventilation and where supply of moisture and all nutrients (other than H_2S) necessary for full activity of organisms producing sulphuric acid is sufficient to make H_2S concentration the controlling factor.

In practice this limiting value would not be reached commonly for such reasons as the following.

First, equation [7.4] is based on the assumption that all of the acid formed on the sewer walls reacts with the concrete, but as noted earlier a significant proportion, up to say 50%, actually gravitates back into the sewage without reacting. Furthermore, aggregates in concrete often are not completely inert to reaction with sulphuric acid, hence some of the acid reacts with aggregate, and the overall depth of the corrosion of the concrete is lessened accordingly. Some ventilation either is provided or occurs naturally. Inevitably, some of the H_2S in the sewer air escapes without producing acid, and to some extent also, ventilation commonly reduces the moisture content of the sewer air which in turn partly dries the sewer walls and reduces bacterial activity.

Computations are frequently based on dry weather conditions, but in systems where stormwater infiltration occurs during periods of significant duration, sulphide concentrations during such periods are much less than those for dry weather conditions, and in fact often are reduced to zero, or practically zero. Consequently, much more data must become available before it will be feasible to correlate average rates of corrosion, as computed by equation [7.4], with observed rates, and it is clear that computed rates will be higher than observed rates. However, on the basis of available information, maximum rates of corrosion in areas just above the sewage T.W.L. may be taken as being roughly approximated by the following equation [7.5].

$$(R_C)_{MAX.} = (6 \times 10^5) \times \frac{K_{sG} \times P_{H_2S} \times A_{Sa}}{M_C \times \rho_C \times A_{aW}} \text{ inches per year} \qquad [7.5]$$

Construction materials' relationships

Attack on sewer materials

Preceding sections deal with the sequences involved in the problem of sulphide attack up to the stage where sulphuric acid is formed on the sewer walls exposed to sewer air. The final stage is the attack by the acid on the materials of which the sewer is constructed. Since sewers are commonly constructed of concrete, this material is considered first.

Concrete

Parker(*45*) has reviewed the information available respecting the attack on concrete due to sulphuric acid formed by bacterial oxidation of sulphur compounds present in the sewer air.

The concrete considered in the first place is that prepared by mixing fine and coarse aggregates, Portland cement, and water, in proportions and using raw materials which, while perhaps generally similar, nevertheless differ for any one sewer compared to another. The mixture used also may differ in detail for different parts of the same sewer, for example for sections in tunnel compared with aqueducts, and so on.

To simplify the discussion it will be taken initially that the fine and coarse aggregates are chemically inert and, consequently, it is only by the hydration reactions between the cement and the water that the concrete develops mechanical and other properties suitable for sewerage structures. To understand how these properties, including strength, are developed initially but can be destroyed by sulphuric acid formed on concrete surfaces, it is necessary to know something of the nature of Portland cement and what takes place during hydration.

Portland cement or a modification is most commonly used, but there are other types of cement sometimes used, such as pozzolanic or aluminous cements. Portland cement is made by heating suitable mixtures of clay and limestone. The "burnt clinker" so formed consists of a mixture of calcium silicates, aluminates and aluminoferrites, with some uncombined lime (CaO), magnesium oxide (MgO) and other minor constituents, which include alkali oxides of sodium and potassium, titanium oxide, phosphorus pentoxide, etc. The principal compounds which "hydrate" are tricalcium silicate (C_3S), dicalcium silicate (C_2S), tricalcium aluminate (C_3A) and tetracalcium aluminoferrite (C_4AF). An analysis of a typical Portland cement used in Australia is shown in Table 7.2.

TABLE 7.2

Analysis of Portland Cement

Analytical Property Reported	Expressed in terms of	Percentage by Weight
Soluble Silica	SiO_2	21.7
Aluminium Oxide	Al_2O_3	5.3
Ferric Oxide	Fe_2O_3	3.2
Calcium Oxide	CaO	64.6
Magnesium Oxide	MgO	1.3
Sulphur Trioxide	SO_3	2.2
Potassium Oxide	K_2O	0.7
Sodium Oxide	Na_2O	0.6
Loss on ignition	—	0.8
Insoluble residue	—	0.4

During mixing and placing of concrete, these various principal cement compounds of the ground clinker react with the water to form hydration products which in the form of hydrated silicates and aluminates, free alkalis, etc., confer on concrete its particular properties.

The reaction of cement with water is in the first instance a reaction of individual constituents. The C_3A reacts extremely rapidly, also the C_4AF. Then follows hydrolysis of the C_3S which soon leads to saturation of the solution with lime. The gypsum in the cement also commences to dissolve rapidly and slows the hydration of the aluminate. Hydrated calcium aluminate and calcium sulpho-aluminate crystallize out, also calcium hydroxide. The product of the tri- and dicalcium silicates may be initially $2CaO.\ SiO_2aq.$ and then $3CaO.\ 2SiO_2aq.$ with the probability of the latter being the final product. It is not certain whether the immediate reaction of C_3A with gypsum leads to the low-sulphate sulpho-aluminate ($C_3A.\ CaSO_4.12H_2O$) or the high-sulphate form ($C_3A.\ 3CaSO_4.\ 31H_2O$), but in any case this depends on the relative rates of solution of gypsum and separation of $Ca(OH)_2$, perhaps also on the rate at which alkalies pass into solution. Once saturated solutions are established, higher sulphate compounds supersede lower compounds. The C_4AF may be expected to combine with the gypsum and lime to form the high-sulphate sulpho-aluminate and sulpho-ferrite.

Cements of different compositions, but also classified as Portland cements, are manufactured by adjusting the relative proportions of lime, silica and alumina, yielding types called, for example, rapid-hardening, low-heat, or sulphate-resistant cements. Other types of cement are prepared from entirely different raw materials, or from quite different proportions of raw materials. These include, for example, trass or pozzolan cements, blast furnace slag cements, aluminous cements, and others.

Low heat cement is a normal Portland cement with a somewhat lower lime content leading to the presence of a higher content of C_2S and a lower content of C_3S. Such cement liberates hydrated lime on hydration both more slowly and in much smaller amounts.

Sulphate-resistant Portland cement is obtained by increasing the iron:alumina ratio, whereby the content of C_3A is reduced and that of of C_4AF increased. Specifications in the U.S.A. require that the C_3A content be less than 5% to justify the name of sulphate-resistant.

Pozzolans may be used as a primary cementing agent in admixture with lime, the two combining in the presence of water to form calcium silicates with cementing properties. Alternatively, pozzolanic material may be added to a normal Portland cement-aggregate mix, when the free lime liberated by hydration of the C_3S reacts with the pozzolan, thus forming calcium silicates also, and reducing the final proportion of free lime. Materials capable of displaying pozzolanic properties are volcanic matter such as pumice, tuffs, diatomaceous earth, certain burnt clays and shales, etc.

High-alumina cement called "ciment fondu", is manufactured by clinkering limestone and bauxite and consists predominantly of CA or of C_6A_4FS. Special conditions must be ensured for proper hydration which inter alia requires some temperature control.

Supersulphated cement is manufactured by mixing ground blast furnace slag (80-85%), calcium sulphate (10-15%) and a small amount of Portland cement (5%). On hydration substantial quantities of calcium sulpho-aluminate are formed.

The reactions of cement products with sulphuric acid apparently have not yet been studied in detail. However, there is a substantial literature on the reaction of concretes with neutral sulphate solutions. Thus in sodium sulphate (aq.) the following reactions are said to occur:

$$Ca(OH)_2 + Na_2SO_4 \xrightarrow{H_2O} CaSO_4.2H_2O(cryst.) + 2NaOH$$

$$3CaO.Al_2O_3 + 3CaSO_4 \xrightarrow{H_2O} 3CaO.Al_2O_3.3CaSO_4.31H_2O(cryst.)$$

$$2(3CaO.Al_2O_3) + 3(Na_2SO_4) \xrightarrow{H_2O} 3CaO.Al_2O_3.3CaSO_4.31H_2O(cryst.) + 2Al(OH)_3 + 6NaOH$$

Magnesium sulphate reacts as follows:

$$3CaO.2SiO_2 + MgSO_4 \xrightarrow{H_2O} CaSO_4.2H_2O(cryst.) + Mg(OH)_2(pptc.) + SiO_2(aq.)$$

It has been shown by Miller & Manson(*48*) that concrete made from Portland cements with low C_3A content (sulphate-resistant cement) is much more resistant to sulphate attack. This appears to be due to the fact that reduction in C_3A corresponds to an increase in C_4AF and with such material a solid solution of C_4AF forms and this increases its resistivity to sulphate as the alumina/ferric oxide ratio decreases.

Steam curing of ordinary Portland cement concrete also tends to alter its reactivity with sulphates. Such concrete may be almost completely unreactive with calcium or sodium sulphates and rates of attack by magnesium sulphate also much reduced, presumably due to calcium monosilicate CSH being formed by reaction between the free lime, e.g. from hydration of the C_3S, with siliceous aggregate.

In the case of H_2S attack on ordinary Portland cement concrete sewers, the sulphuric acid is first neutralised by some of the free lime present, forming calcium sulphate, which then brings about a normal sulphate attack with the formation of sulpho-aluminate. However with increasing acid reaction the pH drops and sulpho-aluminate tends to break down, with the end products calcium and aluminium sulphates. With increasing acidity continual neutralisation occurs, leading ultimately to complete decomposition of silicate to hydrated silica and crystalline calcium sulphate.

Apart from changing the nature of the cement as above, it is possible to produce concrete with different physical properties by varying the relative proportions of cement, fine and coarse aggregate; using suitable gradings of aggregates; by varying the water:cement ratio; using better curing techniques; and ensuring proper ageing of the concrete. However, investigations by exposure of test specimens in sewers under relatively severe aggressive conditions have indicated almost without exception, that such changes in mixing and placing, etc., of Portland cement concretes produced little, if any, significant reduction in overall corrosion rates taken over longer terms. Exceptions occur where the proportion of cement used in the concrete mix is varied. The results of experiments indicate that the depth of attack for a given exposure period was inversely proportional to the proportion of cement in the mix. This suggests strongly that over a given period the amount of acid formed per unit area of concrete wall remains much the same, and the depth of penetration of attack in this period depends upon the proportion of acid neutralizing substance, i.e. the proportion of cement in the concrete. The picture appears to be that whereas lower concentrations of alkaline substances may lower corrosion rates initially, once concentration of such substance reaches some particular level, acid formation proceeds at rates controlled by some other factor, as for example, the available rates of supply of H_2S to the wall surfaces. This rate of acid formation then is stabilized under the control of this second factor, causing different thicknesses of surface concrete to be destroyed in a given period in inverse proportion to the relative amounts of alkaline substances available. The greater the proportion of alkaline substance present in the concrete the lesser will be the rate of corrosion, measured for example in inches per year of concrete destroyed. South African investigators(*47*) have suggested that corrosion rates can in fact be reduced by the use of highly alkaline aggregates such as limestone. Tests in both South Africa and Australia confirm that use of limestone aggregate does in fact reduce the rate of corrosion compared to inert aggregates.

There is little evidence to indicate that stronger, or less permeable, concrete shows materially increased resistance to corrosion. Rates of permeation are usually high compared to the rate of progress of attack and here also rates of acid formation and the relative amounts of alkaline substances present appear to be dominant.

When concrete is exposed to the atmosphere the free lime formed on hydration reacts with carbon dioxide. This carbonation may penetrate for some little distance from the surface (say ⅛in. to ½in) according to exposure, commonly even before the sewers are put into operation. This carbonated layer actually appears to increase the resistance of the concrete surface to acid attack for a period but finally acidic conditions penetrate the carbonated skin and corrosion proceeds normally.

The other types of cement referred to which may now be available and for which higher resistance to sulphate attack is claimed are discussed further below. It is noted here that with such "more resistant" cements the acidity of the wall film will increase more rapidly towards the conditions of maximum activity of the relevant organisms and consequently the amounts of acid formed tend to be adjusted to the added resistance of the particular cement.

Resistance to sulphide attack

The methods for increasing the resistance of concrete to sulphide (sulphuric acid and sulphate) attack thus include many different procedures. Some of these have been suggested and tried experimentally from time to time and from place to place throughout the world. They include:

(i) The use of special cements (e.g. high-alumina cements, super-sulphated cements);

(ii) The use of concretes containing a higher content of alkaline substances so that lesser thicknesses of concrete are corroded by given amounts of acid. Such alkaline substances may take the form of either additional cement or of *alkaline aggregates* such as limestone;

(iii) The inclusion in concrete of slightly soluble substances which may be toxic to the organisms which produce the acid;

(iv) Application to the hydrated concrete surface of such coatings as may act as protective membranes and barriers preventing corrosive acid contact with the concrete; and

(v) The treatment of the concrete surface by chemical reagents such as silicate solutions, soaps, silicon tetrafluoride, which penetrate the surface of the concrete and by reaction with certain hydration products form a resistant layer in or near the surface.

Performance tests of inherently resistant or of treated concrete mixtures have been made, including special tests under unusually aggressive conditions. A short review follows.

Many experiments have been conducted by Australian authorities, and by others, in attempts to assess relative resistance of these various concretes and treatments. Much of the early work, aimed at seeking a concrete entirely resistant to the severe conditions then being experienced in certain sewers, was commenced before the complex chain of processes leading to concrete corrosion was well understood. Though this earlier work did give indications that some concretes were corroded at reduced rates, no fully resistant concrete was found which might be suitable for most ordinary situations. However, the use of more resistant construction might be justified for particular cases.

Over a long period many possible types of coatings have been suggested as economical means of sulphide corrosion prevention or control. Experimental trials almost invariably have been disappointing. For a coating to be effective and provide an economical solution to the problem, it must generally remain impermeable to H_2S and to any acid which may form on or be transferred to it, must be itself resistant to attack and free of significant deterioraion after long exposure to sewer conditions, and must adhere sufficiently to the sewer walls for the effective life of the coating.

Impermeability of coatings is important. Local failure to prevent attack even though only small will initiate a progressive and spreading attack under the coating. A single pin-hole admitting H_2S and oxygen to underlying concrete will become a point of initial break-down. As corrosion proceeds, a blister is formed and the coating is forced from the concrete locally. Eventually it ruptures and so allows progressive development of the corrosion pattern. Although the number of such pin-holes may be reduced by using several coats or layers, it seems practically impossible completely to avoid such effects. Pin-holes appear to be more frequent when drying of the coating requires evaporation of solvent than when drying takes place by cooling of a coating applied at a higher temperature, as in the case of bituminous coatings applied molten, thermosetting resins, polymerization in situ, and so on, but experience even with these has not been encouraging.

Another requirement is that the coating material must not itself be susceptible to attack by sewer gases, moisture or acid. There is also the possibility of attack by particular organisms found on sewer walls or by sewage constituents.

Continued adhesion to sewer walls is very important. Manufacturers using certain additives to paint coatings, claim that they will adhere satisfactorily to dry concrete under ordinary atmospheric conditions, but certainly this is difficult under operating sewer conditions. Where a sewer a constructed under conditions such that the sewer is below the natural water table for long periods, some ground water will percolate through the sewer walls and even though the rate of percolation may be very low, hydrostatic water pressure will build up behind the coating and tend to force it away from the concrete surface. It is of interest to note that failure of coatings usually takes place first just above sewage level and gradually extends upwards towards the crown of the sewer.

A treatment which has been proposed is that of subjecting concrete to penetration by silicon tetrafluoride (SiF_4) gas, the gas reacting with vulnerable components of the set concrete to form acid resistant compounds within the concrete adjacent to the surface. In effect this treatment forms a resistant coating or layer within the concrete near its surface; the thickness of the resistant layer depending upon the depth to which the gas penetrates. The effectiveness of such treatment depends upon the depth of penetration and upon whether or not the resistant layer cracks at any time after formation. As yet the method is applicable only to precast items such as pipes and the depths of penetration are small. Fatty acid preparations and silicate solutions are other examples of surface "conversion" by chemicals. The results of a comprehensive series of tests on various concretes and concrete treatments have been reported recently. For these tests, various concrete pipes of 36in. dia. were prepared and laid to form a continuous test pipe line through which a stream of septic sewage flowed at constant rate and depth. The ends of the test line were sealed by bulkheads carried down below the sewage surface so that the enclosed air showed consistently higher H_2S, CO_2 and moisture content, about as follows:

	Range	Average
Hydrogen sulphide	19-300	140 (p.p.m. by volume)
Oxygen	close to normal	
Relative humidity	78-100%	93%

Since these tests were made for normal sewer pipe conditions, albeit unusually severe and thus producing unusually high rates of corrosion, they presented only an acceleration of the corrosion patterns which normally occur in sewers. The information obtained from them therefore probably is the best yet available on the relative resistance of various materials of construction. The following conclusions were drawn.

For any given sulphide conditions, relative life expectancy for concrete made from more corrosion-resistant cement types would be about as follows:

TABLE 7.2

Relative Life of Different Types of Concrete

Type of concrete tested	Estimated relative life
Using:	
ordinary type Portland cement	1.0
low-heat type Portland cement	1.8-2.0
supersulphated cement	1.5-2.0
aluminous cement (ciment fondu)	2.6
ordinary Portland cement but with limestone aggregate	1.9[1]
ordinary Portland cement concrete but incorporating powdered copper amounting to 1% by wt. of cement used	1.7[2]
ordinary concrete but incorporating locally available[4] pozzolans	about 1[4]
effect of other surface treatments:	
ordinary Portland cement concrete treated with gaseous SiF_4.	see note [3]
most surface treatment showed only temporary delays in onset of corrosion.	
any two changes or treatments were at least partly additive.	
Paint coatings were disappointing even in the short-term. Coatings which persisted for total test period of 76 weeks were:—	
(a) 1/8″ Epoxy-Coal Tar Mastic.	
(b) 1/10″ rigid PVC keyed to concrete.	see note [5]

Note: (1) This ratio is based on high-grade basaltic aggregate.

(2) Various types of bactericide were tested, but only powdered copper showed any significant improvement.

(3) For the first three months SiF_4 treatment gave a ratio up to 4.5. At that stage, however, the more resistant but relatively thin outer layer had been almost penetrated by acid-sulphate, and normal rates of attack then are to be expected. Improved penetration may be found more practicable, in which case this treatment may become worthwhile.

(4) Other pozzolans not available from local sources might prove more useful.

(5) Keyed PVC coatings used overseas have provided good protection for many years at least in some cases.

Ceramic materials, burnt clay products

Ceramic materials especially in the form of burnt clay pipes or bricks have long been, and are still, used for the construction of sewers. Before Portland cement concrete they were the materials in common use; brickwork for larger sewers and preformed burnt clay pipes for smaller sewers. Sometimes also face-tiles were used. Provided these products are properly made from suitable materials they can be very durable in sewer construction. However, experience has shown that unless quality is high, marked deterioration can take place. Two important causes of deterioration are first, where the clay is not suitable, or where not sufficiently burnt, the finished product is likely to swell when exposed to the moist conditions prevailing in sewers. Under sewer conditions, moisture may enter the material from both groundwater and sewage. The progressive expansion of the burnt clay material sets up compressive stress, both longitudinal and circumferential, under the conditions of restraint applying, and this stress may become so great as to cause compression failure within the wall material. Experiments moreover have indicated that this expansion rate increases when the absorbed moisture contains dissolved sulphate.

Secondly, where Portland cement mortar has been employed as the jointing material, the joints expand when attacked by sulphuric acid. Part of the acid also permeates along the interface between the mortar and the ceramic material, and the forces generated by the expanding mortar frequently cause spalling of bricks or stoneware pipes. The sulphates formed by acid attack on jointing material also enter the ceramic material and increase the expansive forces, including those referred to beforehand.

In some cases of spalling, it appeared also that acid carried into the ceramic by capillary and other such mechanisms had formed crystalline sulphate growths in crevices, possibly where ground water and acid met.

In spite of the fact that some failures of ceramic materials both in the form of stoneware pipes and of brick and tile linings have been reported, experience generally has shown that ceramic materials of such forms have given satisfactory service for very long periods. The problems requiring solution appear to be the devising of satisfactory acceptance tests for the ceramic materials and the development of satisfactory joints and jointing materials. Both of these problems are now receiving attention and, provided satisfactory solutions are obtained, these materials may be expected to provide sewers with adequate resistance to sulphide attack for indefinitely long periods. Also adequate provision for drainage and pressure relief where unfavourable situations are anticipated should avoid the spalling and simple compression fracture, root penetration problems, etc.

Unfortunately stoneware pipes of accurate dimensions and forms are not generally available in diameters much above two feet, and it is in sewers of greater diameter that more costly sulphide problems are more likely to occur. The use of brick sewers may justify special acid-resisting mortars which are usually of the sodium silicate type and require the use of special liquids in their mixing and special skills for their application. It might be noted that they require also the use of special quality bricks which may not be readily available, with the result that such sewers usually are costly.

Sulphide corrosion of metals

A detailed discussion of metallic corrosion is beyond the scope of this Manual. There is a voluminous body of literature available covering the subject of metallic corrosion generally but it should be noted that metalwork in sewers is subjected to conditions which tend to increase the intensity of corrosion above what might be expected otherwise. In particular, sewer atmospheres are in most cases so moist that the relative humidity remains at near 100% at all times, and sulphide is commonly present and reacts with most metal surfaces; much of the metalwork may be embedded in concrete or other structural material on which high concentrations of sulphuric acid also are produced continuously. Acidic condensates are likely to drain steadily over the metal surfaces, and highly corrosive conditions ensue. Because the concrete itself is initially highly alkaline, metals such as aluminium and zinc should not be used. The mixed circumstances introduce other complications.

In a similar way, acid and sulphate solutions are transferred to the metallic assemblies which may contain more than one metal. They act as electrolytes and galvanic effects are likely, especially with units such as penstock assemblies. As noted above, hydrogen sulphide directly reacts with the surfaces of many metals of which copper, as an example, is particularly liable to direct H_2S attack and also acidic oxidizing conditions. Most of the copper-based alloys such as brass, monel and even some of the bronzes suffer rapid sulphide corrosion.

The corrosion environment in the sulphide corrosion situation generally is both highly acidic and highly oxidizing. This sort of situation leads to unequal behaviour patterns and comparatively rapid failure of all but a few rather highly characteristic metal types. It should be noted, however, that the term "sewer conditions" in fact covers conditions very different from what is referred to above, and a wide range of metallic compositions could well be durable in many particular circumstances accordingly.

It appears that stainless steels such as 316 grade molybdenum stabilized will give good service. While there are other suitable grades, they may not be so readily available. However, rigid specifications and testing are required if special alloys are to be used successfully.

Because of the lack of data obtained under field conditions, it has not been possible to correlate life expectation of various metallic components with varying sulphide concentrations in sewer air. Considerable corrosion normally occurs under the highly humid conditions which generally prevail in sewers even when little or no sulphide is present, and it is doubtful if such correlations are feasible for any but the more resistant metals, which up to date have not often been tried.

Generally, the problems of metallic corrosion arise only with respect to incidental metalwork used in or about sewers, such as step-irons, ladders, manhole covers, penstocks and vent stacks. Nevertheless in the aggregate, metallic corrosion does constitute a considerable maintenance problem which warrants some properly planned investigation to arrive at the best and most economical applications of metals, and the best renewal procedures. Replacement usually costs more than the initial installation.

Experience has shown that in the design of metalwork, sharp edges should be avoided and round-edged sections used wherever practicable. All parts should be self-draining as assembled. In important manholes where frequent access is required and it is important that metalwork be maintained in good condition, experience has shown that forced induct ventilation can reduce corrosion to a marked extent.

Other materials of construction

There are many materials available for construction in sewerage systems other than concrete, ceramics, and metals. However, when all possibilities are considered, the choice is relatively restricted, both economic and practical factors providing most of the restraints.

The use of plastic brings up so many questions quite unrelated to sulphide corrosion that they must be regarded as beyond the scope of this Manual.

Timbers may be used for many applications throughout sewerage systems. The lighter timbers generally are unsuitable for acidic oxidizing situations, but top grades and classes of heavy timbers show good overall resistance. This applies say to teak woods and many of the Australian hardwoods such as tallow, blackwood, mahogany, mountain ash, box brush and so on, particularly where conditions are not highly variable.

PART TWO

MANUAL FOR DESIGN TO CONTROL THE SULPHIDE PROBLEM

CHAPTER 8

DESIGN DATA

Introductory

In the first part of this book, the occurrence and nature of sulphide problems were examined in detail, showing that sulphide formation in sewers is initiated when sewage becomes "septic". The mechanisms involved include many aspects, among which are the physical chemistry of oxygen absorption into the flowing sewage, hydrogen sulphide evolution, diffusion processes and so on along with the chemistry and biochemistry of oxygen consumption, slime growth on the wetted walls of sewers and generation of sulphides within these slimes, and the formation finally of sulphuric acid on the sewer walls above the normal flow level.

In considering this complex train of inter-related phenomena, the obvious conclusion is: "eliminate sulphide altogether and there can be no problem". In practice, however, the presence of sulphide in sewage within certain limits, and the consequential corrosion, is accepted as not being sufficiently harmful to justify adoption of special anti-corrosion measures. These acceptable limits are set out in Chapter 10, page 111.

There are two quite distinct types of design problems which may confront engineers—the conceptual and detailed design of *new* systems, and correction of sulphide corrosion conditions in *existing* systems. This part of the book sets out procedures which the engineer and associates, faced with these tasks, is recommended to follow—Chapters 9 to 12 being concerned mainly but not wholly with the new systems and Chapter 13 with existing systems.

In this Chapter 8, the principal mathematical formulations of the foregoing Part I are collected into forms suitable for direct application to the design of sewerage systems, and also included are some basic data presented as a guide for Australian conditions.

Economic Aspects of Design for Sulphide Control

An important aspect involves economic considerations. These include not only first costs but also costs of replacement, maintenance and repairs.

For the design of new systems, the alternatives available for economic comparison generally will fall into two or more of the following:—

(1) To provide a system such that the sulphide concentrations which result will have no significant long-term corrosion effects, nor cause other nuisance, e.g. due to odours; or

(2) To provide for renewal (replacement) of any seriously corroded parts of the system from time to time as they approach partial or near failure (including replacement of appurtenances, etc.), or for remedy of any nuisances; or

(3) Similarly, to provide in the original structures for loss of concrete by corrosion with or without provision for subsequent replacement of lost materials from time to time, associated with suitable maintenance and operational procedures; or

(4) Assuming that corrosive sulphide conditions will develop, to provide suitable measures in the system aimed at preventing corrosion by complete "insulation" of the structures from the corrosive environment, e.g. by encasing the exposed concrete surfaces with suitable acid-resisting plastic linings; or

(5) Likewise assuming corrosion will occur, to design on the basis of *reducing* the time-intensity effects of any corrosion, e.g. by using acid-resistant cements or mortars, incorporation of pozzolans, fluoride or silicate hardening, use of carbonate aggregates.

For any of these alternatives which assume that some corrosion or nuisance will develop, allowance also should be made for any fall-off in hydraulic capacity which may result accordingly.

Comparisons of economic costs for these alternatives involve inherent difficulties in ascribing realistic life expectations to various materials, and also in making realistic appraisals of the difficulties in carrying out future repairs or renewals. Development or re-development, additional services or deferment of a start on the repairs or renewals introduce factors which could make the works provided for in the estimates either more difficult than anticipated or even impracticable. The problems of odour, public relations and the overall safety of the system also warrant attention and consideration. Even though economic assessments may not be attainable for all factors, the responsible engineer must provide himself with the soundest possible basis in developing economic comparisons and in exercising his judgment as to the proper principles for the design to be developed.

Under the circumstances, what may appear to be the most economical solution may not be the most advantageous. Regard should be given to orders of accuracy. Comparisons should be made on cost-benefit basis but with allowances for uncertainties. A useful method of displaying the various alternative costs is to set out for each alternative the capital costs and discounted cash flow in which allowance should be made for escalation of costs with time.

Hawken(*49*) presents a sound approach to this task, based on what he calls the vestance sum, the amount of money required initially to cover not only the first costs, but in addition, all subsequent costs of repair, refitting, renewal, etc. If some allowance is to be made for the effects of inflation, Hawken's term vestance may be replace by the term virtual cost and the idea generalized in the mathematical expression of equation [8.1], as follows:—

$$V.C. = \sum_{j=0}^{\infty} C_j \times \left(\frac{1+i}{1+r}\right)^j \qquad [8.1]$$

where V.C. is Virtual Cost of an asset, the amount of money set aside now, invested at r%, to cover at each period j the total capital, maintenance, replacement, renewal and overhead costs, etc., allowing for inflation at a rate of i%.

C_j is equivalent total present costs for period j.

j is start of repayment period (usually years).

r is % investment interest rate % for unit time.

i is % inflation rate % per unit of time interval.

Making the simplifying assumptions first that unit costs remain unchanged or nearly so, and that renewal costs are nearly enough equal to initial or present capital costs, a rather unlikely condition for sewerage works, we have the essence of Hawken's relationship:

$$\frac{V}{C} = \frac{1}{1 - \left(\frac{1+i}{1+r}\right)^n},$$

except that the term $(1 + i)^n$ becomes 1 if inflation be neglected (as by Hawken), V and C being then respectively vestance and capital costs. This simplification leads to the following:—

The maximum percentage increase in original capital cost justified by an extension of renewal period from n years to m years.

$$= 100 \times \frac{V_A \text{ max.} - V}{V}$$

$$= \frac{100 \times \left(\frac{V}{C}\right)_m}{\left(\frac{V}{C}\right)_n} - 100 \qquad [8.2]$$

The following Table 8.1 provides indicative computations.

The data of Table 8.1(a) show for example that if sulphate resisting cement could be relied upon to increase the working life of a section of sewer from 5 years to 10 years, it would be economical to use the special cement if additional first costs did not exceed 75% of the estimates based on ordinary cement. An extension from 15 years to 30 years however would justify only 42% more first cost. Table 8.1(b) sets out the corresponding values obtained when additional allowance is made for the effects of inflation, in this case reckoning 4% per annum for inflation on top of 6% for normal charges. This shows 91% instead of 75% for a life increase from 5 years to 10, and 75% rather than 42% for an increase from 15 years to 30.

Such data are clearly of great significance for designers who must choose between various alternatives; and who must accordingly make careful investigations of the real cost rates for investments, including allowances for probable future trends respecting the economy of money and labour.

It should not be supposed that the accuracy inherent in such arithmetical computations can be transferred to designed reality; these tables merely indicate the apparent relativity of the costs of various alternatives.

Formulation of Equations for Sewer Design

It is believed the principles and recommended practice of this book will do much to bring sewerage sulphide problems nearer to full control. However, these principles have not yet been fully worked out, and further experience has to be gathered to allow useful re-evaluations to be made. There remain therefore, the tasks of perfecting the various criteria, the re-formulation of the rules to provide better predictions at least for most practical conditions and devising better approaches to related problems such as the adequate control of industrial wastes. For these reasons, and because changes in sewage composition over the past years suggest that other changes will occur in the future, it is prudent to assume that sewages generally, but more particularly sewages flowing in larger main sewers, will not necessarily remain fresh throughout, but rather may in places contain dissolved sulphide in significant concentrations. Therefore, under conditions of shock turbulence some H_2S may be emitted from the sewage in sufficient quantities to cause corrosion of the walls of the sewer structure.

In gravity sewerage systems where sulphide generation by biochemical agencies may occur, the sewer length may be considered as a sequence of three different zones characterized according to "sulphide conditions", respectively, an upstream Zone (1) where free dissolved oxygen occurs in appreciable concentrations, then an intermediate Zone (2) into which Zone (1) merges, with this Zone (2) merging downstream into Zone (3) where sulphides appear. A typical illustration of this situation is indicated in Figure 8.1 and is commonly found at those parts of a system where the local reticulation sewers join the carrier or submain system. However, such a situation may be found anywhere along a submain or main sewer likewise where the dissolved oxygen originally present in the sewage has been used up "chemically" or "biochemically", faster than the rate of solution of more oxygen from the sewer air (the re-aeration rate).

TABLE 8.1 MAXIMUM PERCENTAGE WHICH MAY ECONOMICALLY BE ADDED TO ORIGINAL COST TO ACHIEVE LIVES INDICATED

(a) Investment 6%/year Inflation 0%/year

Original Life (years)	Life Extended to .... (years)											
	5	10	15	20	25	30	35	40	45	50	100	inf.
(factor)	3.96	2.26	1.72	1.45	1.30	1.21	1.15	1.11	1.078	1.057	1.003	1.000
5	0	75	131	172	203	227	244	257	267	274	294	296
10		0	32	56	74	87	97	104	110	114	126	126
15			0	18	32	42	49	55	59	62	71	72
20				0	11	20	26	31	35	37	45	45
25					0	8	13	18	21	23	30	30
30						0	5	9	12	15	21	21
35							0	4	7	9	15	15
40								0	3	5	10	11
45									0	2	8	8
50										0	5	6
100											0	0

TABLE 8.1 (b) Investment 6%/year Inflation 4%/year

Original Life (years)	Life Extended to .... (years)											
	5	10	15	20	25	30	35	40	45	50	100	inf.
(factor)	11.01	5.77	4.02	3.16	2.64	2.30	2.06	1.88	1.737	1.628	1.175	1.000
5	0	91	174	249	317	379	436	487	534	576	837	1001
10		0	43	83	118	151	181	207	232	254	391	477
15			0	27	52	75	96	115	132	147	242	302
20				0	20	37	54	68	82	94	169	216
25					0	15	28	41	52	62	125	164
30						0	12	23	32	41	96	130
35							0	10	18	26	75	106
40								0	8	15	60	88
45									0	7	48	74
50										0	39	63
100											0	17

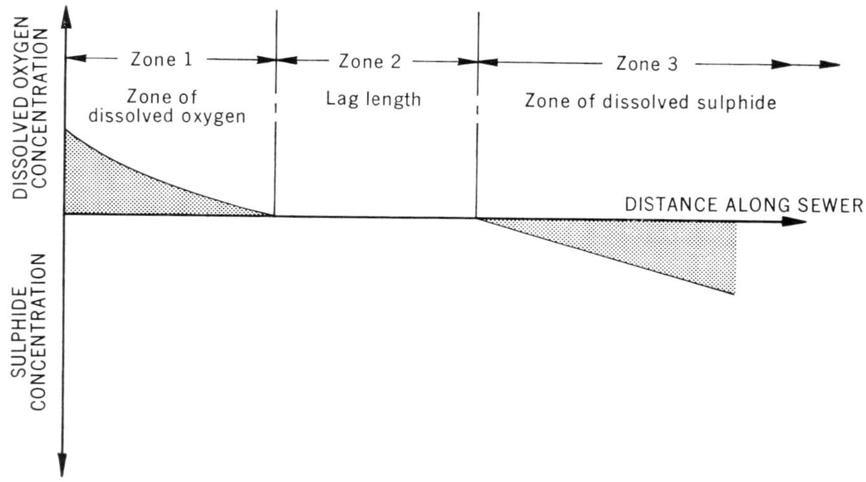

FIG. 8.1 Characteristic zones in sewers.

Many factors are involved in this complex of phenomena, but the designing engineer will find in this chapter a series of mathematical expressions and graphs from which inter-relationships can be reckoned readily and used to predict the consequences of any particular proposal in respect to dissolved oxygen and sulphide balances. While these formulations must not be expected to yield precise estimates of conditions at any one place, experience so far suggests that they may be used with an acceptable degree of confidence for predicting "sulphide" conditions.

Formulation for dissolved oxygen balance along Zone (1): There are many individual factors to be allowed for in any estimation of this balance. A detailed discussion of these factors and the development of appropriate mathematical expressions are presented in Part I, pp. 21-32. The following expression covers the simple concepts of the steady depletion of dissolved oxygen *in new sewers without wall slimes* (eqn. [8.3]).

$$L_{AB} = \frac{O_S \times V_S}{\frac{d}{dt}[BOD_0] - \frac{1}{60} \times O_D \times K_{O_2S}} \qquad [8.3]$$

where L_{AB} is the effective length of Zone (1) under the conditions covered by the following definitions:

O_S is the dissolved oxygen concentration in the inflowing sewage at A, mg/l.

O_D is the average dissolved oxygen saturation deficit over the length AB, mg/l. This normally may be obtained by referring to Design Chart No. 10 or to Table 4.1 p.22, which gives the saturation value for the actual sewage temperature applying, ($(O_S)_{sat.}$, mg/l). When sewage dissolved oxygen is O_S, then $O_D = (O_S)_{sat.} - O_S$.

V_S is the mean sewage velocity, feet per second.

$\frac{d}{dt}[BOD_0]$ is the instantaneous rate of consumption of oxygen by biochemical oxidation within the flowing sewage, mg/l equivalent oxygen per second. It is recommended, making allowances for an increased rate of consumption of oxygen in flowing sewage compared with quiescent laboratory tests, that $BOD_5 \times 6 \times 10^{-6}$ be taken as the normal value for $d/dt\ [BOD_0]$, see Chapter 4, p. 31.

K_{O_2S} is the rate of re-aeration of sewage, mg/l per unit dissolved oxygen saturation deficit per minute. This may be estimated using equations [4.14 and 4.15] and Figures 4.4 and 4.5. The term H = effective depth $\frac{A}{B}$ is brought in with the coefficient K_{O_2S}, other corrections being made likewise as in Chapter 4.

Circumstances of Zone (2): A formulation has not been developed for the progressive changes along Zone (2). It is an indefinite stage of marked differentials in redox potential inevitably involving "lag" phases during which micro-organisms either slowly adapt to the changing conditions or are superseded by new growths of better adapted species. Various new sequences of oxidation-reduction mechanisms replace those established where some free dissolved oxygen was present, which lead finally to microbe-assisted reactions such as carbonate-nitrate-sulphate reduction mechanisms.

Sulphide build-up probably appears in the sewage first not as a direct consequence of these changes in the flowing sewage but as a result of changes in wall-slime character and activity. A relatively high rate of sulphate reduction in these slimes probably will be found some distance upstream of the zone where "full septicity" has developed in the flowing sewage.

Formulation for sulphide build-up along Zone (3): The sulphide concentration at a particular point in the system is the result of the following overall balance:—

(Total Sulphide concentration in sewage at down-stream site Y) = (Sulphide concentration at upstream Site X) + (Build-up due to generation in slimes etc.

along the sewer length X to Y) + (direct discharges of sulphide e.g. with trade wastes) — (Sulphides lost by oxidation within sewage) — (Sulphides lost by incorporation into silt deposits) — (Sulphides lost by emission from the sewage into the sewer air).

Of these items, those of the additions of sulphide in trade wastes etc. and losses by oxidation are not formulated. The oxidation loss might be taken as being close enough to the equivalent oxygen uptake, allowing for other short term biochemical and immediate chemical oxygen demands. Loss of sulphide by precipitation, including accumulation in silt deposits, may be neglected.

Build-up due to generation in slimes

A detailed discussion of all the factors involved in this source of sulphide build-up is presented in Chapter 3. The following equation summarizes the effect of the principal sewage variables:—

$$G_S = \frac{K \times V_S \times [BOD_5]^a \times [SO_4]^b}{a + b} \times C^{\Delta T^\circ F} \quad [8.3a]$$

where G_S is the Sulphide S build-up due to slime generation, lb./1,000 sq. ft./hour.

V_S is the mean sewage flow velocity, ft./sec.

$[BOD_5]$ is the 5-day Biochemical oxygen demand of the sewage.

$[SO_4]$ is the Sulphate concentration in the sewage expressed as mg $[SO_4]$ per litre.

$\Delta T^\circ F$ is the appropriate temperature correction exponent.

K, a, b, and C are constants derived empirically or experimentally. Co-operative investigations (15) of build-up in pressure mains operating under various conditions in different parts of Australia yielded the following correlation equation [8.4] (cf. [3.8] p. 15).

$$G_S = 32.2 \times 10^{-6} \times V_S \times [BOD_5]^{0.8} \times [SO_4]^{0.4} \times 1.139^{(T^\circ C - 20)} \quad [8.4]$$

Note: The conditions applying in these co-operative tests varied between the following overall limits: pipe diameters 1 ft. to 4 ft.; sewage velocities during pumping 1 ft. per second to 3 ft. per second; temperatures 67°F (19°C) to 81°F (27°C); BOD_5 (overall range) 90 mg/l to 800 mg/l; sulphate sulphur concentrations 14 mg/l to 200 mg/l. Wall shear values lay between about 0.004 and 0.056 p.s.f.

It is to be expected that the rates of generation would be practically the same for slime areas of pressure mains as for continuously submerged areas of gravity sewers along Zones (2) and (3). However, portion of the slime areas in gravity sewers is subject to the effects of the diurnal rise and fall of the sewage surface. Boundary shear forces also vary over the submerged periphery of the sewer, and may reach values high enough to affect the thickness and activity, or even adhesion, of the slimes. Accordingly, we may write:—

$$(G_S)_{eff.} = \frac{F_1 \times P_T \times G_S}{1{,}000} \quad [8.5]$$

where $(G_S)_{eff.}$ is the effective generating rate for a gravity sewer, lb. per sq. ft. per hour.

P_T is the total submerged perimeter,

F_1 is a factor covering the overall effects of intermittent submergence, wall shear, etc. on slime growth and activity.

G_S is the generating rate given by Eqn. [8.4] above.

Three cases need to be considered (see Figs. 8.2(a), (b) and (c):

(a) where shear rates are too low to affect slimes and slime activity,

(b) where slimes are removed or affected by shear forces only at levels below the sewage surface level at minimum flow, and

(c) where shear may affect slime also at levels above minimum flow.

Figures 8.2(a), (b) and (c) respectively depict these circumstances. Where P_T, P_i, P_c and P_S give respectively the total submerged area, the intermittently and continuously submerged areas and the total area of slimes, per foot run of sewer, ($P_S = P_i + P_c$), and taking F_S as the proportion (P_S/P_T) of the area on which the slimes develop; and where $(G_S)_i$ is the effective contribution to the sulphide build-up of the intermittently submerged slimes expressed as the equivalent rate in lb of sulphide sulphur per 1,000 sq. ft. per hour, we have from Fig. 8.2:—

$$\text{(a)} \quad F_1 = \frac{P_c}{P_T} + \frac{(G_S)_i}{G_S}\left(1 - \frac{P_c}{P_T}\right) \quad [8.6a]$$

$$\text{(b)} \quad F_1 = F_S - \frac{P_i}{P_T}\left(1 - \frac{(G_S)_i}{G_S}\right) \quad [8.6b]$$

$$= \frac{P_S}{P_T} - \frac{P_i}{P_T}\left(1 - \frac{(G_S)_i}{G_S}\right)$$

$$\text{(c)} \quad F_1 = F_S \times \frac{(G_S)_i}{G_S} = \frac{P_S}{P_T} \times \frac{(G_S)_i}{G_S} \quad [8.6c]$$

From Eqn. [8.5] we have $(G_S)_{eff.}$

$$= \frac{F_1 \times P_T \times G_S}{1{,}000} \text{ lb./hour}$$

$$= \frac{F_1 \times P_T \times G_S}{3.6 \times 10^6} \text{ lb./second}$$

∴ Build-up per ft. of sewer, due to generation,

$$B_S = \frac{L}{V_S} \times F_1 \times \frac{P_T \times G_S}{36 \times 10^5} \text{ lb. per ft. per second}$$

$$= \frac{L}{V_S} \times F_1 \times G_S \times \frac{P_T}{A} \times \frac{10^6}{62.5 \times 36 \times 10^5} \text{ mg./l per second}$$

Substituting from eqn. [8.4],

$$B_S = 1.43 \times 10^{-7} \times V_S \times [BOD_5]^{0.8} \times [SO_4]^{0.4} \times 1{\cdot}139^{(t^\circ C-20)} \times \frac{F_1 P_T}{A} \text{ mg./l/sec.} \quad [8.7]$$

Build-up due to generation in silt deposits

There is no information, based on direct observations, as to the rate of sulphide generation per unit area of silt. Counts of sulphide-producing bacteria per unit volume indicate that silt is not important as a source of sulphide compared with wall slimes, and accordingly, it is recommended that sulphide contributions, by generation in the silt be ignored unless the design engineer knows from particular local experience that some allowance is desirable. For such cases, it is suggested that the contribution of the silt be taken in proportion to the relative counts of sulphide-reducing bacteria in silts compared with slimes. In very sluggish sewers, a normal slime layer may develop also on the surface of silt deposits, and comparable unit rates of build-up then may well occur.

Build-up due to generation in the body of the sewage

This aspect has not been formulated but is discussed in Part I, pp. 19, 20, where it is concluded finally that in most cases this source of build-up also is relatively small and may safely be ignored. The design engineer is warned again, however, that this may not be so where special circumstances arise. In one case of a

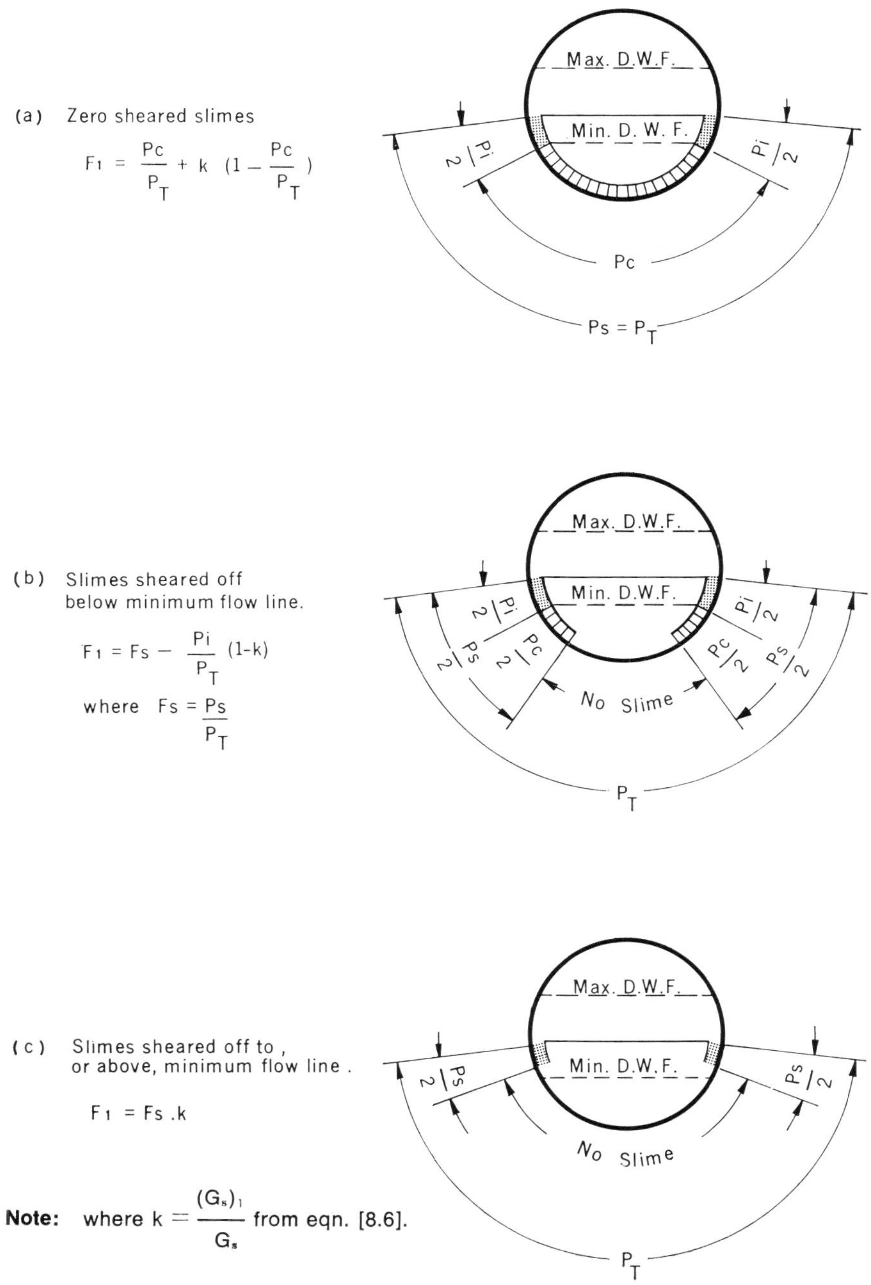

FIG. 8.2 (a), (b) and (c) Slime distribution diagrams.

sewer near Sydney, for example, a sedimentation pit installed in a dyehouse was particularly favourable for the development of sulphate-reducing bacteria. This resulted in comparatively low redox potentials and comparatively high rates of generation of sulphides in the body of the receiving sewage. These circumstances were brought under control by reducing the effective detention time in the sedimentation pit and also instituting improved cleaning and maintenance schedules for the pit.

Contribution of sulphides in trade waste discharges

In addition to various indirect effects on the oxygen demands and redox potentials of sewages, trade or industrial waste discharges come from a wide variety of factories operating many different processses, handling many different raw materials, by-products and finished products, and particular wastes may carry significant concentrations of total sulphide. Such wastes therefore require special attention. The question of which industries may require such special consideration, and the control of trade waste discharges, are discussed elsewhere, including a discussion of the forms in which sulphides occur in sewage as suspended or dissolved sulphide or as free dissolved molecular H_2S. Only the dissolved sulphides are significant for sulphide problems in sewerage systems, this dissolved portion normally being about 40% to 60%, or say one-half, of the total sulphide. Experience suggests that sulphide problems are not likely to be significant if the *dissolved* sulphide content is less than about $\frac{1}{4}$ mg/l or about $\frac{1}{2}$ mg/l equivalent total sulphide.

It is recommended that the design engineer take account of possible trade waste problems as follows:—

Step 1: If a preliminary review of the area to be sewered suggests that some industry is or may be established in the drainage area, an experienced sewage or trade waste chemist should be asked to investigate the circumstances, giving special attention to BOD and sulphide concentrations and also suspended matter. From the report, the design engineer should determine whether special consideration will be given to the area on this account, including provisions for particular types of junctions, or for special ventilation requirements.

Step 2: These decisions made, appropriate notes recorded and the necessary design provisions incorporated in the relevant detailed local designs, the design engineer should then proceed as outlined below in preparing the remaining detailed designs of the sewers etc., without allowance for these trade waste possibilities; bringing them again into account at the stage of final assessment.

Step 3: This subsequent and, in most cases, final step should be provided for *quite clearly* in the design engineer's preliminary reports and in any negotiated contracts arising therefrom. When the system ultimately is in operation, with constructions and connexions for the particular stage of the systems' development practically completed, the consulting chemist or other such agent should be asked to investigate the operating system and to report on sulphide conditions in relation both to the detailed and overall intentions of the design.

Should the contributions from trade and industry be found or judged significant, appropriate action should then follow so as to bring the sulphide contributions from these sources into line with the conditions aimed at, or otherwise to remedy the situation. The advice of a sewage chemist or chemical engineer then should be sought again.

Forms of sulphide

The following diagram summarizes the interrelationships of the various forms of sulphide which the total sulphide content of sewages normally represents.

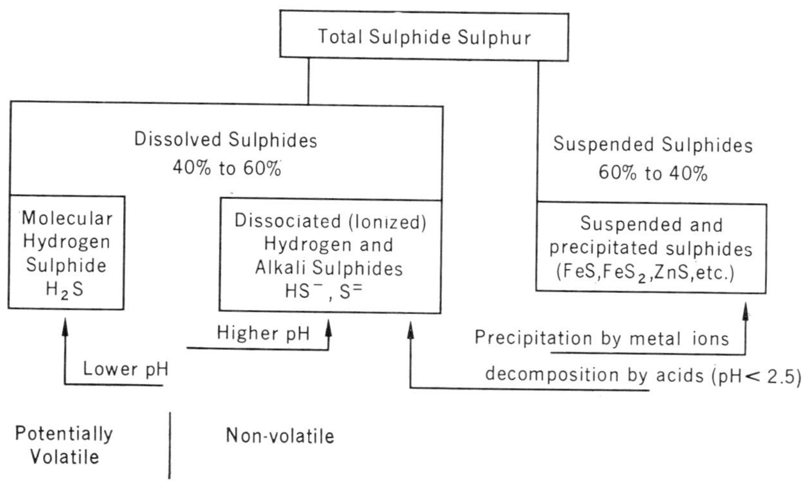

The occurrence and distribution of the sulphide sulphur content of sewage.

The proportion of total sulphides which are in the form of dissolved sulphides varies from sewer to sewer and from time to time, the amounts and nature of trade waste usually also having an effect. Until more is known of the cause and magnitude of these variations, it is recommended that a value of 0.5 be applied to the sulphide prediction calculations, that is, Dissolved Sulphide = 0.5 × Total Sulphide, see pp. 34, 35.

Sulphide sulphur balance

Losses due to oxidation: Sulphides are subject to oxidation in sewage only if the redox potential be high enough. Generally oxidation occurs wherever free dissolved oxygen persists in the sewage, but some oxidation will occur also even in stale sewage, at the expense of the oxygen which is dissolving at the sewage surface. Some of the oxygen will be used also in the oxidation of other substances. The overall effect may be formulated as follows:—

For clean water, the rate at which O_2 is absorbed at the surface $= K_2 \times O_D$

where O_D is the oxygen Saturation deficit (mg/l)

K_2 is the rate of oxygen absorption (re-aeration) for clean water, mg/l per min/unit dissolved oxygen deficit

$$= 0.163 \times \frac{(V_S \times S)^{0.408}}{f_d^{0.66}} \quad \text{see [4.7] p. 24}$$

where V is the water velocity ft./sec.

S is the slope of pipe or channel

f_d is the hydraulic mean depth $\left(\frac{A}{B}\right)$ ft.

The absorption of oxygen into sewage surfaces is more or less reduced by the presence of absorbed films, so we write

The rate of absorption of oxygen by sewage,

$$K_{O_2S} = F_{dt} \times K_2$$

The factor F_{dt} may be taken as being correlated with the organic matter of sewage, which in turn is correlated with the sewage BOD, Fig. 4.4, p. 26, also eqn. [4.14].

As noted above, the oxygen taken up will be used both in the oxidation of sulphides and to satisfy sewage oxygen demand. In Chapter 4, sewage oxygen demand is considered in detail and the normal instantaneous demand at any particular place in the system was reckoned at about $6 \times 10^{-6} \times [BOD_5]$ mg. O_2 per l. per sec. However, a better practical value appears to be

$$\frac{d}{dt}[BOD_0] = 6 \times 10^{-6} \times [BOD_5] \text{ mg. } O_2 \text{ per l. per sec.}$$

It has been suggested that sulphide oxidizes either to sulphate or to tetra-thionate. For the first alternative, the theoretical oxygen demand of the sulphide formed by the slimes would be equal to twice the rate of sulphide generation, $2 \times G_S$. For the second alternative, the corresponding multiplier for G_S would be 1.125. Measurements in sewers indicate that the multiplier approximates to 2, consistent with the proposition that sulphide is oxidized to sulphate.

Allotting the oxygen taken up in proportion to the rate of "total" instantaneous demand, reckoning on both a practically instantaneous oxidation of sulphide and a continuing "instantaneous" biochemical demand, we have

O_2 consumed in oxidation of sulphide

$$= \frac{K_{O_2S} \times O_D}{60} \times \frac{2 \times G_S}{\frac{d}{dt}[BOD_0] + 2G_S}$$

and the equivalent rate of loss of sulphide due to oxidation

$$\triangle[S]_{Ox} = \frac{K_{O_2S} \times O_D}{60} \times \frac{G_S}{\frac{d}{dt}[BOD_0] + 2G_S} \quad \text{mg. } O_2/\text{l/sec.} \quad [8.8]$$

Losses by emission: These losses are discussed in detail in Chapter 6, where formulations are developed primarily from the viewpoint of the sulphide balance in the sewer air, the main determinant of the rate of corrosion. Normally, the proportion of the sulphide present in sewage which is lost by emission is not a large proportion, whether one reckons the proportion in terms of the total amount present in the sewage, or from the rate of generation. If this item be ignored, designs for sulphide control will be conservative accordingly. It is recommended that the design engineer bring this item into account only where he judges that the circumstances are rather unusual and favour relatively high rates of emission.

Note: *The effect of drops.* Equation [8.8] is intended to cover only the conditions to be found in sewers corresponding to Zone 3 (Fig. 8.1) or to the adjacent regions of Zone 2. Both oxygen uptake and H_2S emission are greatly increased by drops, the effects of which are dealt with in detail in Chapter 9. Where drops come into question, an entirely new regime is almost certain to result, so that the foregoing argument and formulation then will not be directly relevant.

Final design equation for build-up along Zone 3

This may now be summarized to formulate the following design equation. From above,

$(G_S)_{eff}$ = effective rate of build-up of sulphide sulphur mg./l per second

$$= G_S \times \left[1 - \frac{K_{O_2S} \times O_D}{60\,(\frac{d}{dt}[BOD_0] + 2G_S)}\right] \quad [8.8]$$

where $G_S = 1.43 \times 10^{-7} \times V_S \times [BOD_5]^{0.8} \times [SO_4]^{0.4} \times 1.139^{(T°C-20)} \times \frac{F_1 P_T}{A}$ [8.7]

$$K_{O_2S} = F_{dt} \times 0.163 \times \frac{(V_S \times S)^{0.408}}{f_d^{0.66}} \text{ mg. } O_2/\text{l}$$

per minute per unit oxygen deficit

O_D is the oxygen saturation deficit, see Figure 10.8

$\frac{d}{dt}[BOD_0]$ is the instantaneous rate of sewage BOD, mg. O_2/l per second = say $6 \times 10^{-6} \times [BOD_5]$

V_S is the mean sewage velocity ft., per second

S is the sewer slope (dimensionless ratio)

f_d is the depth of flow $\left(\frac{A}{B}\right)$, ft.

$[SO_4]$ is the sewage sulphate content, mg. (SO_4)/l.

F_1 = slime activity correction factor, see Table 6·1.

P_T = total submerged perimeter, ft.

F_{dt} = correction factor for surface active films see Figure 4.4.

Determination of Slope required for zero sulphide build-up: From the above equation [8.8] we see that for zero build-up, $(G_S)_{eff.} = 0$,

then $$1 - \frac{K_{O_2S} \times O_D}{60 \times \left(\frac{d}{dt}[BOD_0] + 2G_S\right)} = 0,$$

or $$K_{O_2S} \times O_D = 60\left(\frac{d}{dt}[BOD_0] + 2G_S\right)$$

Substituting from above,

$$F_{dt} \times 0.163 \times \frac{(V_S \times S)^{0.408}}{f_d^{0.66}}$$

$$= 3.6 \times 10^{-4} \times [BOD_5] + \left(1.43 \times 10^{-7} \times V_S \times [BOD_5]^{0.8} \times [SO_4]^{0.4} \times 1.075^{(T^\circ F - 68)} \times \frac{F_1 \times P_T}{A}\right) \qquad [8.9]$$

V_S and S are sewer design values for zero sulphide build-up.

It is emphasized that build-up of sulphides will not always occur with velocities below those determined using this equation, but some build-up as predicted is likely. The equation tends to be conservative, and it is on the contrary quite unlikely that any build-up at all would occur with velocities equal to or greater than any velocity V_S which is consistent with the formulation of this equation [8.9].

The details of its formulation can be found from equations [3.8] to [3.13] and [4.6] to [4.17], see pp. 15-18, 23-28.

Approximate methods for predicting sulphide conditions also have become available. In Chapter 10 it is suggested that such methods be used to obtain preliminary assessments of the likelihood that undesirable levels of sulphide build-up will occur. Thus in cases where there is not likely to be any significant build-up, more precise assessments based on the relatively complicated complete reckonings of equation [8.9] become unnecessary, and much time is saved accordingly. Such preliminary assessments on the other hand indicate in what areas particular attention is required, also the areas where conditions may be marginal and so on. Three such methods are illustrated by example in Chapter 10, pp. 111, 112, one of which is the empirical formulation due to Pomeroy, as set out in the following.

Pomeroy's Formulation for the parameter Z_p: Based on his own work and of Davy (*15*), Pomeroy has developed the equation (*50*)

$$Z = \frac{(E\ BOD) \times P}{S^{\frac{1}{2}} \times Q^{\frac{1}{3}} \times b}$$

This has been modified to give equation [8.10]

$$Z_p = \frac{(BOD_5) \times 1.038^{(T^\circ F - 68)}}{Q^{\frac{1}{3}} \times S^{\frac{1}{2}} \times f\left(\frac{d}{D}\right)} \qquad [8.10]$$

where Zp is Pomeroy's sulphide corrosion parameter

$f\left(\frac{d}{D}\right)$ is $1.4 \times \frac{b}{P} \times \left(\frac{Q}{Q_f}\right)^{0.064}$

S is slope of sewer

b is surface width

P is wetted perimeter

Q is part full discharge

Q_f is full discharge

If Camp's variation of Manning's n be used the following relationship between $f\left(\frac{d}{D}\right)$ and $\frac{d}{D}$ may be derived:

$\frac{d}{D}$	1	0.9	0.8	0.7	0.6	0.5	0.4	0.3	0.2	0.1
$f\left(\frac{d}{D}\right)$	0	0.34	0.50	0.63	0.74	0.84	0.92	0.98	1.02	1.01

Pomeroy (loc. cit.) relates this parameter Z_p to the possible occurrence of sulphides in sewers, considering conditions during the least favourable quarter, as follows:

Condition

Z_p below 5,000
sulphide rarely present and then only in very small concentrations; silver and copper alloys and perhaps lead-bearing paints may be darkened; no other nuisance.

Z_p around 7,500
peak concentrations of a few tenths of a mg/l of dissolved sulphide may be reached; slight corrosion of concrete and masonry in structures may occur, especially where there is high turbulence.

Z_p around 10,000
sulphide sometimes may develop in sufficient proportion materially to increase odours, and concrete and masonry structures again especially where there is high turbulence may suffer substantial damage, perhaps with slight attack of concrete and asbestos-cement elsewhere.

Z_p around 15,000
odour of the sewage will increase markedly at times, also chlorine demand and oxygen require-

ments, e.g. at secondary treatment plants. Some rapid attack of concrete structures is to be expected at points of turbulence, with some serious attack elsewhere also. With concrete pipe of 1 in. wall thickness there is a strong possibility of failure within 25 years.

Z_p at or above 25,000
dissolved sulphide will be present most of the time, and small concrete pipes possibly will fail in 5-10 years. (If made with calcareous aggregate, they may last longer, see p. 71).

Thus according to Pomeroy sulphide build-up occurs rarely if ever where Z_p is less than 5000, and at worst only marginal build-up where Z_p = 5000 to 10,000.

This concept therefore provides a simple means by which the designer might first make preliminary assessments of a particular sewerage system; so that he might first determine which areas are unlikely to cause sulphide build-up and which areas on the contrary require more detailed and more accurate investigations, based on equations [8.8] and [8.9], as illustrated in the example appended to Chapter 10.

Figure 8.3 gives an idea of the relativity of Pomeroy's simplified approach and the more detailed procedure.

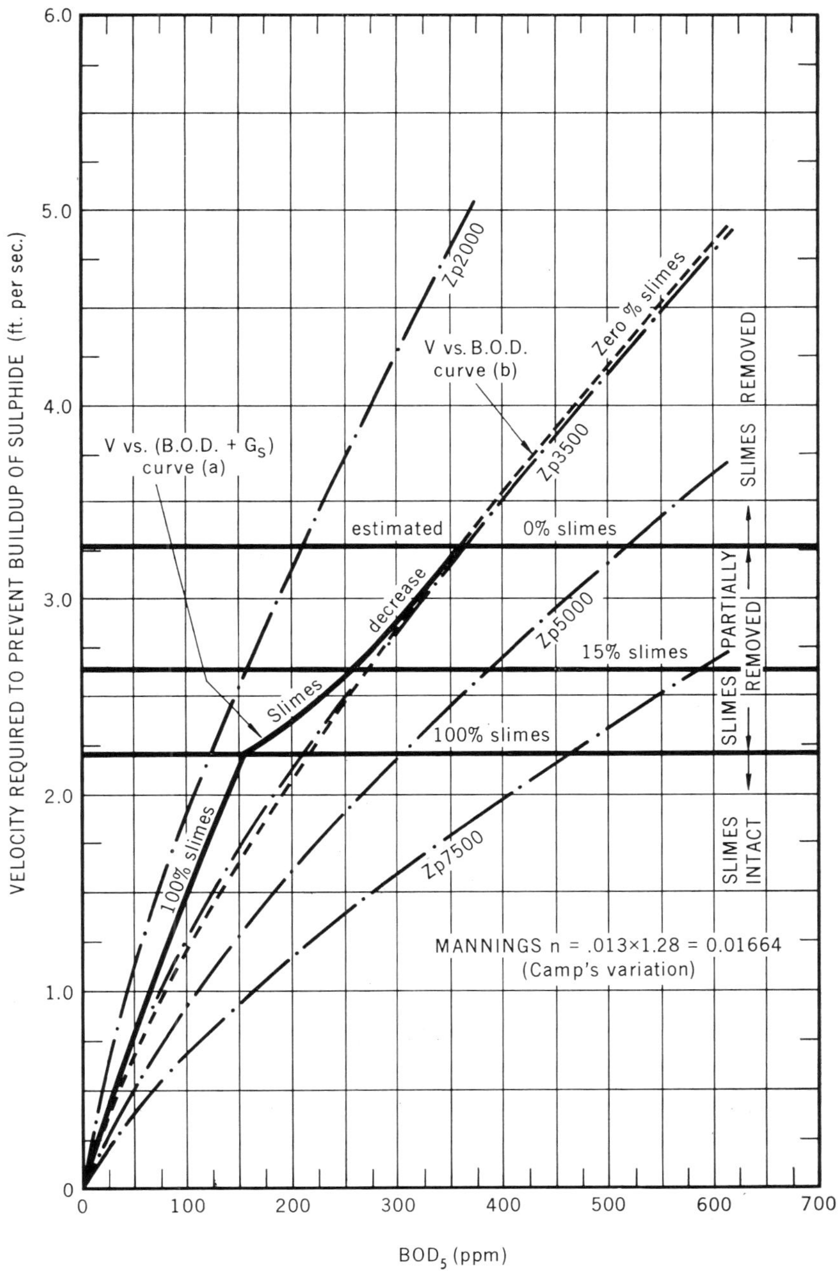

FIG. 8.3 Critical sewage velocities to obviate sulphide buildup.

Figure 8.3 compares Pomeroy's Zp criterion with equation [8.9], for a sewer in which D = 2ft. 6in., d/D = ¼, Temp. = 68°F, $[SO_4]$ = 60 mg/l. The respective values of (V_S) crit., the sewage velocities for no build-up, are plotted against different values for sewage $[BOD_5]$.

The effect of various factors on sulphide build-up

The following Figures 8.4 to 8.9 have been prepared from computations using the foregoing Sewer Design Equations so that design engineers might perceive the overall general effect of the separate variables of the equations [8.8] and [8.9]. Each figure shows graphically the effect of varying one only of the six main factors, the remaining five factors being held constant to the values corresponding to the following assumed sewer and sewage flow dimensions and characteristics (reference values):—

(1) Reference Diameter of (Circular) sewer D = 2ft. 6in.

(2) Reference Grade of sewer S = 1:1000

(3) Reference Proportion depth of flow (d/D) = ¼

(4) Reference Sewage temperature T°F = 68°F

(5) Reference BOD of sewage $[BOD_5]$ = 400 mg/l

(6) Reference Sulphate content of sewage $[SO_4]$ = 60 mg/l

Note: For this purpose it has been assumed that the diurnal rise and fall of the sewage (for any of the values of d/D) is negligible, and the Factor F_1 is taken to mean the percentage proportion of the wetted perimeter supporting active slimes ("percent slimes").

The effect of changing of diameter (other variables held constant): This effect is shown in Figure 8.4. The effect of increasing the diameter with d/D constant is to decrease the rate of build-up. This effect however is proportionately less and less up to about 5ft. diameter for d/D = ¼ until a dramatic fall-off commences due to the shearing of slimes from more and more of the wetted sewer walls.

It will be appreciated that the change in diameter means corresponding increases in total sewage discharge rates.

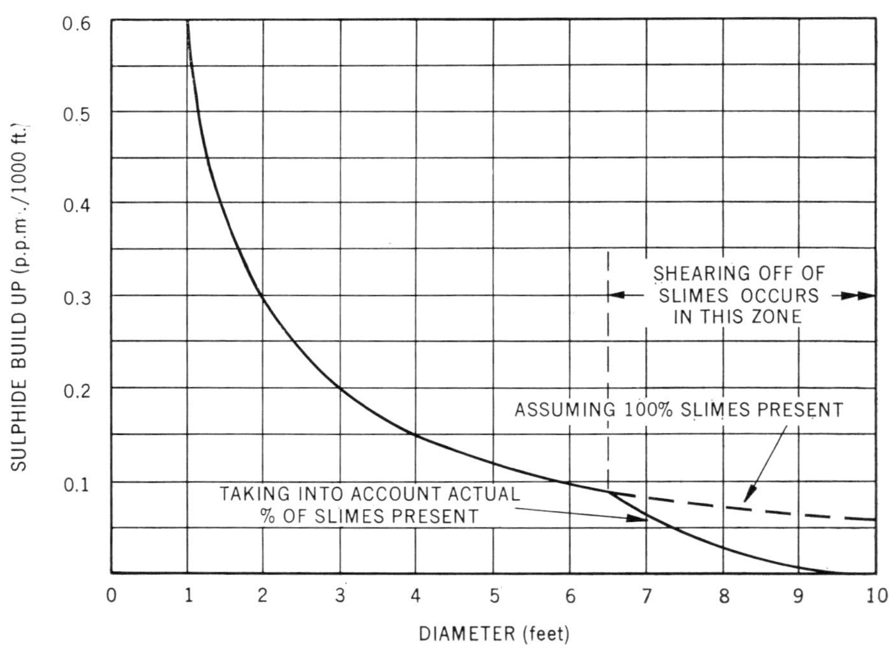

FIG. 8.4 Effect of changing sewer diameter ($\frac{d}{D}$ = ¼, etc.).

The effect of slope is shown in Fig. 8.5 (other variables held constant). Until grades are reached where the shearing off of slimes commences—for 2ft. 6in. diameter at about 1:500 slope—build-up decreases directly with increasing slope, but relatively more and more slowly until shearing conditions are reached, when the differential effect of slope changes dramatically.

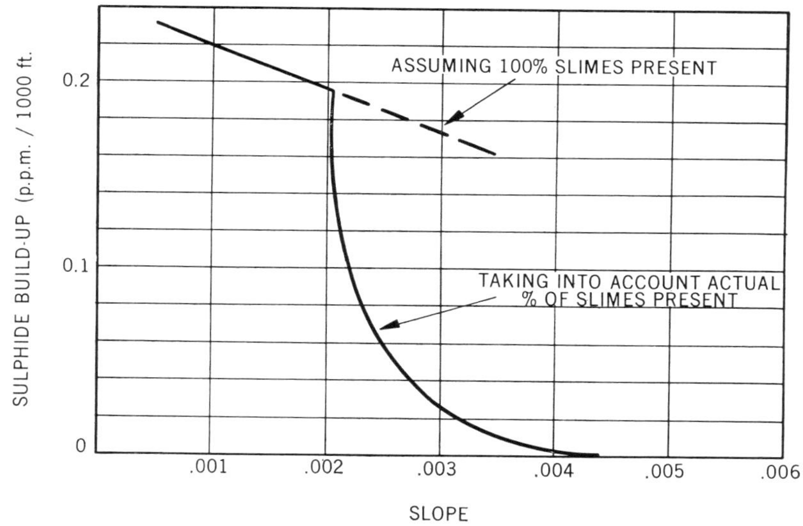

FIG. 8.5 Effect of changing sewer grades (D = 2½ ft., $\frac{d}{D}$ = ¼, etc.).

The effect of proportional depth of flow is shown in Fig. 8.6 (other variables held constant). Note that this particular case corresponds nearly to the changing conditions which occur progressively after construction and commissioning, as the contributing population increases with development of the area served. Sulphide build-up decreases with increasing flows (corresponding to increasing d/D), but since the grade has been predetermined and fixed at 0.001 slope, there is no shearing off of slimes, and % slimes also remain substantially unchanged.

When Fig. 8.6 is examined, there appears at first sight to be an anomaly in that a lesser build-up of sulphide occurs with the pipe running full, than at any proportional depth up to about 1/3 full, the proportional depth at which most sewers operate under dry weather conditions and at which most sulphide problems occur. This apparent anomaly however is cleared away when it is considered that if the sewer had been designed to take the equivalent dry weather flow running full, the diameter of the pipe would have been not 30in. diameter but only 12in. diameter and the corresponding build-up per 1,000ft. under these conditions would be about twice as much as with the 30in. pipe operating at quarter depth.

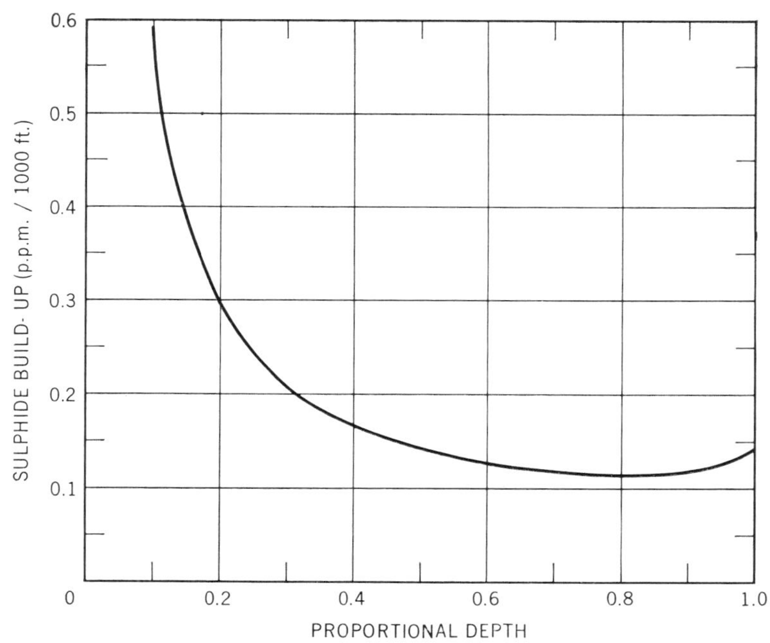

FIG. 8.6 Effect of changing proportional depth of flow (S = 1:1000 etc.).

Effects of sewage properties (sewer dimensions and slope and sewage flow held constant) are shown in Figs. 8.7 to 8.10, respectively showing effects of changes of temperature, and BOD_5 and sulphate contents, varied one at a time only in turn, and other properties held constant along with the sewer conditions. All these changes have marked effects on sulphide build-up, which increases more and more as temperature increases at least up to 90°F, practically directly with increasing BOD_5, and increasing at first very rapidly, but above about 60 mg/l progressively much less, with increasing sulphate content.

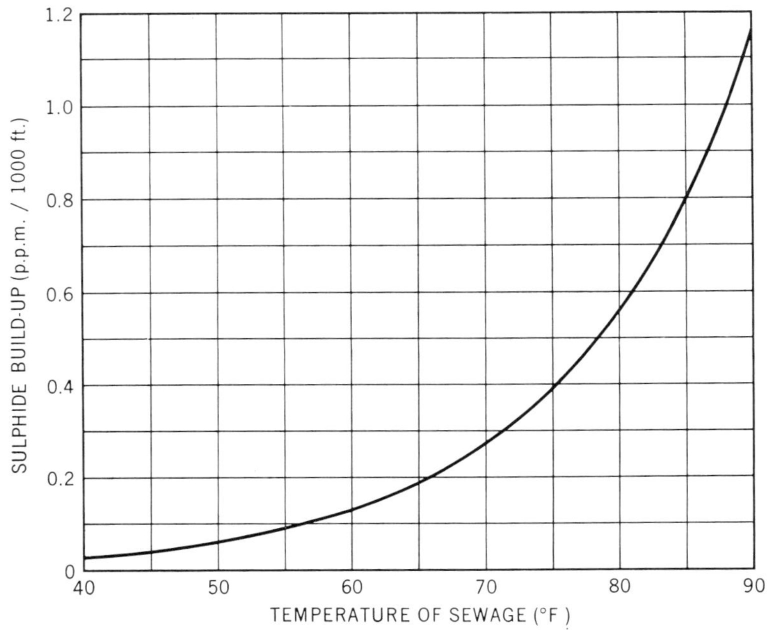

FIG. 8.7 Effect of changing sewage temperature (other conditions and sewage composition constant).

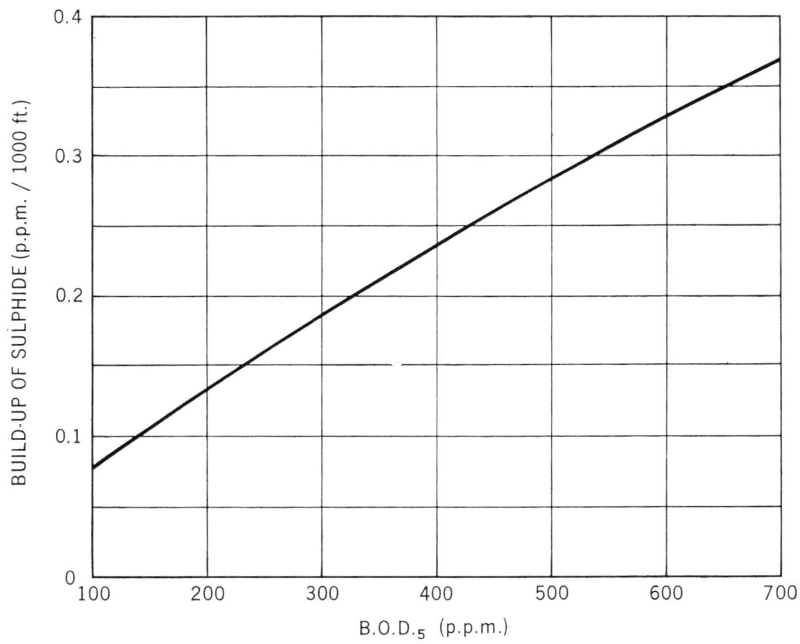

FIG. 8.8 Effect of changing BOD (other conditions and sewage components constant).

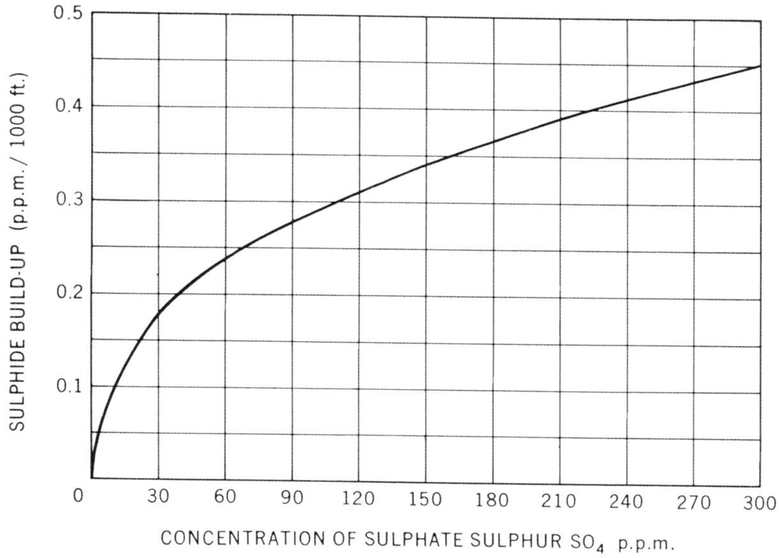

FIG. 8.9 **Effect of changing sulphate concentration (other factors constant).**

Data Guide for Australian Conditions

The designer aiming at preventing or limiting sulphide build-up must determine as nearly as he can the volumes and character of the sewage flows which now occur or will occur from time to time in the sewerage system in question. For existing systems, adequate flow measuring procedures combined with appropriate arrangements for sampling and testing will be practicable. For proposed new systems assessments must be made, and preferably should be based on studies of comparable systems, if feasible from the same or a similar geographical region. In either case, the assistance of chemists already experienced with sewerage systems should be sought. Such assistance should be extended to questions of industrial or other abnormal waste inflows, including groundwaters and significant aspects especially of temperature effects, metal-ion concentrations, acidity or alkalinity and so on.

The following data provide only broad guide-lines for planning such investigations and making basic assessments. There are no particular conditions *typical* for Australia, but these data indicate overall ranges to which most cases would be found to conform. Unfortunately, data are collected regularly in the cases of only relatively few sewerage systems in Australia. Data from associated water supply systems may also be relevant, especially relating to sewage temperatures and sulphate contents, but these also are often deficient.

Similar guide-lines should be sought by designers working in other countries, where appropriate local data should likewise provide the basis for determining sewerage conditions and sewage character.

The specific data referred to in this book include weather conditions and sewage flows and temperatures, water supply dissolved oxygen contents, and the concentrations in sewage of sulphides, sulphates, BOD, and detergents; and sewage pH levels. Other factors are referred to in more general terms.

Sewage flows in Australia are distributed widely both above and below a median range of around 40 to 50 gallons per capita per day (g/c/d). In country towns where piped water supply and sewerage have not long been operating, sewage flows may be only about 30 to 35 g/c/d or even as low as 25 g/c/d, but subsequently water consumption and wastage to the sewers tends to increase steadily, reaching or even exceeding the above median levels. Flows in cities or larger country towns now not uncommonly exceed 60 g/c/d, although at least one of the larger cities still records flows of less than 50 g/c/d including a substantial proportion of industrial wastes. This may be due partly to low rates of groundwater infiltration. Flow patterns suggest that infiltration usually reaches at least 10 g/c/d/ average in dry weather. In the more industrialized areas, trade or industry waste flows amount to between 15% and 30% of the total sewage flow, the figure for most cities being at least about 20%. Two of the larger cities now have total sewage flows of 90 g/c/d or more in dry weather, perhaps including say 55 essentially domestic, 25 industrial and 10-15 infiltration, dry weather.

Wet weather conditions are highly variable, usually very much dependent on rainfall patterns, which tend to be relatively very "flashy" throughout the southern two-thirds or more of the Australian continent.

Table 8.2 presents available representative data.

TABLE 8.2 DWF AND PEAK WWF RATIOS FOR AUSTRALIAN SEWERS

System \ Flows and Ratios	Spring-Summer DWF g/c/d Average	Spring-Summer DWF g/c/d Peak	Spring-Summer Peak WWF/DWF	Autumn-Winter DWF g/c/d Average	Autumn-Winter DWF g/c/d Peak	Autumn-Winter Peak WWF/DWF
A large	70	100	3.1	66	80	3.7
B large	66	92	1.0	?	?	1.3
C large	60	70	3.2	60	70	3.2
D moderate	39	?	?	41	?	?
E small	28	?	1.5	?	?	abt. 5

Note: The ratios given for WWF/DWF probably are low because of overflows upstream of the Flow Recorder Stations.

Sewage temperatures depend mainly on the temperature of the carrying (flushing) water, but also are influenced to a variable extent by ambient air temperatures, relative depth of sewage flow and relative air-sewage flow patterns. The significant temperatures are rather the weighted weekly mean temperatures than daily mean temperatures. For any one area, the minimum sewage temperature is at least slightly higher than the minimum weekly mean air temperature; while maximum sewage temperatures lag behind the weekly mean maximum air temperatures. Data from North Queensland show that in areas where daily average air temperatures might range between 95°F and 100°F or even more for 1-2 weeks together, sewage temperatures would not exceed about 85°F.

In Table 8.3, normal ranges of temperatures for Australian sewages are collected under three broad geographical regions for two seasonal periods. Note that temperatures in much of Tasmania or for some mainland mountain regions would be rather lower, and for the more northerly areas of the continent, rather higher, than those of the table.

TABLE 8.3 TEMPERATURE RANGES FOR TYPICAL AUSTRALIAN SEWERS

ZONE	COMMON TEMPERATURE RANGES °F			
	Spring-Summer		Autumn-Winter	
	Minimum	Maximum	Minimum	Maximum
Colder Temperate	about 65	about 75	about 55	about 65
Warmer Temperate	about 65	about 82	about 55	about 68
Sub-Tropical	about 70	about 85	about 60	about 75

As stated, actual values should be checked against local records of mains water temperatures. Modifying factors include ventilation-evaporation and storm-water-infiltration effects.

Few records of sewer air temperatures are available, but they indicate that sewer air generally is slightly warmer than the flowing sewage.

Mains water properties and composition: Most waters from piped supplies normally carry dissolved oxygen at levels between about 50% and 80% of the saturation value, but some groundwater supplies carry much lower levels. Probably 50% is a satisfactory value to assume for designs.

In some areas, the mains waters are relatively highly mineralized, with sulphate contents of up to 60 to 100 mg SO_4 per 1 or more. Such concentrations are highly significant for purely domestic areas. Again for purely domestic sewerage areas mains waters of low mineral content are associated generally with relatively low sewage sulphate content, in some cases perhaps only about 6 mg SO_4 per 1 of sewage, sulphide problems then commonly being of limited intensity.

Water treatment processes using aluminium

sulphate preparations for coagulation normally increase the sulphate concentrations in mains waters by up to about 10 mg SO_4/1.

Sulphate concentrations in sewages in domestic areas do not differ much from the sulphate levels in the water supply but individual wastes and ground and estuarine water infiltration may add relatively large amounts of sulphate to municipal sewers. Thus the water supply to one large Australian city contributes normally only about 6 mg/l of SO_4 to the sewage flow, but sulphate levels in the main sewers reach about 100 mg SO_4/1. Often these sources of extraneous sulphate are heavily infected with sulphate reducing anaerobic bacteria and justify special attention. The highest SO_4 concentration reported was 2000 mg/1, the lowest about 9 mg/l.

The general character of sewages cannot be typified, but it appears that, in general, sewages in Australia tend to be more often "stale" or "septic" than is commonly the case, for example, in, Europe or the United Kingdom. pH values also tend more often to be "acid", with pH levels centring around say 6.5, rather than alkaline, pH levels say 7 to 7.5. The former is broadly characteristic of soft water areas, the latter for moderately hard or hard. pH taken along with sulphide concentrations is a very important determinant of sulphide conditions, and designers are advised to seek assistance from chemists before adopting a design pH value. Trade wastes may be specially important in this connexion.

Other specific characters of Australian sewages also cannot be typified. *Suspended Solids* contents appear mostly to be low rather than high, and for most cases do not justify special consideration. Some special cases may arise because of industrial waste discharges. *BOD_5 values* for Australian residential sewages show a trend towards high rather than low per capita values. For suburban areas mean values normally are between about 0.13 and 0.15 lb/capita/day, but some values as low as 0.07 lb/c/d, even less, have been recorded. However, such values are too low even to match standard physiological norms for human body wastes alone, and probably such values reflect the problems of sampling and testing sewages from reticulation sewers. It is recommended that when values appear to fall below say 0.11 lb/c/d BOD_5 they be re-investigated. All values must of course be combined with carefully determined flow data to reach appropriate estimates of sewage composition. The weighted and simple mean values for a wide range of Australian data are respectively 0.146 and 0.145 lb/c/d. The range for 70% of all reported data, excluding the lowest and highest 15%, is from 0.10 to 0.17 lb/c/d BOD_5, the overall range being from 0.06 for a very small virtually unattended plant to 0.5 for a large city. Actual recorded BOD_5 contents have varied from about 20 mg/l to 1200 mg/l, but the general pattern reported is about as shown in Table 8.4.

Detergent contents in sewages in Australia appear to be following trends overseas. Levels of 20 mg/l or more have been recorded, but most values are well below 10 mg/l. Most manufacturers now are marketing compounds which are largely bio-degradable.

TABLE 8.4 BOD_5 CONTENTS OF MOST AUSTRALIAN SEWAGES

BOD_5, measured as mg O_2/1 of sewage		
Systems say to 15,000 persons	Say 15,000-50,000 persons	Greater than 50,000 persons
200 to 300 (mostly 220 to 260)	280 to 400 (mostly 300 to 360)	350 to 600 (mostly 400 to 500)
NB. See text above respecting sewage flows and lb/c/d values. 50 g/c/d and 0.14 lb/c/d gives BOD_5 equal to 280 mg/l.		

Grease contents of sewages generally appear to be between about 50 and 100 mg/l, but in some local areas trade wastes raise this perhaps to around 200 mg/l or more.

Nitrogen and phosphorous levels have not normally been determined, but it appears levels are always high enough to maintain biological activity.

The chlorine demand of raw sewages in Australia varies normally between about 20 mg Cl_2/1 and about 50 mg Cl_2/1. The main factors are sewage strength and the stage or extent of decomposition.

Chapter 9

DESIGN OF DROPS AND JUNCTIONS

Introductory

The design principles set out in the following pages are derived directly from those discussed in Part I of this book, including the mechanisms of gas transfer between sewage and sewer air. A short section is included which emphasizes that *all* details of sewer layouts should take appropriate account of the hydraulics both of the flowing sewage and the sewer air. The remainder of the chapter shows how the design of sewer drops and junctions in particular can be used as positive control measures. Sewer appurtenances generally should be considered from the same points of view. The design basis relates directly to the following two simplified ideas:—

(i) If the sewage entering a sewer is fresh, then it should be kept in that state as long as possible in its further travel along the sewer, and this aim will be helped (but only while freshness is maintained) by drops and other causes of turbulence which will tend to increase the rate of absorption of oxygen into solution.

(ii) If the sewage is septic, then turbulence must be avoided because although the absorption of oxygen by the sewage still may be desirable, the very factors which favour oxygen uptake will at the same time encourage the emission of hydrogen sulphide from the sewage into the sewer air, and other undesirable gases and vapours also.

These two opposing propositions must remain in the designer's mind not as unrelated ideas, but rather as two integral but interwoven entities always to be considered and taken together, to be kept constantly in mind by the designing engineer while developing both concepts and detailed designs.

"Fresh" sewage is defined as sewage containing *some* free dissolved *oxygen*—i.e. in an aerobic condition.

"Septic" sewage is sewage containing *some* free dissolved *sulphide*.

"Stale" sewage is sewage in a condition somewhere between these states of "freshness" and "septicity". It is an anaerobic state, but at an higher redox potential than would characterize "septicity".

These important ideas are shown diagrammatically in Fig. 9.1.

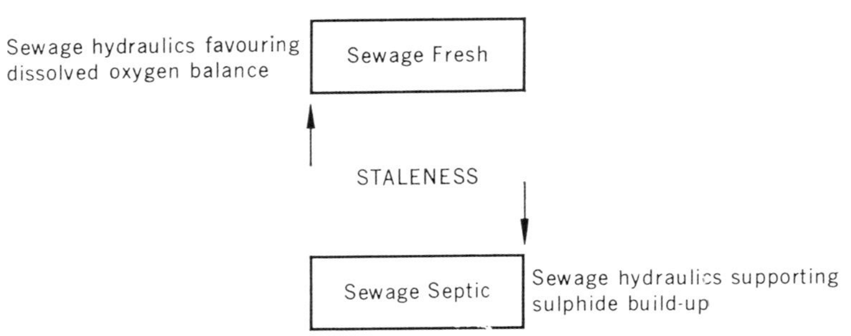

FIG. 9.1 Fresh and Septic Sewage relationship.

Streamlining of sewers

One of the main principles inherent in the foregoing is that for sulphide control the hydraulic regimes of the flowing sewage must be kept under a corresponding strict control. The application of this principle will be apparent generally from the following treatment of the design problems of drops, junctions, etc., but the designer should keep it always to the fore, applied to the whole length and all parts of each sewer.

Stated briefly, it is recommended that for sewer designs generally, except as qualified herein, projections into the sewage flow, rapid changes in direction including sharp-angle bends of small radius inside manholes, sudden enlargement or contraction of section, drops and weirs, should all be avoided. Where it can be established as unlikely that significant concentrations of dissolved sulphide will be present, less than say 0.25 mg/l, some relaxation of these requirements may be desirable and would be permissible—having regard for example to construction costs or to land acquisition difficulties—provided other important factors relating to siltation or other accumulations of solids, oils, etc., are taken into account and allowed for.

The idea of strict general application of this principle requires some qualification at least for sewered areas where sewage freshness is assured, as for example, by the topography. However, the above warning regarding the accumulation of solids may be important even for the smaller sewers which might be used for such areas. Apart from this, sewer hydraulics is likely to be important also for control of surcharge conditions in wet weather.

The designer should keep in mind, without attempting too much long-range planning, that areas now to be entirely residential may change character later, and sewage composition likewise change accordingly.

Although the influence on corrosion of shock turbulence in the air stream above the sewage may not be so great as when such turbulence occurs in the sewage it appears desirable also to streamline the air-way. It has been observed that larger projections into the air-way, such as roof beams, cause areas of increased corrosion on their downstream side and therefore should be avoided. The rates of transfer of hydrogen sulphide from sewer air to the roof and wall surfaces are promoted by high air velocity or shock turbulence in the air-stream, and it is to be expected that corrosion will be increased at and near projections as a consequence.

Sewer junctions

Sewer junctions require special attention because they occur throughout all sewerage systems, in many cases involving a considerable variety in circumstances, such as changes in direction and elevation and in the types and character of the sewered areas in question, the sewers involved, and the respective sewages to be merged.

In the following, simple junctions, where large energy changes due to drops are not involved, are considered as a general design problem, whereas drops and drop junctions are dealt with subsequently as special cases.

Design of junctions

A primary distinction is first made between sewers which are both small and aerobic, and those where conditions are more critical, more nearly on balance.

Junctions of small sewers carrying fresh sewage:— At the upper ends of a sewerage system, where in general the sewage is fresh, stream-lining of junctions is not generally important except in so far as it lessens head losses and "dead-water" areas where silting may occur, or in special cases such as where trade wastes occur. Stream-lining also assists the free flow of sewer air.

Crowns of inflowing sewers should not be lower than the crowns of outflowing sewers. The construction of dead-end channels for future connexions to manholes should not form pockets where detritus may collect, but might be filled with weak mortar until required. Likewise, channels in manholes generally should be uniformly graded from inlet to outlet and smoothly finished to avoid any projections which may catch detritus and reduce water-way. Such restrictions also produce slack water zones upstream, thus reducing velocities and leading to siltation and sulphide formation.

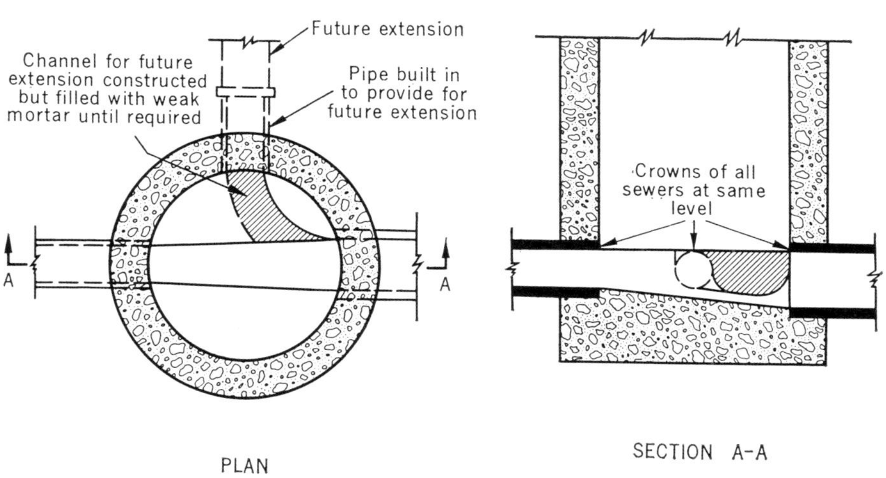

FIG. 9.2 Simple junctions—small 'aerobic' sewers.

Junctions of sewers which may carry sulphides:—At junctions of main sewers and other sewers which may carry septic sewage, it is important that the confluent sewages merge with a minimum of turbulence (Fig. 9.3). It will normally be impossible to design a sewer junction which will give ideal merging conditions at all times of the day and for all stages of development of the contributing areas. The best compromise must be worked out for each case. The following generalizations provide an approach to the problem:—

Excessive turbulence during periods of storm flow is of little significance, but for normal dry weather flow, when the sewage is stronger and might even be septic, excessive turbulence should be avoided by appropriate design.

It is more important to prevent excessive turbulence at and near peak dry weather flow than for flows near the minimum, when the sewage would be weaker and the sulphide problem lessened or absent.

Excessive turbulence during the early years of development of the areas served by the sewers may appear to be more important than during the later years, because, excepting for trade waste contributions, sulphide concentrations usually are lessened as the flows increase. The later conditions usually apply for the greater part of the life of the sewer and therefore are to be preferred as the design basis. Nevertheless, possible bad or undesirable conditions in the earlier years justify attention to both circumstances.

Minimum turbulence occurs when the energy loss in the junction is a minimum. For practical purposes, this will be achieved when:—

(i) the angle of convergence within the junction zone is as small as practicable;

(ii) velocity changes at the junction are gradual, or preferably avoided altogether;

(iii) there is a "balanced lateral momentum" condition within the junction zone.

This last requirement is necessary to avoid conditions such as where the flow from one incoming sewer sweeps across the junction zone, producing a standing wave and considerable turbulence on the far wall.

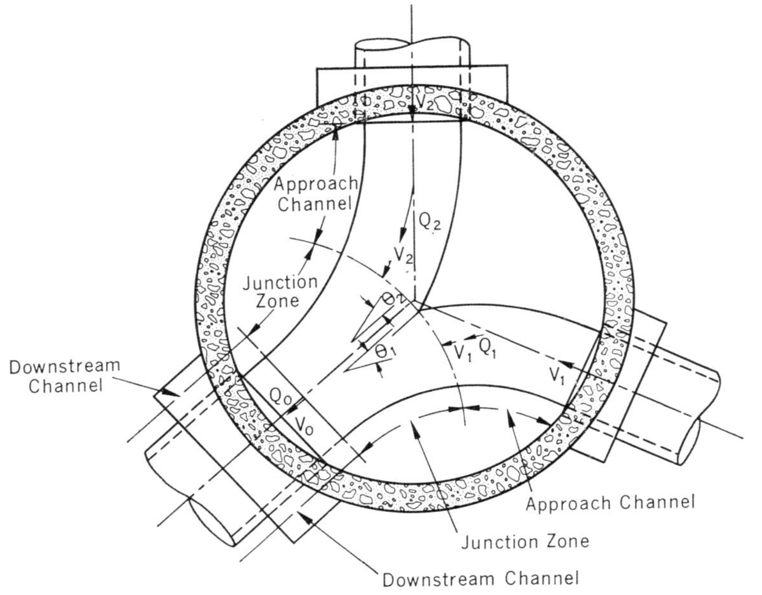

FIG. 9.3 Stream-lined junction.

Procedure for design of junctions:—Assuming that all sewers are of sufficient size, with sufficient discharge to have sub-critical flow, the following procedure is recommended (see Fig. 9.3):

Step 1. First determine the invert levels as follows:—

(a) For the peak dry weather flow selected compute the depth and velocity (V_0) at the downstream end of the junction. This will be governed by the hydraulics of the outgoing sewer only.

(b) Assuming that the water surface within the junction is horizontal, fix the invert levels of the incoming sewers so that the velocities (V_1 and V_2) in these sewers will be of equal magnitude. This will generally produce a minor drawdown or backwater in at least one sewer, but is the most important requirement in the design to prevent turbulence. If it is feasible to make these velocities equal to V_0, the subsequent design work will be simplified.

Step 2. Fix the shape of the approach channels and the junction zone as follows:—

(a) Arrange the geometry of the sewers and the approach channels so that the angle of convergence within the junction zone is as small as practicable, and so that $Q_1 V_1 \sin \theta_1 = Q_2 V_2 \sin \theta_2$, which ensures that the lateral momentum balance is obtained under design conditions.

(b) Where the condition $V_0 = V_1 = V_2$ has resulted, shape the junction zone so that the cross-sectional areas normal to the stream lines are maintained correctly throughout the junction zone.

(c) Where V_0, V_1 and V_2 are not equal, shape the junction zone so that there are *no abrupt changes* in the time rates of change of velocity at any stage.

(d) For either case (b) or (c) ensure that the shapes are such that there is no flow across the cusp at the invert.

(e) If it be not possible to arrange for $Q_1 V_1 \sin \theta_1$ to equal $Q_2 V_2 \sin \theta_2$ as recommended in (a), raise the cusp so that it becomes a deflecting wall between the two incoming sewers.

(f) If the Froude Number parameter for the outgoing sewer,

$$\frac{V_0{}^2 b_0}{g A_0}$$, be greater than 0.2, it is suggested

that the design be verified by hydraulic model experiments.
[b_0 = surface width, A_0 = area of cross-section of flow]

Design of sewer drops (including simple drops and drop junctions)

The factors determining the general design and structural details of drops include the following:—

(i) the condition of the sewage—whether fresh or septic;

(ii) the volumes of flow to be dropped, and the relative volumes of any associated sewage flows;

(iii) the differential height of the drop;

(iv) other practical limitations such as manhole dimensions.

Such considerations make it convenient again to divide drops into two separate categories, (a) those belonging essentially to smaller sewers carrying fresh sewage, and (b) those which are likely to carry sewage which may or may not be septic from time to time, and including larger size drops involving substantial potential energy changes.

Beneficial aspects of vertical drops

Before proceeding with these two categories of drops, it is convenient here to consider the deliberate use of drops, as a means of preventing sulphide generation, in systems wherein any need for drops otherwise would not be considered.

Such a question arises as a possibility when designing a sewer system for areas where slopes are not very steep, viz whether moderate grades with occasional vertical drops should be used in preference to steeper grades without drops. It is therefore of interest to consider a particular example of the overall absorption and emission characteristics of a length of sewer with and without a drop, assuming that the available head loss over the section is to be the same in either case. These alternative methods of grading a length of sewer are shown in Fig. 9.4. Estimates (Table 9.1) of the comparative amounts of oxygen absorbed for these alternative designs have been made assuming 100% dissolved oxygen deficit of the influent sewage, the case where the sewage entering upstream carries no dissolved oxygen.

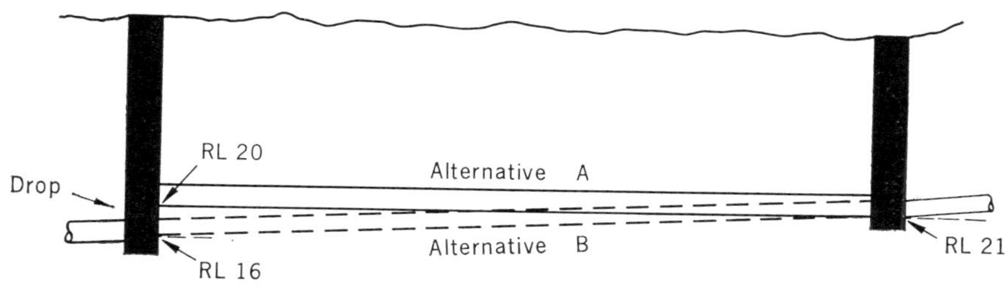

FIG. 9.4 Alternative ways of grading sewers (refer to Table 9.1).

For these particular conditions Alternative A, using the drop, shows nearly 50 times as much oxygen uptake as for the sewer without drop (Alternative B). If there were some dissolved oxygen present in the sewage this ratio would remain much the same, i.e. about 50, but the relative amounts of oxygen absorbed would be reduced in proportion to the actual oxygen saturation deficits. Whether or not the use of such a drop would be worthwhile or could be justified requires consideration of the expected dissolved oxygen concentrations upstream of the influent sewage and dissolved oxygen balances downstream.

It will be noted that Alternative A shows a positive oxygen balance for the sewer itself over the distance Manhole 1 to Manhole 2 (Fig. 9.4). In general it would be undesirable to use grades with an unfavourable oxygen balance, but certainly the sewer downstream of Manhole 2, where the use of the drop has boosted the dissolved oxygen by more than 3 mg/l, could be laid to an even flatter grade for considerable lengths without exhausting the dissolved oxygen. Such a layout might well be much cheaper taking the whole sewered area into account.

Drops generally are not favoured where the influent sewage already is septic, because the relative emission rates for sulphide will parallel the relative absorption rates for oxygen and therefore more sulphide will be emitted under the conditions of Alternative A.

The data of Table 9.1 show that most of the oxygen absorption and sulphide emission will occur at the drop. Thus it is clear that the use of drops is beneficial when sewage does not contain dissolved sulphides, but may be detrimental when significant concentrations of sulphides are present. In the latter case the sulphide emitted at the drop is likely to be drawn into the downstream section of the sewer with the prevailing

TABLE 9.1 COMPARATIVE UPTAKE OF OXYGEN ALONG A SEWER, A WITH AND B WITHOUT DROP

	Alternative A	Alternative B
Dry weather flow (max. 6 hour)	2 Cusecs	2 Cusecs
Diameter of pipe	2 ft.	2 ft.
Grade	1 in 1000	1 in 200
Velocity dry weather flow	1.74 ft./sec.	3.15 ft./sec.
Oxygen absorbed assuming saturation deficit of 10 mg/l (i.e. no dissolved oxygen in sewage)		
Whilst flowing in sewer	0.038 mg/l[1]	0.069 mg/l[1]
At drop	3.18 mg/l[2]	—
Total	3.218 mg/l	0.069 mg/l

1 Based on equation [4.7] (see pp. 23 et seq.) taking sewage BOD_5 at 300 mg/1.

2 Based on formulation of Gameson.

air flow. Sulphide emission would be increased also along the turbulent zone which would persist for about 10 to 20 diameters downstream.

One measure to control possible corrosion at drops, particularly at sites where odour nuisance would not be expected, is shown diagrammatically in Fig. 9.5. The idea is to induce or force a continuous local air current upward in the drop manhole of sufficient strength and volume to ensure that all the sulphide gases emitted from the sewage because of the drop are continuously moved out of the drop manhole and the downstream section of sewer.

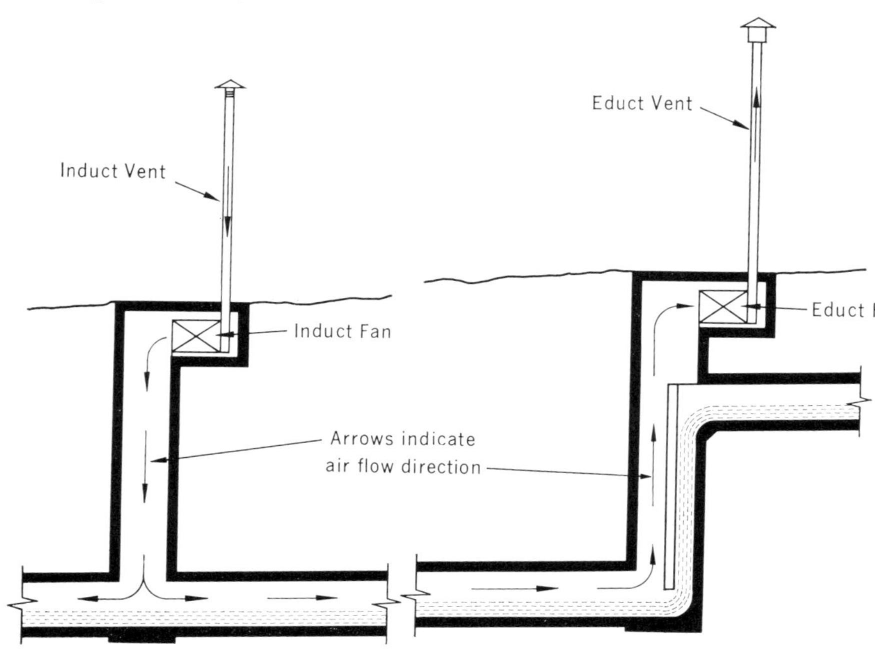

FIG. 9.5 Special Ventilation at Drops.

Care must be taken so that the induct fan is of sufficient capacity to provide for the maintenance of the air stream interrupted by the exhaust system at the drop manhole, as well as for any displaced upstream air current between the two manholes shown. The final setting of fan capacities and the general efficacy in functioning of the system should be checked, when the sewer is in operation, by air velocity determinations in manholes and sewers with manhole covers in place, and by determinations of moisture and sulphide concentrations in the discharged air.

Drops in small sewers or at small tributary sewer junctions, in cases where sewage is expected always to be still fresh and turbulence generally to be beneficial. All that is required for this are layouts which, while simple, nevertheless do not encourage accumulation of debris, etc.

Figures 9.6, 9.7 and 9.8 illustrate satisfactory practical designs for such drops of various heights. As the height increases, the bottom of the drop is made more resistant to erosion and a water cushion is used for the same purpose.

Drops and tributary junctions where the sewage condition may not always be fresh (including larger sewers and main sewers): The final discharge from all tributary drops on main sewers should take place at levels below the lowest level of the sewage surface in the main sewer, and the outlet stream should be directed in a downstream direction, as shown in Fig. 9.9.

Where the beneficial effects are assured, as discussed above, an arrangement such as that shown in Fig. 9.10 is typical (see p. 103).

Tributary sewer junctions where sewage conditions are septic: The problem of discharging a high level tributary sewer into a low level carrier or main sewer is met frequently—especially so if the main sewer is an interceptor. If the sewage in the main sewer is septic, minimizing its disturbance should be the main aim. The above suggests however that the use of drops upstream, where the sewage still may be fresh, perhaps should be extended as required to maintain freshness. If the tributary sewage is septic, special ventilation arrangements also may be desirable, even necessary.

Where the height of the drop is great, say 20 feet or more, the problem is to dissipate harmlessly as much as possible of the potential energy which the tributary stream possesses due to its elevation. One method of achieving this is by inducing a wall-hugging spiral flow in the vertical drop pipe, largely to dissipate the excess energy by friction. This spiral flow is set up by means of a circular or volute shaped chamber at the top concentric with the drop pipe, in which a vortex is maintained by discharging the tributary flow tangential to its periphery, as shown in Fig. 9.11. The design theory for such drops is to be found in the references(*51, 52*). If the tributary sewer has been properly designed the inflowing sewage should be free from significant concentrations of sulphides and therefore the effect of the spiral flow drop should be to re-aerate the sewage fully. If however for any reason the tributary sewage does or is likely to contain significant amounts of sulphides then emission rates would be high and probably some corrosion will result above sewage level in the vortex chamber. This chamber therefore should be specially vented to the outer air, also the drop pipe and adjacent sewer.

Figures 9.12 and 9.13 contain the design parameters and criteria for such vortex drops. Fig. 9.14 shows a typical drop arrangement for a 24″ diameter pipe.

Especially where larger discharges are to be dropped comparatively great heights, vortex drops are more beneficial (and more satisfactory generally) than "ordinary" drops, for a number of reasons, which include the following:—

(i) the maintenance of a continuous air core down the shaft ensures stable operation;

(ii) the continuous air core, combined with the spiral flow regime, provides excellent conditions for oxygen uptake;

(iii) because there is free fall at all times, scum and oil, etc., are removed continuously;

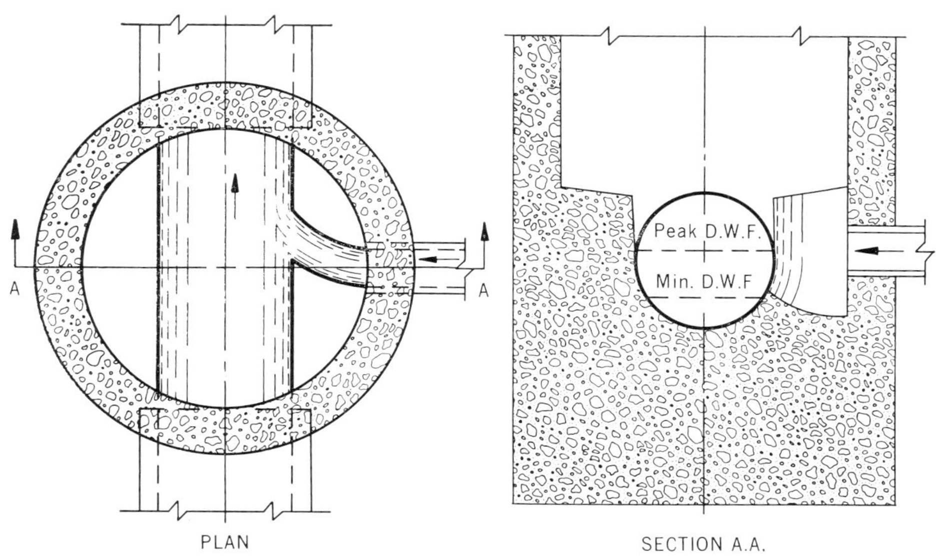

FIG. 9.6 Small Tributary Drop.

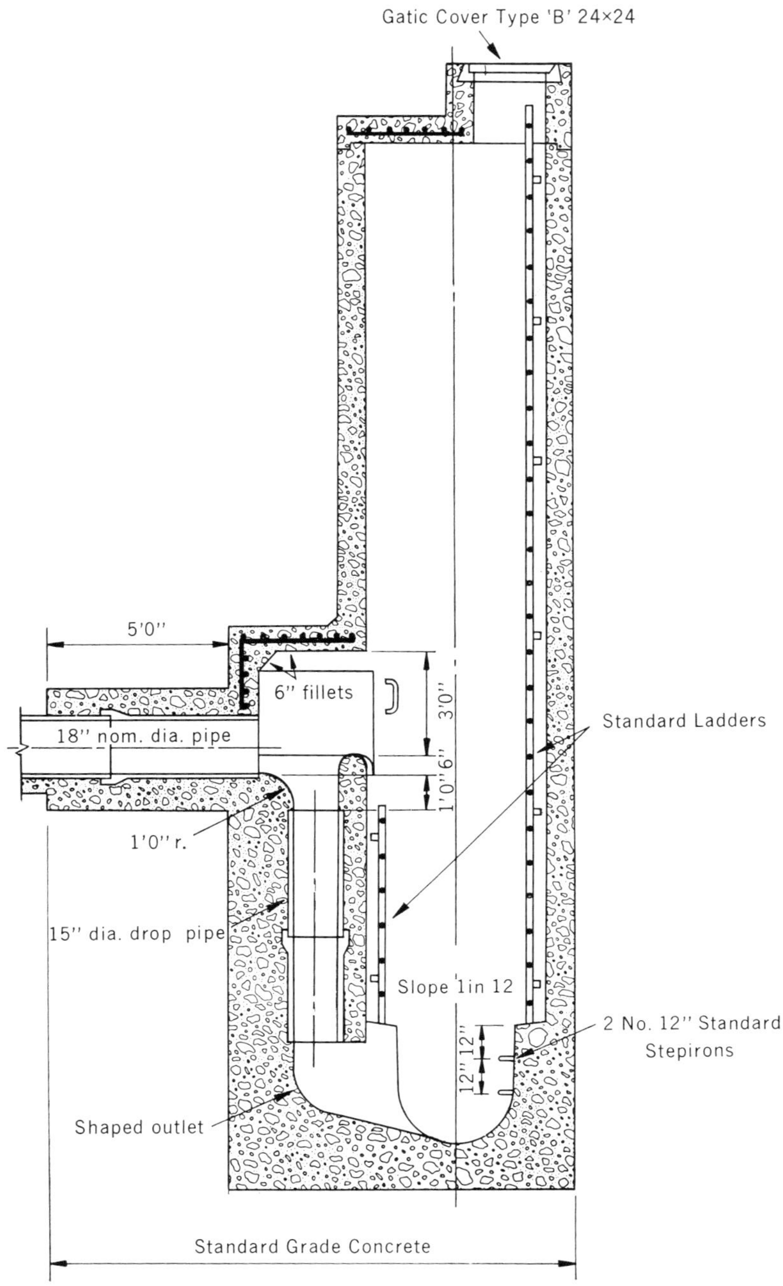

FIG. 9.7 Typical Manhole Drop Structure—Low drop.

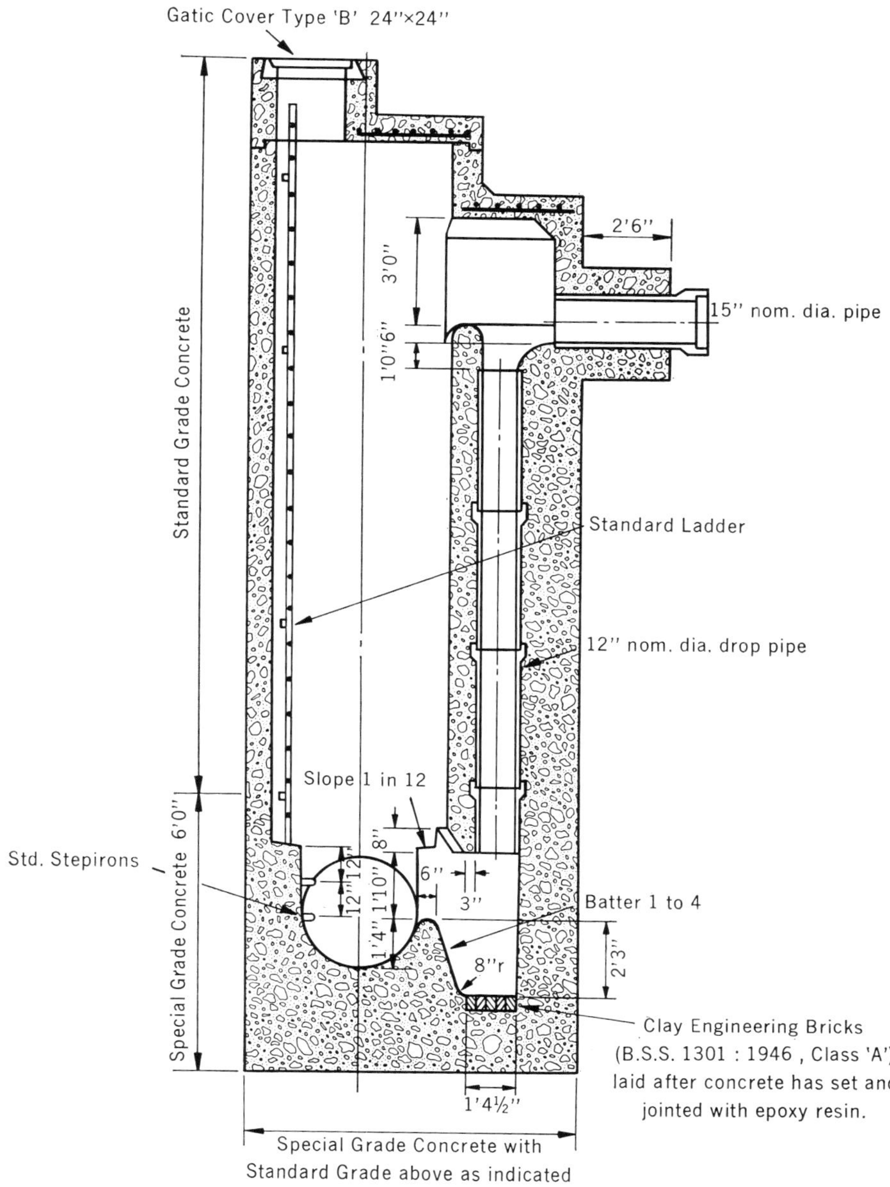

FIG. 9.8 Typical Manhole Drop Structure—Intermediate drop.

(iv) the vortex arrangement also facilitates the passing of solids, thus obviating obstructions under almost any circumstances. Even pieces of timber whose lengths are appreciably greater than the diameter of the drop will enter and pass the drop;

(v) because the dissipation of potential energy is achieved to a much better degree by vortex drops than by others, the bottom entry problems are alleviated substantially.

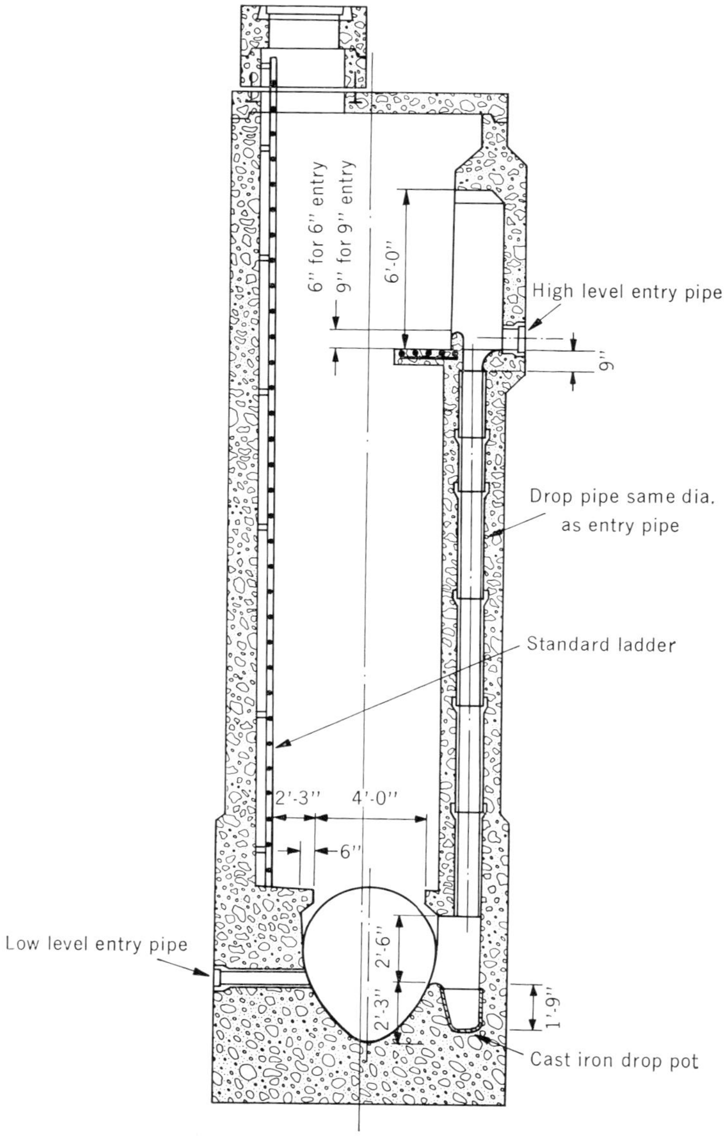

FIG 9.9 Typical Manhole Drop Structure — High drop.

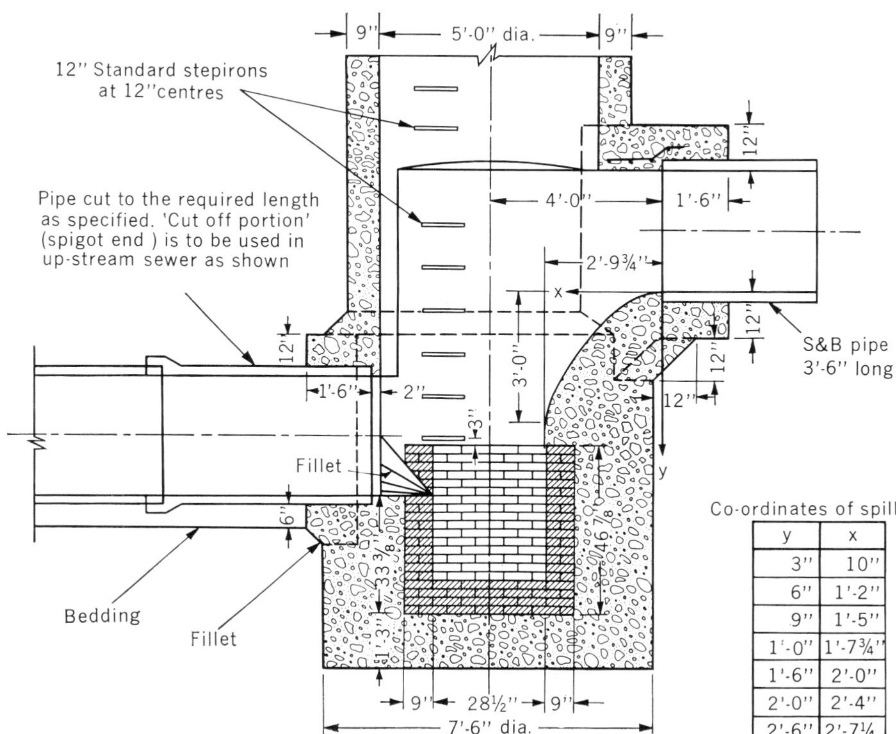

Co-ordinates of spillway

y	x
3"	10"
6"	1'-2"
9"	1'-5"
1'-0"	1'-7¾"
1'-6"	2'-0"
2'-0"	2'-4"
2'-6"	2'-7¼
3'-0"	2'-9¾"

FIG. 9.10 Typical small Drop incorporated in sewer solely to improve re-oxygenation of sewage.

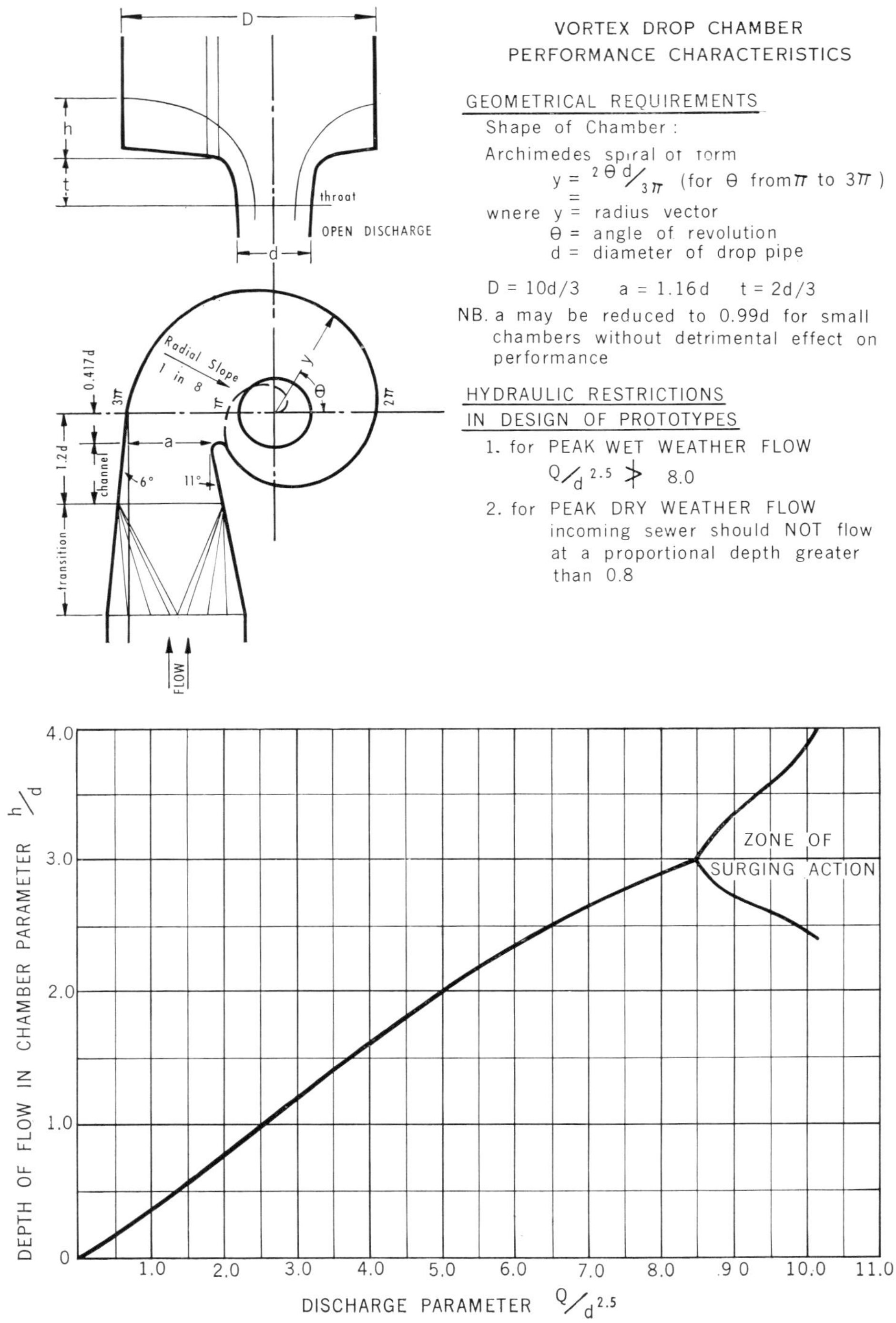

FIG. 9.11 Vortex Drop.

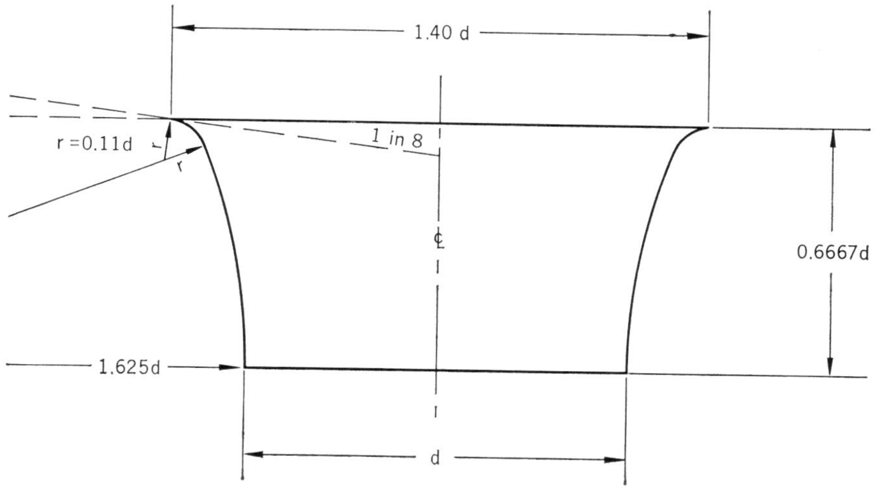

FIG. 9.12 Bellmouth for Vortex Drops.

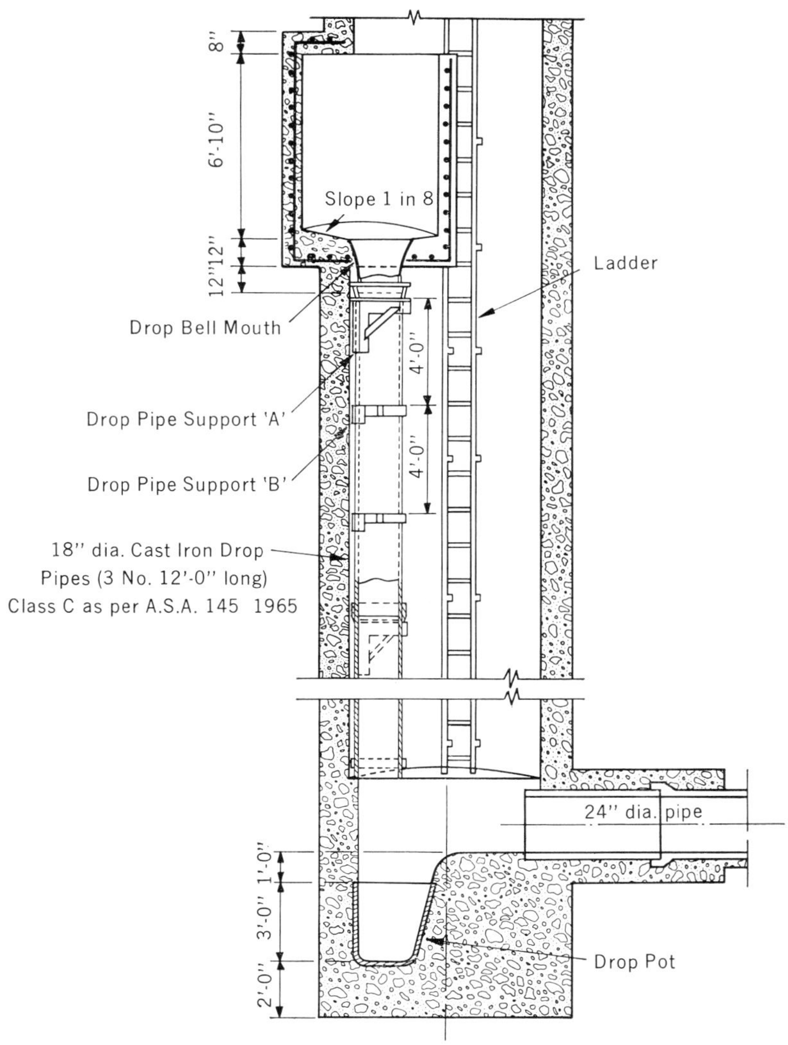

FIG. 9.13 Typical Drop Arrangement showing Bellmouth (Fig. 9.12) and Drop pot.

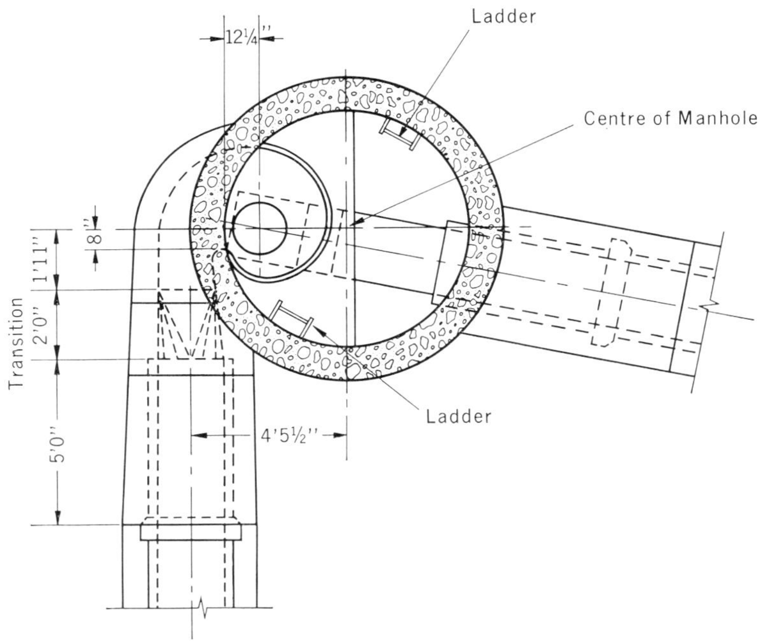

FIG. 9.14 Plan of Drop Layout for 24" pipe.

CHAPTER 10

DESIGN OF GRAVITY SEWER SYSTEMS

Introductory

The detailed consideration of all aspects of sewerage sulphide problems in Chapters 2 to 7 of Part 1 show clearly that such problems are entirely avoided in sewerage systems which operate to keep the flowing sewage in a "fresh" condition, always containing some free dissolved oxygen, however little.

The whole sequence of sulphide corrosion processes, shown in the diagram behind the front cover, is discussed in Part 1, including not only sulphide build-up and corrosion, but also toxicity hazards and associated odour nuisance. At the same time the derivation of appropriate mathematical relationships is pursued, based partly on fundamental physico-chemical and biochemical considerations, partly on empirical expressions derived from experimental observations. Formulations applicable directly in terms of the designers' units of measurement are developed finally in Chapter 8, p. 85. Chapter 8 also deals with sewer characteristics and sewage composition, including some illustrations of their effects on sulphide problems.

These formulations are intended to facilitate the determination of sewerage system layouts and dimensions detailed so that sulphide build-up will not normally occur. Special consideration is required regarding possible direct additions of sulphide-bearing wastes, also other undesirable possibilities such as blockages should be taken into account. There could be situations where the desirable sewer grades determined in this way are not practicable—perhaps where sewage temperatures are unusually high, or because of the nature and profile of the terrain of the area to be sewered. The need for some pumping also may make it difficult to achieve a system approaching freedom from sulphide problems. In such cases the build-up of sulphides in at least part of the system may be practically inevitable, and some corrective measures may be desirable if not essential (see Chapter 13).

In this Chapter basic design procedures are presented for designing a system of sewers with a minimum of sulphide problems. The foregoing Chapter 9 discusses the significance of drops, junctions, etc. Chapter 11 deals with the design of pumping systems; 12 with ventilation systems; and 13 with remedial measures.

Outline of design procedure

The formulations of Chapter 8 involve six main "factors", three of which broadly are characteristic of the sewer, the other three of the sewage, all six being directly relevant to the overall aspects of the dissolved oxygen "balance" and sulphide control.

The sewer factors are size, slope and proportional depth of flow; the sewage, Biochemical Oxygen Demand (standard 5-day test), sulphate content, and temperature. Stated simply, the design procedure results finally in determining the slope required for a sewer of the selected size to prevent sulphide build-up, this size and the proportional depth of D.W.F. being related to both D.W.F. and W.W.F. of the sewage to be provided for.

The procedure is presented as comprising two stages:—

(i) a preliminary assessment, and

(ii) the detailed design.

The idea briefly is first to make a preliminary assessment of the order of value of the slopes which will correspond to only a negligible build-up of sulphide, if any, progressively along the sewers and main sewers. These sewers are then laid out to best suit the topography of the subject area. From this, economic and practical feasibility studies would follow.

At the next stage, these preliminary designs and the lay-out would be examined in detail, section by section, in turn, to yield more reliable predictions of sulphide and dissolved oxygen conditions from place to place. Where the prescribed limits are not achieved the designs then must be adjusted repeatedly until successful.

Design Parameters: Data for typical sewered areas in Australia are included in Chapter 8. From these data it should be feasible to make satisfactory assessments of sewage characteristics; per capita values providing some check. A further check may be made if the area to be sewered already is supplied with mains water, so that per capita water consumption data are available, but special allowances may be needed where relatively large industrial undertakings come also into question.

Economic factors also are discussed in Chapter 8, where suitable techniques for comparing effective real costs are outlined.

Fig. 10.1 illustrates the procedure.

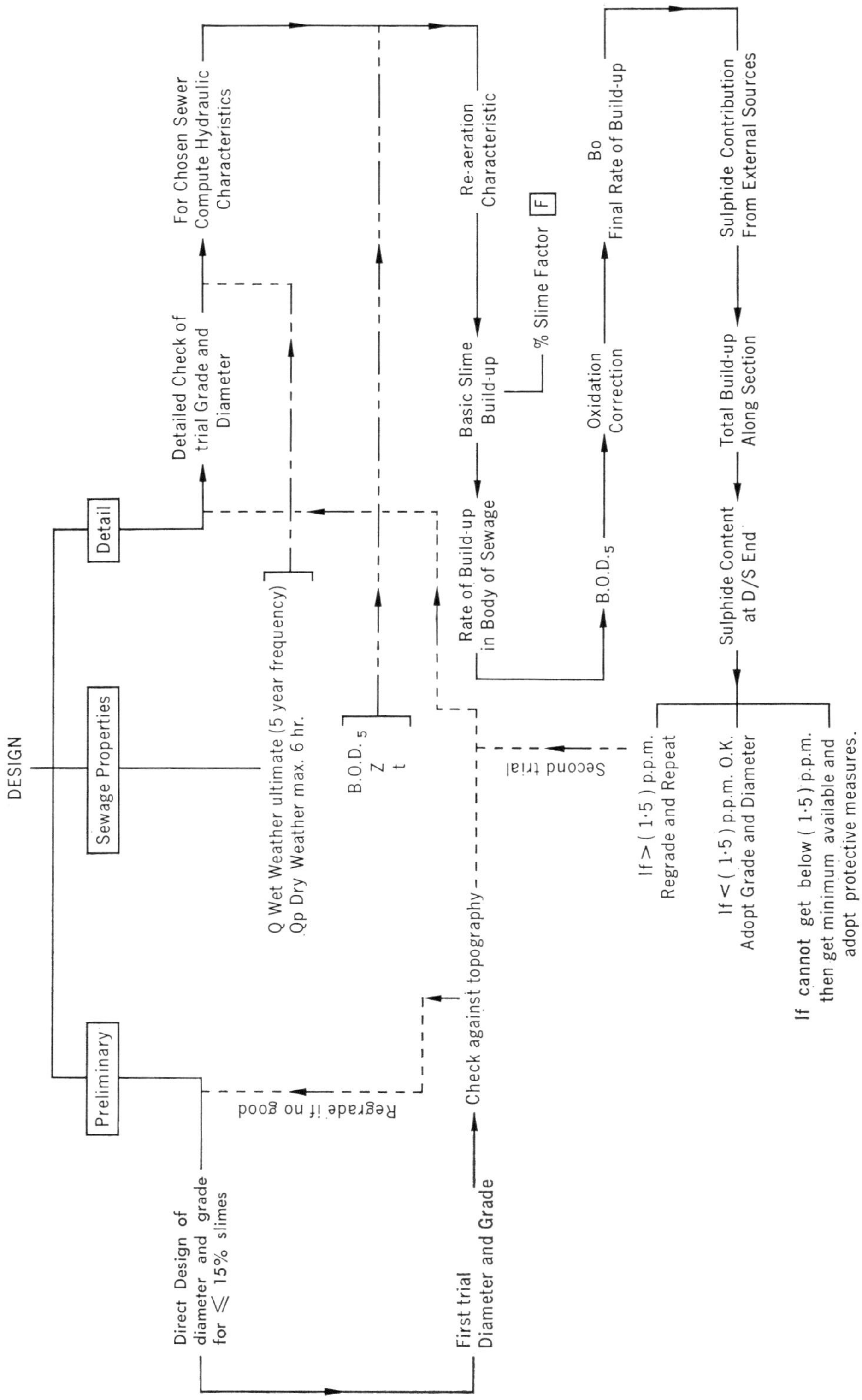

FIG. 10.1 Procedure for Detailed Design of Sewers for Sulphide Control.

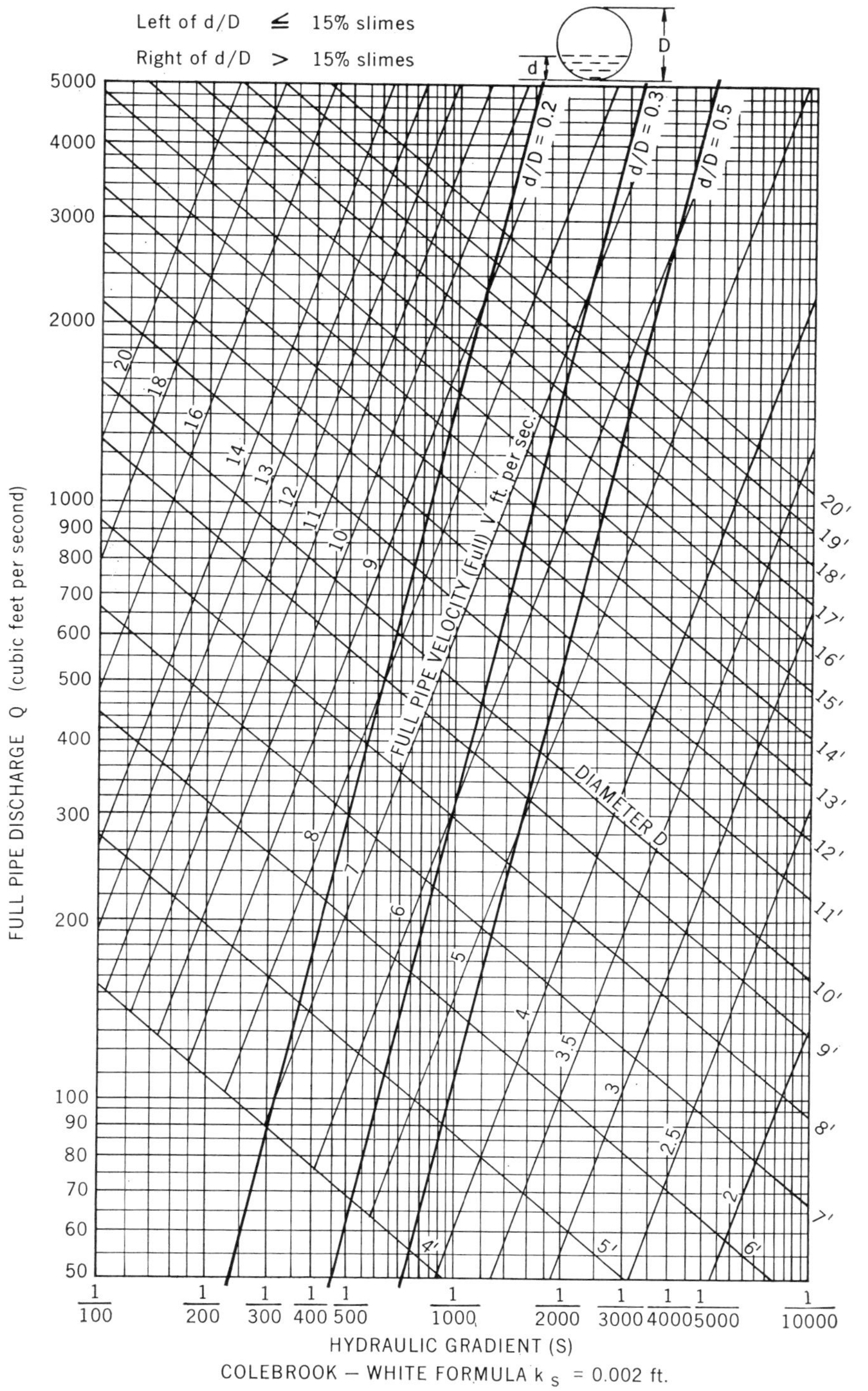

FIG. 10.2 Direct Design Chart No. 1 (Design Chart 1(a)).

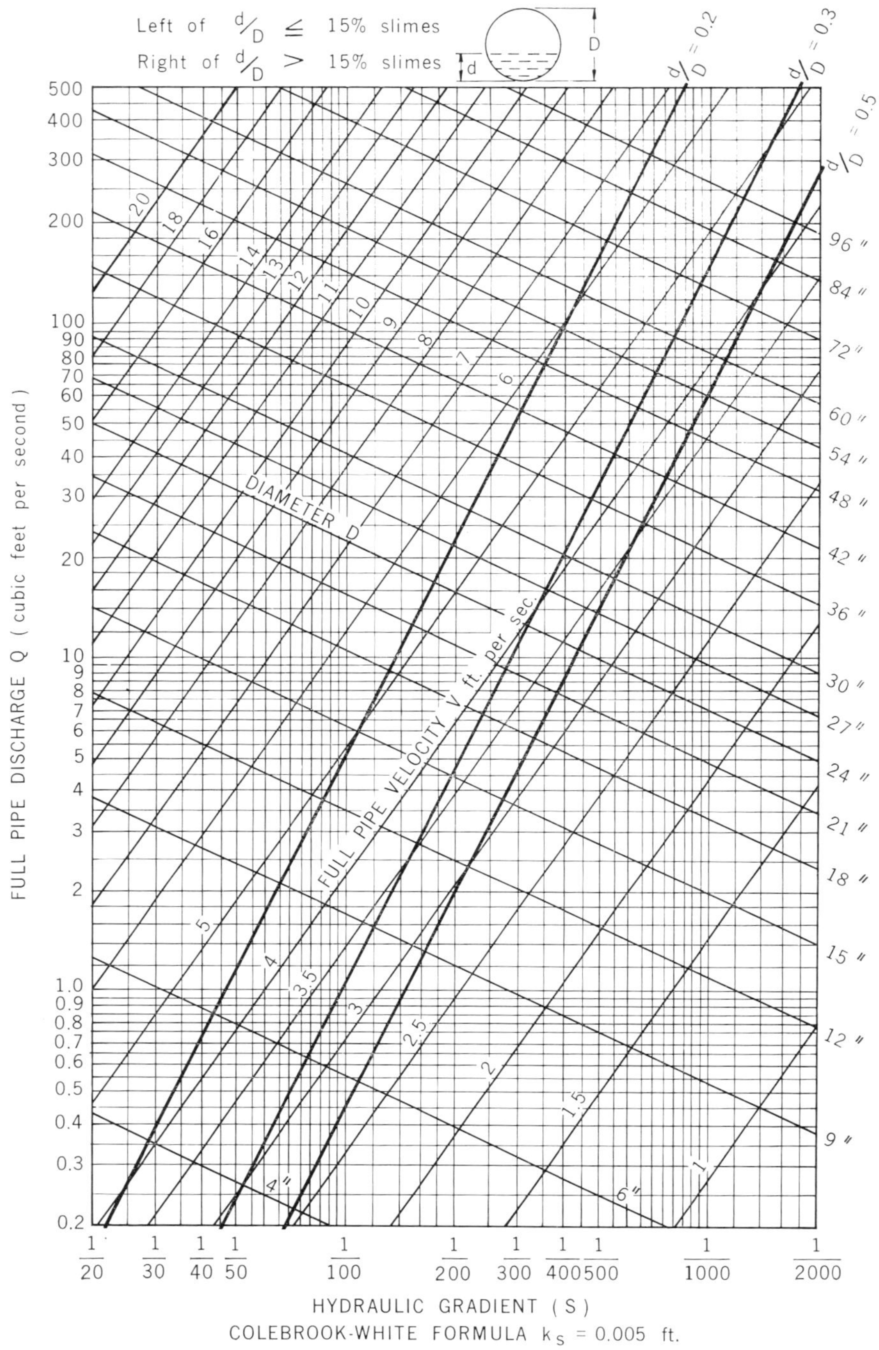

FIG. 10.3 Direct Design Chart No. 2 (Design Chart No. 1(b)).

Permissible concentration limits for sulphides in sewage: Before proceeding, however, it is necessary to consider what levels of sulphide concentrations should be regarded as tolerable and unlikely to cause problems of corrosion or other significant nuisance. Although sulphide concentrations are not the only factors determining the intensity of the sulphide problem at any point, some working guide is required respecting the limits for concentrations of sewage sulphide, under normal conditions of sewage pH and sewer ventilation, such that no significant corrosion of concrete will result. Considering the large fluctuations which occur in the sulphide concentrations from time to time, as discussed in Chapter 5, and possible variations in the many other factors involved, choice of any precise value for such a permissible concentration is difficult. Experience up to the present suggests that a mean concentration over the year of 1 mg/l of *total* sulphides would generally correspond to about 0.5 to 0.6 mg/l of dissolved sulphide, and in turn about 0.3 mg/l molecular hydrogen sulphide in the sewage. At such a level significant sulphide problems are not normally to be expected.

To estimate the average concentration of sulphides in sewage over the year and day however involves a substantial amount of computation. It is therefore suggested that the criterion for sulphide concentration be stated rather in terms of the average for the worst six hours of the day over the quarter of the year of maximum sewage temperatures. If this latter criterion be used the above average concentration of 1 mg/l would need to be increased to 2.0 to 2.5 mg/l. However, 2.0 to 2.5 mg/l appears to be high having regard to known corrosion experience. A better basis seems to be a value around 1.5 mg/l, and therefore the following criterion is proposed: Significant corrosion of structures of good quality concrete is unlikely to occur where the average concentration of total sulphides in the sewage, over the day-time six hours of maximum flow in the quarter of the year of maximum sewage temperature, does not exceed 1.5 mg/l.

Such permissible average concentrations however have little relevance for aspects of odour or toxicity where concentrations over short periods are more significant.

Preliminary design for sulphide control:

Estimates of the correct order of gradings and sizes of all sub-main and main sewers in question may be obtained quickly by one or other of the following means:

(i) Using Pomeroy's Z_p criterion;

(ii) An assessment based on the percentage of active slime; and

(iii) Using Direct Design Charts.

Using Pomeroy's Z_p Criterion for Sulphide Concentrations: Pomeroy(*50*) has developed an empirical expression, based on his experience and observations in California, from which a criterion Z_p can be calculated, such that

If Z_p is less than 5,000,	it may be assumed with safety that, at worst, sulphide conditions will be under satisfactory control
If $7{,}500 < Z_p < 10{,}000$,	conditions are unlikely to be more than barely significant
If $Z_p > 10{,}000$,	severe corrosion problems become more and more likely, the greater the value of Z_p

$$Z_p = \frac{[BOD_5] \times 1.038^{(T^\circ F - 68)}}{S^{\frac{1}{2}} \times Q^{\frac{1}{3}} \times f\left(\frac{d}{D}\right)} \qquad [10.1]$$

The values of $f\left(\frac{d}{D}\right)$ are given in Chapter 8 for $\frac{d}{D}$ from 0.1 to 0.6.

Assessment based on percentage of active slimes: On the basis that sulphide build-up and the corresponding sulphide problems can be controlled proportionately by shearing off more or less of the potentially active slimes (pp. 11, 12), it may be assumed that limitation to say 15% slimes will approach adequate control (that is 85% of the wetted perimeter should be free of active slimes).

Proceed as follows: Taking maximum W.W.F. off the area upstream, choose diameter and slope to match this flow at 90% full. Take the corresponding worst D.W.F. conditions to assess percentage slimes, using equations (p. 11) and (p. 129). If this value exceeds 15%, select a somewhat steeper slope and re-assess.

Flow conditions consistent with a maximum 15% slimes will also ensure self-cleansing conditions.

Using Direct Design Charts: This method of preliminary assessment in essence is that of the foregoing determination of "percentage slimes", but uses the two charts, Figures 10.2 and 10.3, respectively labelled also Direct Design Chart No. 1 and Direct Design Chart No. 2. They are simple chartings of the Colebrook-White discharge and velocity formula for k_s values of 0·002 and 0·005 respectively, onto which appropriate $\frac{d}{D}$ lines have been drawn to correspond to the hydraulic separation limits for 15% slimes, for $\frac{d}{D} = 0{\cdot}2, 0{\cdot}3$ and 0·5. Use of the charts avoids a series of trials and retrials.

Proceed as follows: Enter the Direct Design Chart with $Q_{W.W.F.}$. From this, fix a particular $\frac{d}{D}$ to suit the particular D.W.F. required for design. Any charted combination of D and S lying on the left hand side of the $\frac{d}{D}$ line will correspond to a maximum of 15% slimes or less.

As noted above, these flow conditions will correspond not only to the 15% slimes limitation but will ensure self-cleansing velocities.

Direct Design Chart No. 1 is drawn up for use with in-situ concrete conduits such as tunnel sections where some slimes normally are found on the wetted wall surfaces. Equivalent "roughness height" k_s is taken at 0.002 feet.

Direct Design Chart No. 2 is for use with jointed circular pipe sewers, equivalent roughness height k_s taken at 0.005 feet.

Example using Direct Design Charts. Taking 280 cusecs W.W.F. at 90% full, and designing for D.W.F. = 0·4 × W.W.F. (ratio $\frac{1}{0{\cdot}4} = 2{\cdot}5$), we have $\frac{d}{D} = 0{\cdot}5$. Entering Direct Design Chart No. 1 with 280 cusecs for $k_s = 0{\cdot}002$, $\frac{d}{D} = 0{\cdot}5$ gives $S \approx 1:1500$ and D = 8 ft. 6 in.

Flow conditions for $S = \frac{1}{1500}$ and $\frac{d}{D} = 0{\cdot}5$ corresponding to 112 cusecs (0·4 × 280), shows percentage slimes equal to or less than 15%, which probably would be satisfactory in practice.

Detailed procedure for sewer design and sulphide build-up prediction

The procedure set out and exemplified in the following is essentially a step-wise procedure for the detailed application of equations [8.8] and [8.9] in the form of equation [10.2], to predict sulphide build-up along any particular sewer length.

$$(B_S)_{eff.} = [(B_S)_{sl1} + (B_S)_{sew}] \times \left[1 - \frac{(O_D) \times K_{O_2} \times F_{dt} \times [1 + 0.015^{(t^\circ C - 20)}]}{60 \times \left(\frac{d}{dt}[BOD_0] + 2(B_S)_{sl1} + 2(B_S)_{sew}\right)}\right] \quad [10.2]$$

The detailed part of this procedure consists almost entirely of the arithmetic and algebra involved in evaluating the many factors contained in equation [10.2]. These factors are discussed at length in Part 1 of the book, and the designer should acquire a background knowledge of this discussion in all its parts, including the graphical relationships and formulations presented in Chapter 8.

These data then should be plotted, say as shown in Fig. 10.4. Sections requiring further detailed attention and re-design then are detected readily, and a new trial design re-calculated in the same way, and so on (see Fig. 10.1), until the conditions predicted are judged acceptable both for the particular length in question and for downstream conditions also.

The calculations are greatly simplified by use of the various Design Charts which are appended. These are described briefly in the following pages, and in-

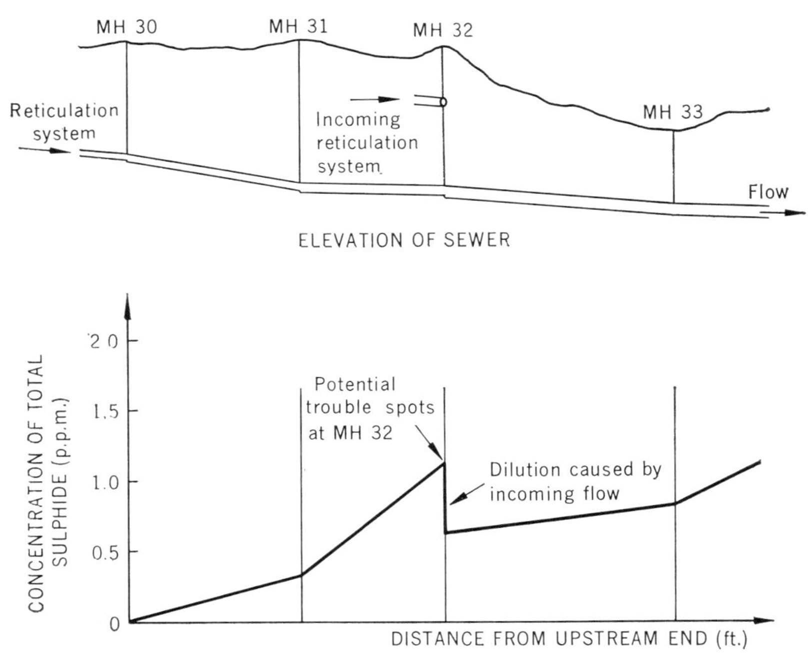

FIG. 10.4 Diagrammatic Plot of Sulphide Conditions.

cluded also, in pp. 128-130, is a tabulation for the whole sequence of the computations. Designers actually making these computations first should prepare an enlarged copy of the tabulation for working. It provides for recording each step towards the final answer, which is the required prediction of the estimated rate of probable sulphide build-up, section by section, for plotting as in Fig. 10.4.

Adoption of this systematic approach, together with the appropriate use of the Design Charts, makes the procedure essentially simple and routine. Accumulation of the data progressively, from small sewer catchments to larger in turn, is facilitated by entering all the calculations one by one along the columns first for the upstream sewer of the first row, then the second sewer, etc. and so on section (row) by section (row).

Before describing each step in detail the following points are emphasized:—

The phenomenon of sulphide build-up along a sewer is essentially restricted to dry-weather flow conditions, hence the flows taken for the reckonings are for dry weather conditions. On the other hand the size of sewer is determined for wet weather flow and consequently the depths of flow for the reckoning usually are below half diameter, that is, $\frac{d}{D}$ commonly will be less than 0·5.

Individual sewerage authorities usually apply their own particular separate methods and data for determining sewer sizes under any given conditions, and it is not within the scope of this manual to lay down rules for determining sewer diameters or dimensions generally. Therefore the type tabulation presented as an example (pp. 128-130), takes the sewer diameter as being determined from predicted wet weather flows. Other methods may be substituted if so desired. The conditions most conducive to sulphide build-up usually occur during the early stages of development within the catchment area to be served by the sewer, in dry summer weather. Also generation of sulphide in the body of the flowing sewage usually does not become significant until the average age of the sewage reaches about 10 to 12 hours.

In the tabulation sheet referred to, the working of the various steps is in most cases made clear by the explanation included above the column numbers. However, some of the steps require a more elaborate explanation than was feasible on the tabulation itself, and these are now given in order of the column numbers.

Column No. 1.—From Manhole to Manhole. This defines the section for reference purposes. Each manhole should be numbered or lettered to enable the section under examination to be fully identified on preliminary and final sewer design plans.

Column No. 2.—Length of Section. The total length of the proposed sewer to be examined in any one trial should be divided into sections, for each of which the slope and diameter remains constant throughout and in which no vertical drop or substantial confluence occurs, allowing that current practice sometimes provides for increased grades in the short length of channel within the manhole. While such minor departures from an uniform grade may be ignored in sulphide calculations, no substantial change of size or grade should be permitted except for deliberate attempts to increase re-aeration rates, which then must be considered as separate problems.

Column No. 3.—Estimated maximum wet weather flows.—Though sewers may be designed on the basis of strictly separate systems for waste waters, some stormwater usually does in fact gain access into the house drains and sewers. Except where methods of design and construction or other circumstances are such as to prevent such stormwater entry, the capacity of sewers usually is made larger to carry these additional flows in wet weather. The wet weather flow in a sewer may amount to several times the dry weather flow, the actual amount varying with soil type, pipe-jointing design, and so on. Even if rainwater could be completely excluded from sewers it would be undesirable, from the point of view of sulphide control, to operate sewers much more than half full under ultimate dry weather flow conditions.

Columns Nos. 5 and 6.—From the peak wet weather and peak dry weather flows, using the direct design charts, the diameter and slope for the first trial may be obtained directly along with preliminary assessments of probable sulphide conditions.

Column No. 7.—Estimated dry weather flow.—As stated before, the build-up of sulphides in sewage is essentially a dry weather flow phenomenon and the associated problems usually diminish or disappear under conditions of increased wet weather flow. The ability to predict dry weather flows is therefore important, and individual sewerage authorities should have as a basis for this their own experience of local conditions. The flow used for design should be the peak 6-hourly Dry Weather Flow (p. 111), but it may be desirable also to examine other conditions.

Time after construction: Experience has shown that the critical period in the life of a sewer in so far as sulphide corrosion is concerned is often the first few years before much building development or connection of properties has taken place. In this period, particularly with flat grades, serious damage to sewers can occur. Accordingly the tabulation under this heading provides for inclusion of dry weather flows as predicted at the end of five years after construction as well as for the ultimate condition (p. 44).

The dry weather flow at the end of the first five years after construction is in many instances more difficult to predict than the ultimate flow except where the sewer in question is of the nature of an interceptor or is a duplication. In such instances the sewer may have a substantial and measurable initial flow and the five year flow then may not be important. Unless better information is available to him, the designer may adopt the rough working rule that the average daily dry weather flow at the end of the first five years of a sewer's life will be 0.2 of the design ultimate dry weather flow. However, the first trials should be on the basis of ultimate conditions which are generally less severe than the five year basis. Only when trials indicate that satisfactory control will be reached for ultimate flows is it worthwhile to examine the five year situation.

Column No. 8.—Capacity of sewer running full.—This is the capacity, running full, of a sewer of the diameter stated in Column 6 laid to the trial slope of Column 5. This can be obtained from Design Chart No. 1A based on the Colebrook-White formula. Similar charts may be produced for other values of "n" if the design practice of the particular authority requires the use of values other than n = 0.012.

Column No. 18.—Season.—Provision is made in the tabulation for insertion of the seasons, summer, autumn, winter and spring, so that average sulphide for the year can be obtained as required for the criterion of acceptable average sulphide concentration. Also the season of highest temperatures should be examined in detail. The main factor relating to season usually is the temperature, but in some localities rainfall also may be a factor through its effect on sewage flows. First trials may be made for spring to autumn conditions and when satisfactory grading is being approached, the full seasonal details can be computed (p. 44).

Column No. 19.—Assumed temperatures.—The temperatures referred to are the sewage temperatures. Individual sewage authorities should have the information necessary to fill in the average sewage temperature for each season. Alternatively mains water temperatures may serve as a basis, with minor corrections based on atmospheric and ground temperatures (p. 92).

Column No. 20.—Oxygen saturation concentrations.—This can be obtained directly from Design Chart No. 4 or Table 4.1 p. 22, the temperature being that from Column 19 and the barometric pressure being taken to be 760 m.m. of Hg for localities not more than 100 ft. above sea level. For localities at a greater height, the barometric pressure to be used can be taken from the following Table 10.1.

Column No. 21.—Assumed dissolved oxygen concentration.—It will be on the safe side to assume that the dissolved oxygen concentration is zero. However when the calculations for a preceding section indicate the presence of dissolved oxygen the appropriate value for the downstream end of that section should be inserted. An acceptable general assumption for this system of calculations might be that all dissolved oxygen present in the upper reaches of the reticulation sewers is just on the point of disappearing at the lower end of all 9 in. reticulation sewers. For this to be valid, reticulation design must be generally sound.

Table 10.1

Barometric Pressures At Various Heights Above Sea Level

Height above sea level	Barometric pressure m.m. of Hg
100	757
200	755
400	749
600	743
800	738
1,000	733
1,200	727
1,400	722
1,600	717
1,800	712
2,000	707

Column No. 25.—Assumed BOD.—The BOD referred to is the five day BOD as determined by the standard test.

(i) *Domestic Sewage.*—The output of five-day BOD per head of population served may be taken to be roughly constant for any given area. Some values are given in Chapter 8 (p. 93), with an average value for Australian conditions about 0.14 lb./head daily. The average BOD of domestic sewage will therefore depend on the quantity of sewage discharged per head per day. Since we are here concerned with the design of new sewers it is reasonable to assume that the dwelling served will be of a modern standard and the water consumption and sewage discharge will be fairly high. The flows in sewers will also contain some groundwater infiltration even in dry weather. The actual total per head dry weather sewage flows vary according to the locality and its climate and to the standard of housing involved. Individual authorities should have their own data on per capita flow.

The domestic sewage BOD can be predicted by the following formula—

$$BOD_5 = \frac{0.14 \times 10^5}{Q_D}$$

Where Q_D = Average daily D.W.F. gals. per capita per day (g.p.c.d.).

(ii) *Industrial Wastes.*—It is extremely difficult to accurately predict the effect of industrial wastes on a particular sewage flow and BOD. If local town planning ordinances prohibit the use for industrial purposes of any portion of the catchment area of the sewer under consideration the effect of industrial wastes may be ignored, but where no such prohibition exists (as is the common case) some allowance should be made depending on the size of the catchment involved. The allowance should be relatively greater the smaller the catchment. It would not be economical to provide in advance for the possibility of the worst type of waste being discharged from every potential industrial site. In fact most industrial establishments do not discharge harmful wastes, hence when such wastes do come into question some time in the future, the effect on the system should be considered then and the appropriate protective and control measures, pre-treatment and so on, instituted. For the ordinary cases where industry is of lesser significance, some overall allowance nevertheless should be made, but if the local authority concerned has no special practice in the control of this aspect, an experienced sewerage chemist or trade waste expert should be consulted.

Column No. 28.—Assumed sulphate concentration. —This can best be determined from sulphate concentrations in existing neighbouring sewers. Two main sources of sulphates are the town water supply and permanent groundwater infiltration. Industrial wastes may also be a contributor particularly when sea or estuarine waters are used by the industry. Assumptions made respecting sulphate concentrations should be made likewise in conjunction with experienced chemists.

Column No. 29.—

(a) *Shear Stress Parameter* (p. 11).

$$\tau_{op} = \gamma_S \times S \times \frac{A_S}{P}$$

This is plotted on Design Chart No. 10.

(b) *Percent slimes.* This is obtained from Design Chart No. 10 for the appropriate proportional depth of flow (p. 124).

Column No. 30.—Rate of sulphide generation in body of sewage.—As an approximation generally of sufficient accuracy this can be obtained from Design Chart No. 8 for the appropriate mean age of sewage. It will be found that this rate normally is not significant until the mean age of the sewage exceeds about 10 hours.

Column No. 41.—Sulphide concentration at upstream end of Section.—This is the downstream end concentration obtained from the trial computations for the next sections upstream. For the upstream end of the uppermost section, if all tributary sewers discharging to this point are reticulation sewers, the value of the sulphide concentration may be taken as zero. If further extension of the branch sewer under examination is likely at some future date, it will be advisable to make preliminary investigations for this future section by these trial methods and hence obtain an estimate for possible future sulphide concentrations.

For systems where dissolved oxygen is present initially in the sewage as discharged from properties, it is normally safe to assume as already stated that for reticulation sewers carrying domestic sewage only there will be no dissolved sulphide at all in the sewage at the upper end of the branch mains.

Main sewers of diameter greater than 12 in. should be graded in accordance with these procedures. Experience has shown that at concentrations up to around 0.25 mg/l of total sulphides, practically all the sulphide will occur in the innocuous insoluble form. It will then be safe and may be advantageous to use vertical drops in main sewers, that is where the predicted total sulphide concentration does not exceed 0.25 mg/l, unless industrial wastes may bring sulphide into the sewage directly.

The following presents a review of the Design Charts. These charts generally are self-explanatory. Their significance has been discussed already in the foregoing, and their use is illustrated by the computed example which follows.

Design Charts 1A and 1B are prepared for box section and circular pipe sewers respectively.

Design Charts 2A and 2B are based on Camps' variation of "n" giving the flow characteristics and geometrical properties of circular pipes flowing part-full, 2A giving the overall range and 2B the detail for shallow depths of flow.

Design Chart No. 3 presents graphically the relationships between $\frac{d}{D}$ and the hydraulic mean depth and radius.

Design Chart No. 4 comprises a Nomogram for the relationships between oxygen dissolved at saturation equilibrium and the total atmospheric pressure and temperature (see also p. 22).

Design Chart No. 5 gives the clean water re-aeration rate constant K_{O_2} (mg./l per min. per mg./l of oxygen deficit) directly for various slopes and velocities (eqn. [4.7], p. 24).

Design Chart No. 6 gives the correction factor F_{dt} for the above K_{O_2} values, allowing for surface active substances correlated with BOD_5.

Design Chart No. 7 allows direct reading of the values of $\frac{d}{dt}[BOD_0]$, $K_{O_2} \times F_{dt}$ and sulphide build-up rate corrected for different sewage temperatures.

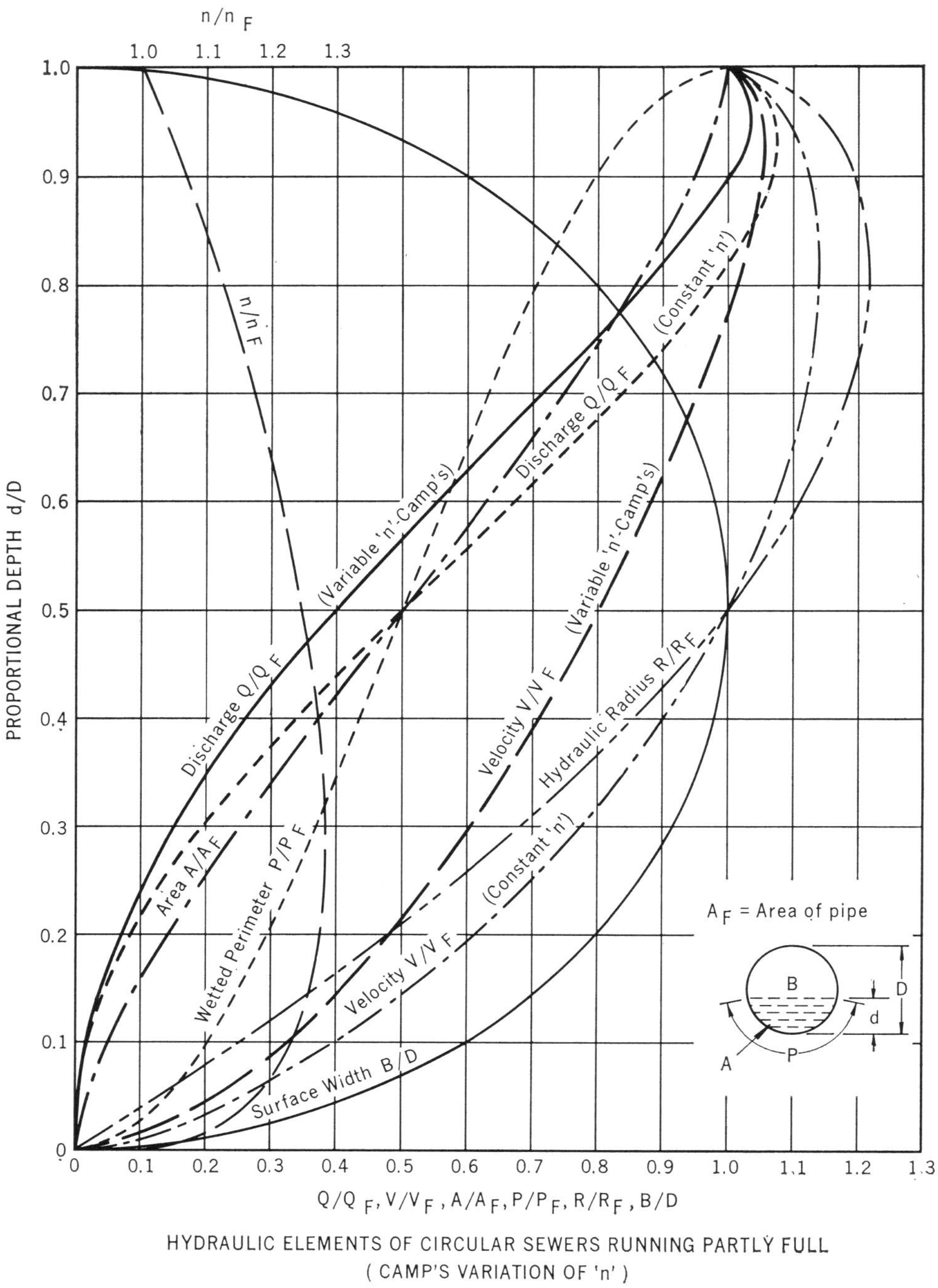

FIG. 10.5 Design Chart No. 2A.

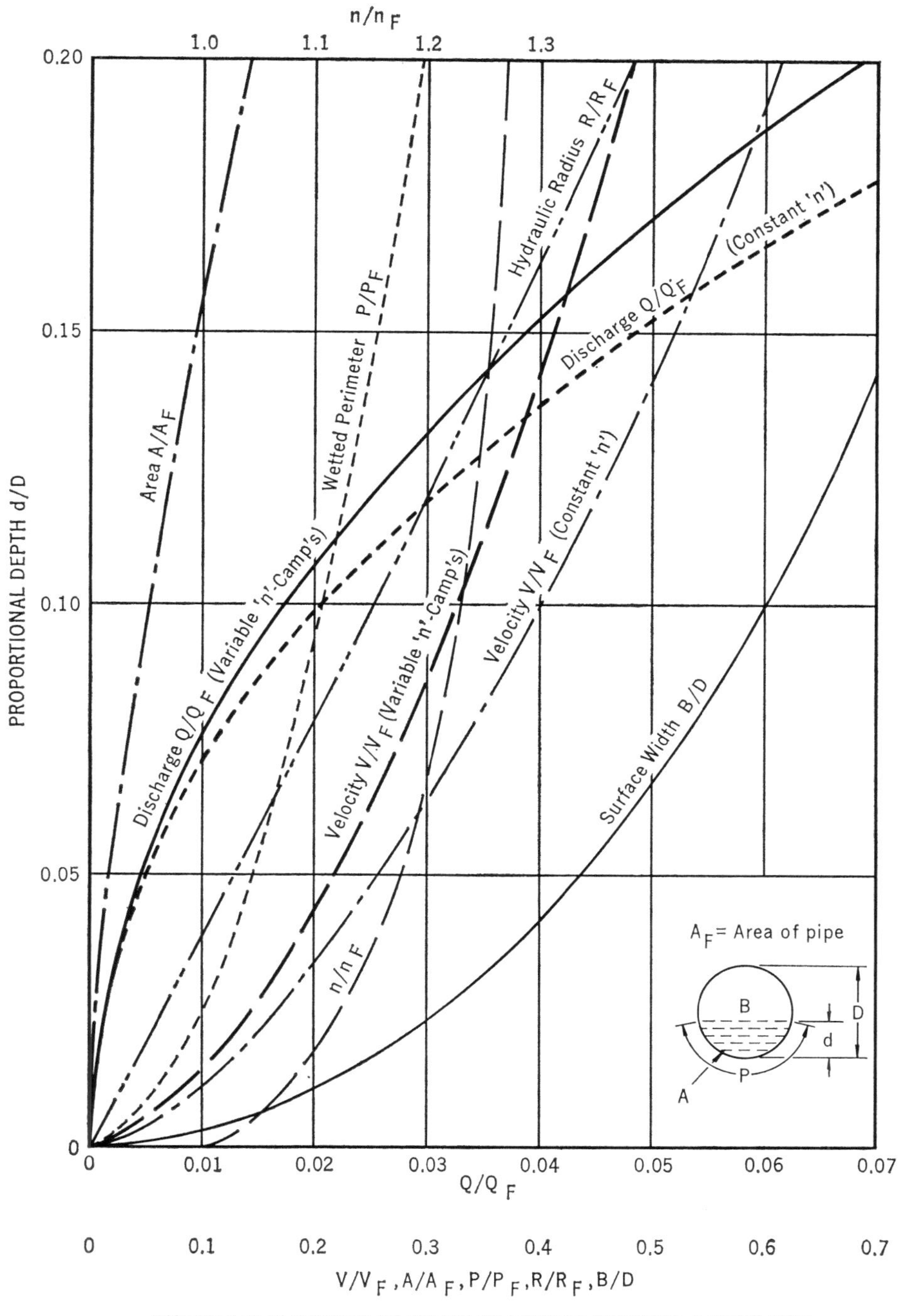

HYDRAULIC ELEMENTS OF CIRCULAR SEWERS RUNNING PARTLY FULL
(CAMP'S VARIATION OF 'n')

FIG. 10.6 Design Chart No. 2B.

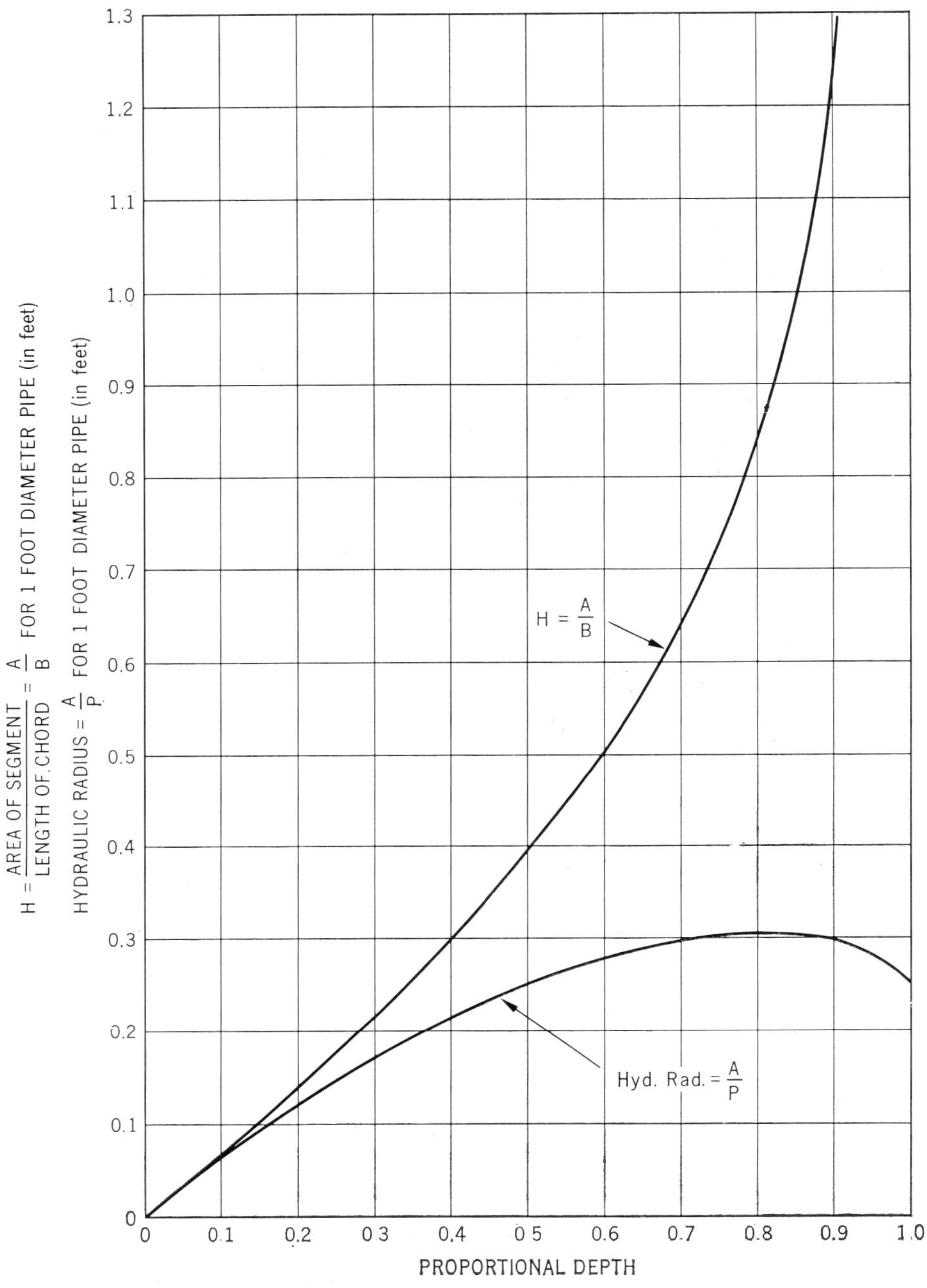

FIG. 10.7 Design Chart No. 3.

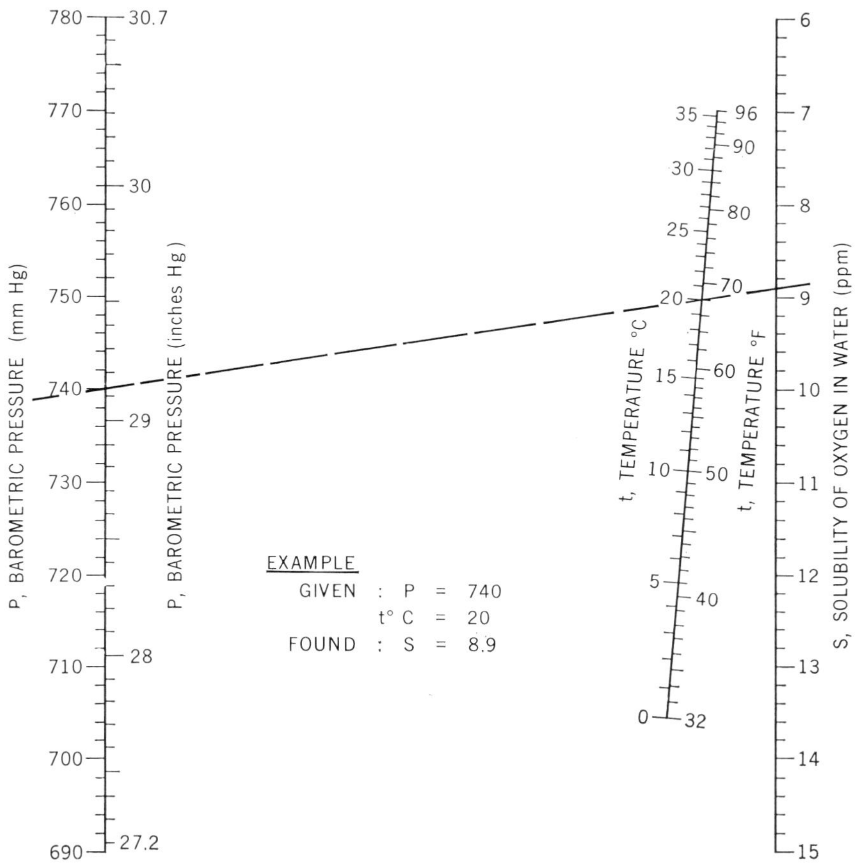

FIG. 10.8 Design Chart No. 4.

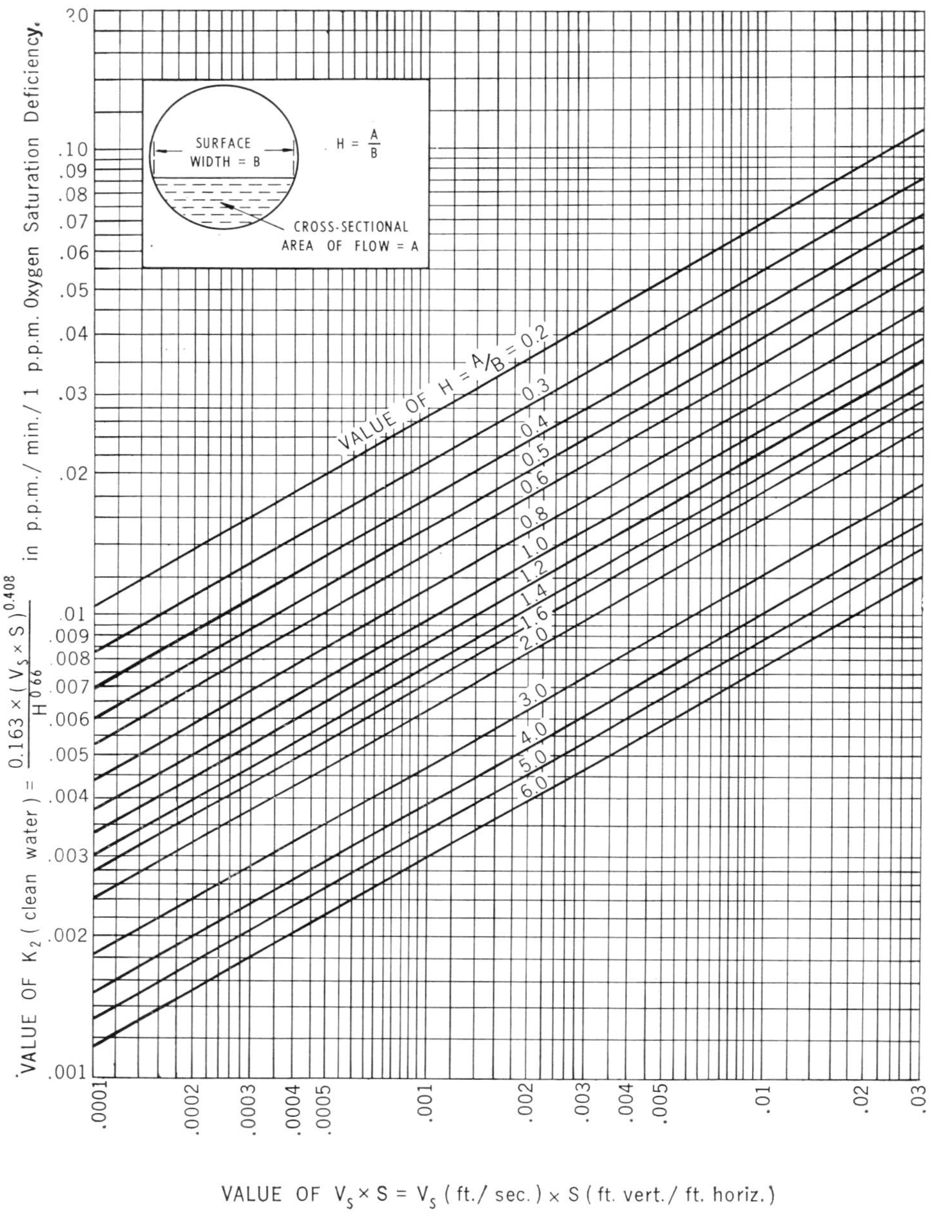

FIG. 10.9 Design Chart No. 5.

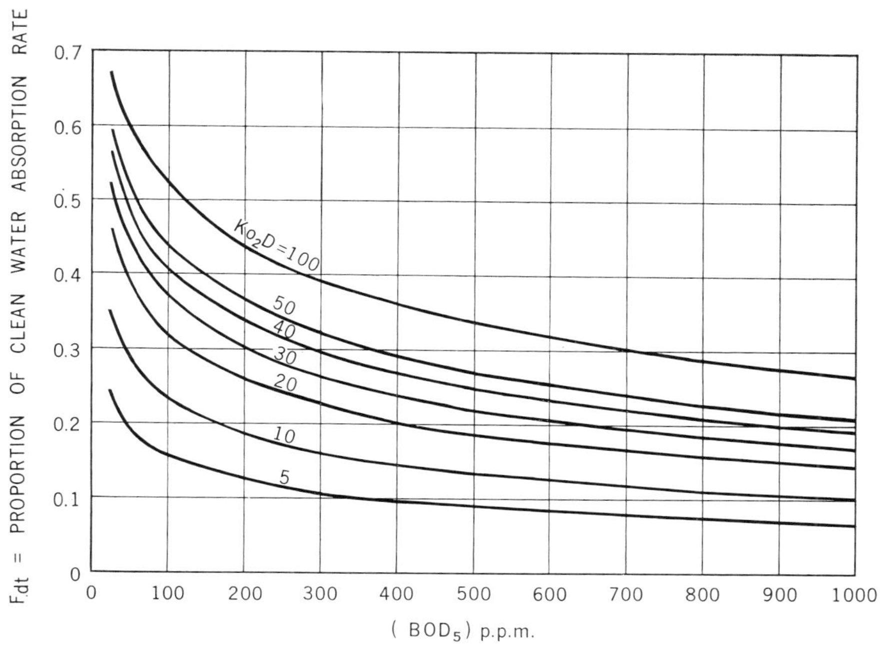

FIG. 10.10 Design Chart No. 6.

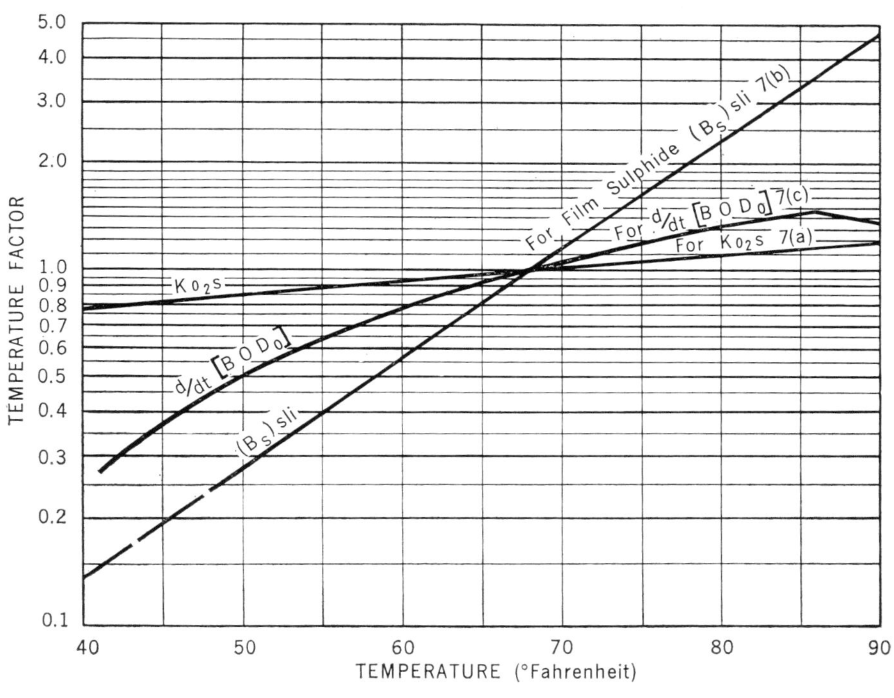

Fig. 10.11 Design Charts Nos. 7(a), (b) and (c).

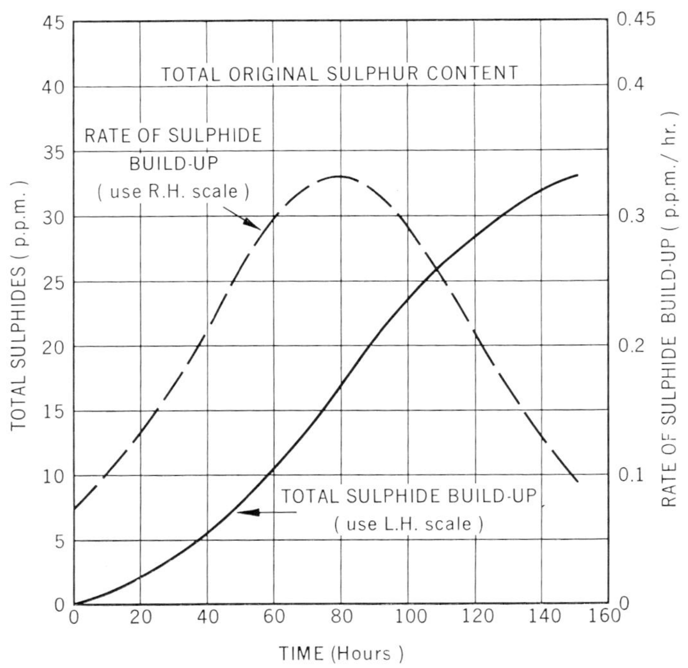

FIG. 10.12 Design Chart No. 8.

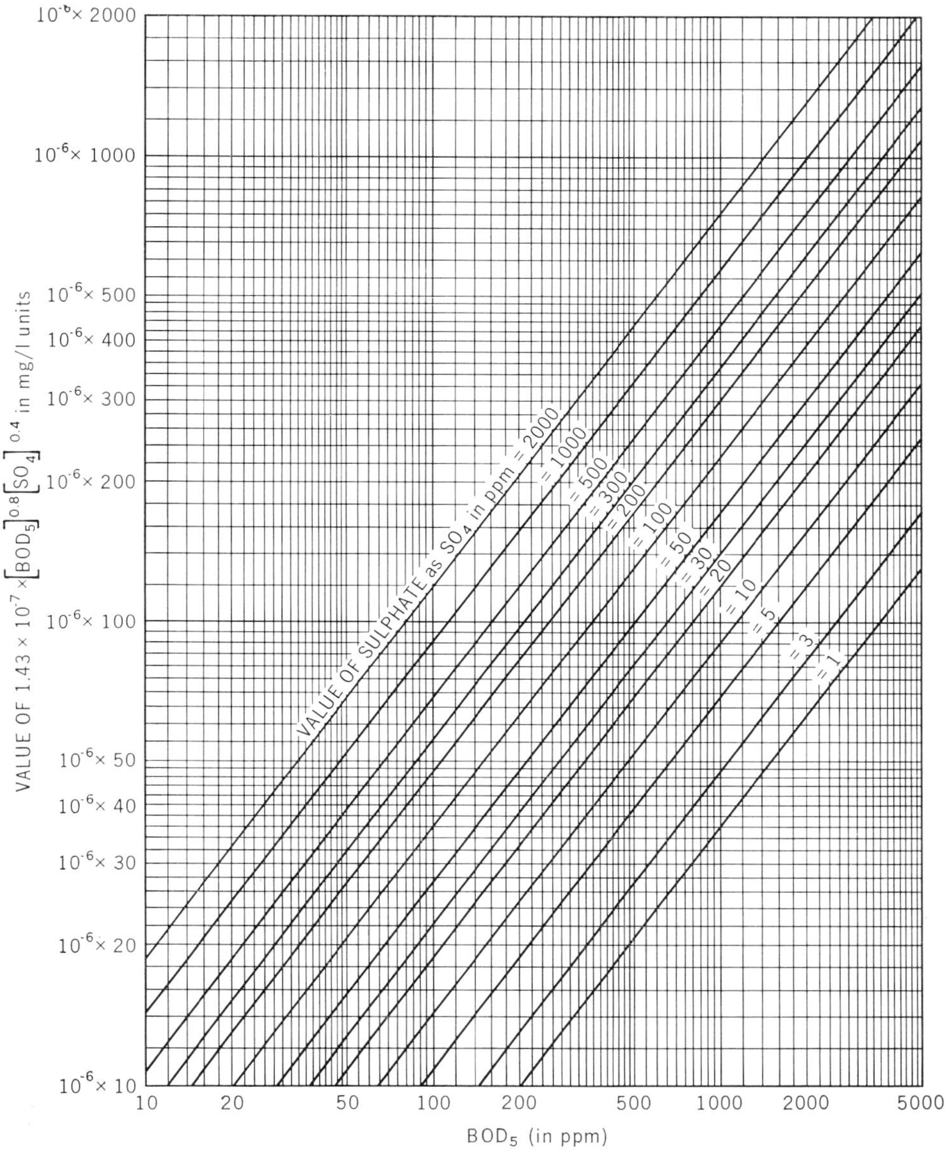

FIG. 10.13 Design Chart No. 9.

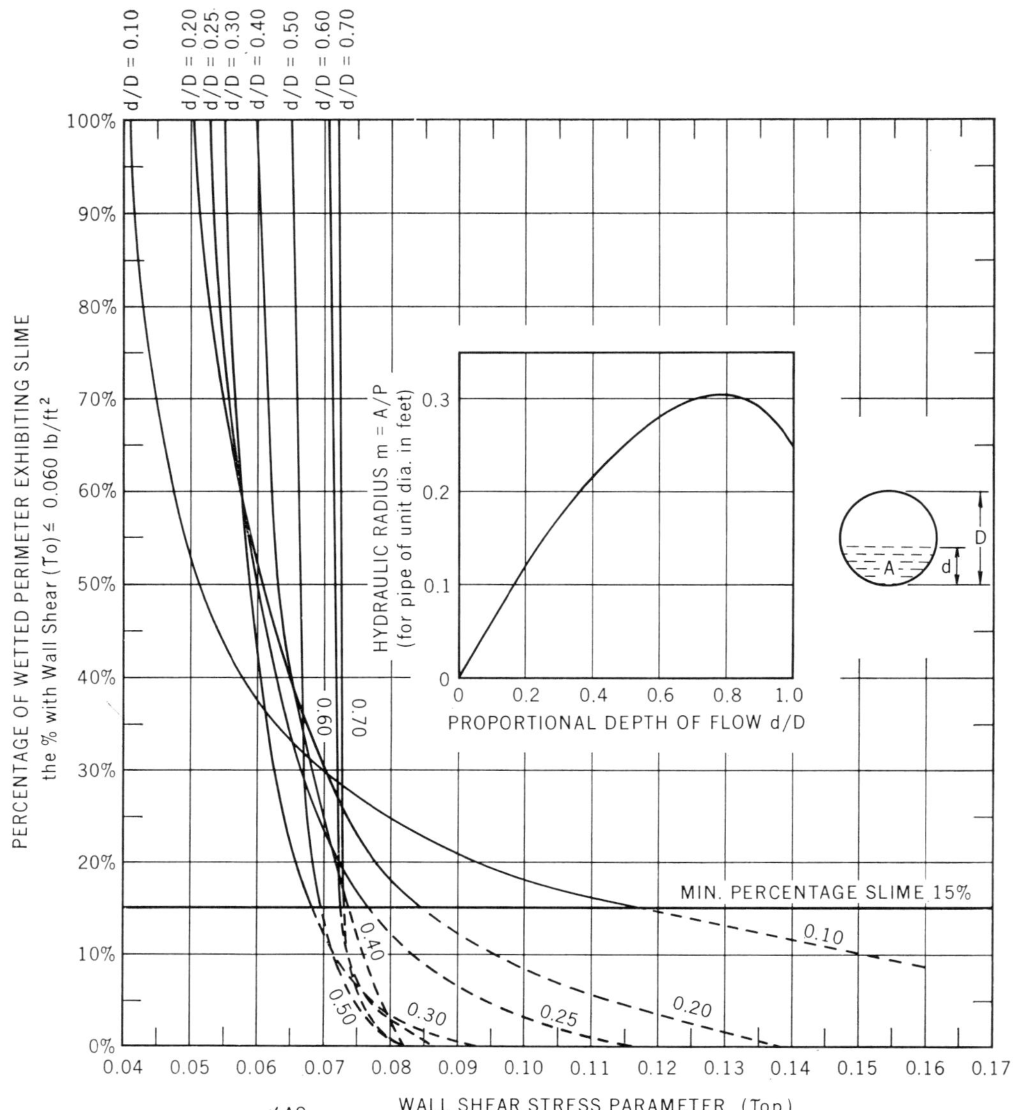

$$\tau_{op} = \frac{\gamma A S}{P}$$

Where
τ_{op} = Shear stress parameter
γ = Specific weight of sewage lb/ft^3
S = Slope of pipeline or tunnel ft/ft
P = Wetted perimeter in feet
A = Area of flow in ft^2
D = Diameter of the pipe in feet
d = Depth of flow in feet

m = A/P for unit diameter pipe at proportional depth d/D

N.B.
Max. 6 hourly = appropriate flow to be considered when using the plots.
Max. 6 hour flow ≈ 1.4 Av. daily dry weather flow.

It may be noted that when the point of maximum velocity occurs at the surface of the flow then this shear stress parameter becomes equal to the average wall shear stress.

SLIME PREDICTION CHART FOR VARYING FLOWS IN MAIN CIRCULAR SEWERS

FIG. 10.14 Design Chart No. 10.

Design Chart No. 8 gives the rate of build-up of sulphide in the body of the sewage $(Bs)_{sew}$ according to the age of the sewage.

Design Chart No. 9 provides for direct reading of the slime sulphide build-up $(0.143 \times 10^{-6}[BOD_5]^{0.8} \times [SO_4]^{0.4})$ for given BOD_5 and sulphate values.

Design Chart No. 10 is a plot of the wall shear stress parameter-percentage slimes-proportional depth relationships.

Self-cleansing grades: If practicable, at reasonable cost, the sewer gradients and resulting velocities should be sufficiently high throughout the system to hold average concentrations to total sulphides below 1 p.p.m. If this is not practicable at a reasonable cost, then comparisons should be made of the total costs using other gradients but providing for supplementary means of sulphide control such as those described in Chapter 13.

In the designing of sewers the engineer must provide the capacity necessary to carry the anticipated dry and wet weather flows in such a way that there is no undue risk respecting either blockage or the deposition of solids which may develop into a blockage, or of sulphide problems developing either in the section of sewer being designed or in the system downstream.

The prevention of siltation in sewers is important not only because of the risk of the deposition building up to a blockage but also because the silt itself can give rise to sulphide generation. Such deposition is also generally a sign that velocity and slope conditions may be such as to reduce the rate of surface absorption of oxygen by sewage to an unduly low value. To maintain sewers free from silt, it is practically necessary that velocities do not fall at any time below velocities needed to prevent siltation. If siltation does occur, then an appreciably higher velocity will be required periodically to entrain and transport the settled material. A common practice then has been to provide for minimum velocities of from 2 to 2½ feet per second for at least one hour daily at the stage when the sewer is carrying the estimated ultimate dry weather flow. During the earlier stages in the development of an area this basis may be inadequate and the normal sewage flow then may have to be augmented by flushing water.

Recent research into bed load movement in watercourses and the application of this research to sewers has shown that requirements to ensure "self-cleansing" properties for a sewer depend upon sewer size, slope, depth of flow, maximum size of particles in silt forming material of sewage and quality of joint alignment *(53-57)*. Jointing quality and true alignment have been found very important. When jointing and bedding of pipes is of very good quality it has been found *(56)* that comparatively flat grades, for example 8″ pipes at 1 in 400, are adequate to prevent deposition. However, even if the necessary quality of jointing and trueness of grading can be achieved to permit such flat grades, it would not be wise at this stage of our knowledge to adopt such a general practice because of the risk of increased sulphide production.

Referring to a further effect of low velocities, Swab *(58)* points out that, on the basis of the work of Reid and Tien-Cheng Yang *(10)*, low velocities in sewers permit the more rapid build-up of slimes on the walls resulting in further decrease in velocity due to the increased surface roughness produced by such slimes. Thus an original velocity of 1.38 ft./sec. in a clean sewer decreased by as much as 35%, whereas with an original velocity of 1.59 to 1.70 ft./sec. the reduction was only 12%.

Reticulation Sewers: For reticulation sewers it can be accepted that generally the necessary sulphide control will be achieved provided that such sewers are laid to the requirements of the following Table 10.2.

Table 10.2

Minimum Satisfactory Grades For Reticulation Sewers

Sewer Dia.	Minimum Grade
4″	1 in 60
6″	1 in 120
9″	1 in 300
12″	1 in 400

Corrective Remedial Measures: It may not be practicable finally to lay out the sewerage system as designed—for example, because of some special factors of topography or geography and consequential high costs—and some corrective measures may become necessary, if only for the first ten to fifteen years of development. This need may be related to corrosion problems, or to odour problems, or to both, or the question may be rather one of safety from toxicity or explosion hazards. Not only are sewage velocities and depths practically uncontrollable, at least as a matter of routine, once the sewerage system has been constructed and commissioned, but the whole interrelationship of the ventilation system must be very different during the first years from those for which the designer must provide as the ultimate.

The action required under such circumstances is in effect covered fully in the detailed discussions in Chapter 13 of the remedial measures applicable to existing systems. However, difficulties arise in considering economic factors, particularly where the problems to be remedied or controlled are expected gradually to be alleviated as the population approaches that designed for.

One of the important recommendations of Chapter 13 is that before anything is done at all, the particular needs for some remedial action should be determined first. This requires adequate surveys of the conditions actually to be found from place to place, in sewers, manholes, chambers etc. respecting both the physical structures and sewage sulphide and dissolved oxygen contents. The data should be plotted on diagrams. It is pointed out that these data subsequently allow direct trials and testing of any measures which appear most promising.

Reference to Chapter 13 will disclose a wide variety of possible control measures, and it must be expected that they will bring also a variety of mechanisms, some more efficacious in one situation, others more suitable for a different situation. It is recommended that in general the simpler techniques be considered preferentially, particularly those which tend to "displace" the circumstances under which the system is operating closer to the designed circumstances. These "simpler" techniques include those which demand only operational or managerial changes, or relatively simple provisions such as, for example, the institution of flushing schedules. Flushing as a means of siltation and slime control by natural scour and shear must be operated at relatively short intervals, especially in prolonged dry weather, and may require mobile tank cars or the use of simple, although substantial, apparatus to dam and pump from local watercourses. However, such procedures are not only essentially simple but also are more positive in character, as well as easier to control and operate, than techniques, for example, requiring precise chemical control. Portable or otherwise temporary mechanical aeration apparatus appears in an intermediate category because it may involve nuisances of odour or ventilation also. More detailed discussions will be found in Chapter 13.

Sample Calculations of the Detailed Design Procedure

The following computations illustrate the DETAILED DESIGN PROCEDURE. For this purpose, the design of the sewer length between manholes C and D in the hypothetical system shown in Figure 10.15 is examined in detail. It is assumed that sewers AC and BC have already been designed, giving basic design data and predicted sulphide concentrations as shown also in Figure 10.15.

The requirement is to design the sewer CD so that the average concentration of total sulphides, during the daily period of ultimate peak 6 hourly flow over the quarter of the year in which sewage temperatures are a maximum, will certainly be less than 1·5 mg/l. For the purposes of this example, the peak 6 hourly dry weather flow is taken as 1·4 times the mean dry weather flow.

Properties of Sewage over length C—D.

The magnitude of flow and values of sulphide concentration $[SO_4]$, biochemical oxygen demand $[BOD_5]$, sewage temperature (T) and sulphides concentration (B_e) required for the design of sewer CD are first determined.

Flows are reckoned as follows:—

ULT. PWWF	= 24 + 32 =	56 cusecs
ULT. PDWF	= 12 + 16 =	28 cusecs
ULT. AV. DWF	= 6 + 8 =	14 cusecs
ULT. MIN. DWF	= 3 + 4 =	7 cusecs
ULT. MAX. 6 HOURLY DWF	= 1·4 × 14 =	19·6 cusecs.

Sulphate concentration. The sulphate concentration may be determined by direct proportion based on maximum 6 hourly dry weather flows. In the following the subscripts 1, 2 and 3 for various measures refer to the sewers AC, BC and CD, as indicated in Figure 10.15.

$$[SO_4]_3 = \frac{[SO_4]_1 \times Q_1 + [SO_4]_2 \times Q_2}{Q_3} = \frac{45 \times 8{\cdot}4 + 80 \times 11{\cdot}2}{19{\cdot}6} = 65 \text{ mg/l}$$

Biochemical oxygen demand, $[BOD_5]$, is reckoned in the same way:

$$[BOD_5]_3 = \frac{[BOD_5]_1 \times Q_1 + [BOD_5]_2 \times Q_2}{Q_3} = \frac{300 \times 8{\cdot}4 + 400 \times 11{\cdot}2}{19{\cdot}6} = 357 \text{ mg/l}$$

Sewage temperature.

$$T_3 = \frac{T_1 \times Q_1 + T_2 \times Q_2}{Q_3} = \frac{73 \times 8{\cdot}4 + 75 \times 11{\cdot}2}{19{\cdot}6} = 74°F$$

Sulphide concentration at upstream end of sewer, at C.

$$[S]_3 = \frac{[S]_1 \times Q_1 + [S]_2 \times Q_2}{Q_3} = \frac{0{\cdot}25 \times 8{\cdot}4 + 0{\cdot}5 \times 11{\cdot}2}{19{\cdot}6} = 0{\cdot}39 \text{ mg/l}$$

Trial Diameter and Grade for Sewer C—D, columns 4 and 5.

The columns referred to are those of the calculation chart see below, p. 128

As a first trial for design purposes a diameter and grade can be selected from Direct Design Chart No. 2 (based on a critical wall shear stress, $\tau_{0crit.}$ of 0·06 psf. and a minimum percentage slime growth of 15%.

$Q_{FULL} \geqslant$ 56 cusec (ULT. PWWF)
Q = 19·6 cusec (ULT. MAX. 6 HOURLY DWF)

$\therefore$ Proportional discharge $\frac{Q}{Q_F} = \frac{19{\cdot}6}{56} = 0{\cdot}35$

From Design Chart No. 2A using Camp's variation of "n"

proportional depth $\frac{d}{D} = 0{\cdot}46$

Hence from Direct Design Chart No. 2 a 48 in. diameter pipe at a grade of 1 in 600 would appear to be satisfactory.

Detailed Design of Sewer C—D.

The detailed calculation of sulphide build-up in the length C—D is the next step. This requires working stepwise through the appropriate sequence columns of the Calculation Chart.

The data for Columns 1 to 10 have been assembled in the foregoing, and are summarized here as follows:—

Columns 1 to 3: From manhole C to manhole D, length L = 8,000 ft, Col. 2, and estimated ULT. PWWF = 56 cusecs.

First trial for diameter and slope, Columns 4 to 7(a) and (b): trial No. 1, trial slope 1 in 600, from Direct Design Chart No. 2, hence required diameter "D" to accommodate ULT. PWWF, 48 ins. = 4·0 ft., from Design Chart No. 1B; and the design flow period after commissioning—ultimate development. Design flow Q = MAX. 6 HOURLY DWF = 19·6 cusecs.

Data for Columns 8 to 10 also are given above, and are repeated here for inclusion in the chart:

Capacity of sewer flowing "just" full Q_F = 56 cusecs, Col. 8; proportional discharge

$\frac{Q}{Q_F} = \frac{19{\cdot}6}{56} = 0{\cdot}35$, Col. 7(b) ÷ Col. 8 = Col. 9,

and required proportional depth, using Camp's variation of "n",

$\frac{d}{D} = 0{\cdot}46$, Col. 10.

Columns 11 to 12(a) and (b): Depth of flow $d = \frac{d}{D} \times D = 0{\cdot}46 \times 4 = 1{\cdot}84$ ft., Col. 10 × Col. 6 = Col. 11; hydraulic radius for 1 ft. dia. = 0·24 ft., from Design Chart No. 3 take 12(a), thence hydraulic radius for D = 4·0 ft. gives 0·24 × 4 = 0·96 ft., Col. 12(a) × Col. 6 = Col. 12(b).

Columns 13(a), (b): mean depth $H = \frac{A}{B}$, across flow for 1 ft. dia. = 0·36 ft., from Design Chart No. 3, then mean depth H,

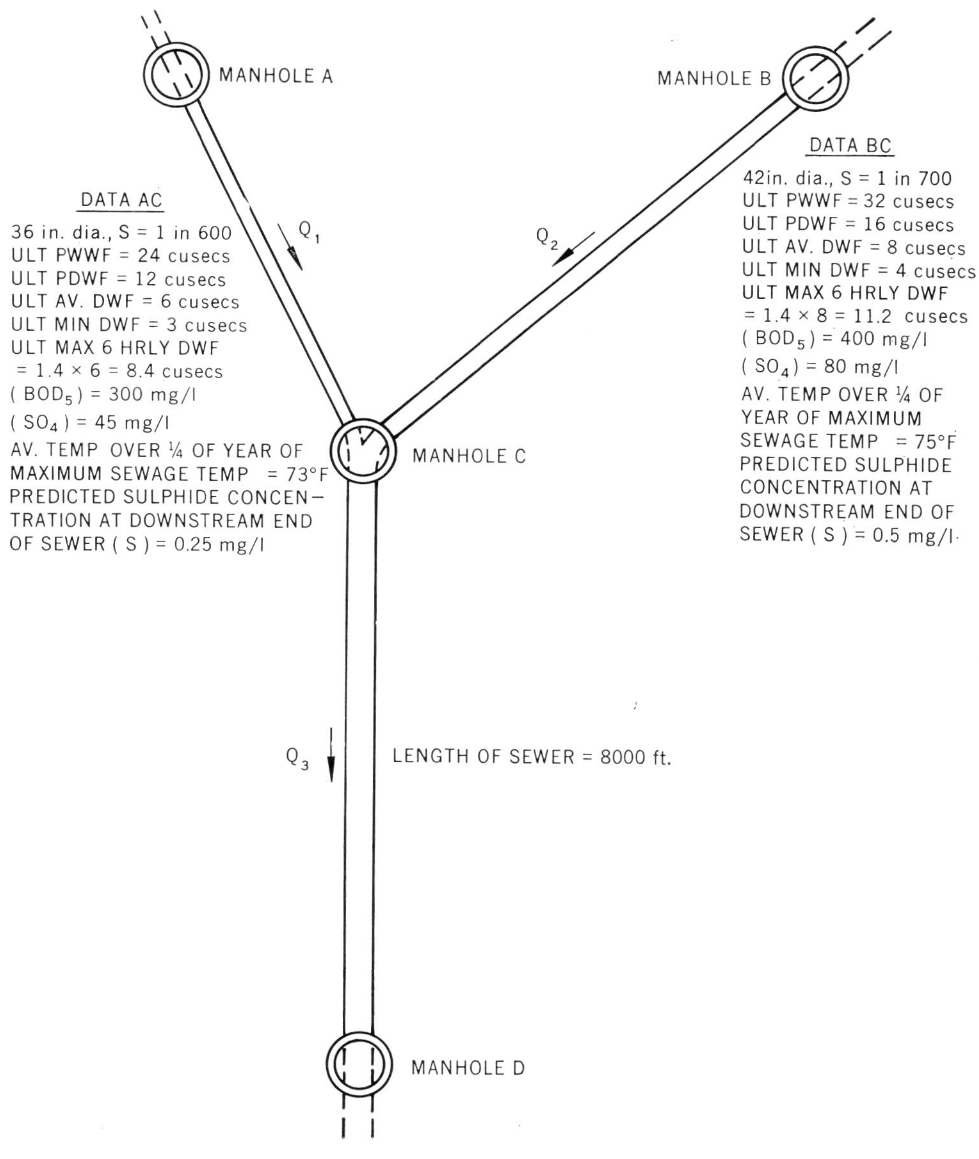

FIG. 10.15 Schematic Plan of Hypothetical System.

1	2	3	4	5	6	7 (a)	7 (b)	8	9	10	11	12 (a)	12 (b)
				BASIC TRIAL	DESIGN CHART No. 1 (NEXT STANDARD SIZE ABOVE)				7 (b) ÷ 8	DESIGN CHART No. 2	10 × 6	DESIGN CHART No. 3	6 × 12 (a)
PARTICULARS OF SECTION						ESTIMATED MAX 6 HOUR DRY WEATHER FLOW DURING QUARTER OF YEAR OF MAXIMUM SEWAGE TEMPERATURE						HYDRAULIC RADIUS $\frac{A}{P}$	
TO MANHOLE No.	L = LENGTH (FT.)	ESTIMATED MAXIMUM WET WEATHER FLOW (cusecs)	TRIAL No.	TRIAL SLOPE 1 IN –	DIAMETER "D" REQUIRED FOR WET WEATHER FLOW	PERIOD AFTER COMMISSIONING	DRY WEATHER FLOW = Q cusecs	CAPACITY OF SEWER FLOWING FULL = Q_F cusecs	VALUE OF $\frac{Q}{Q_F}$	REQUIRED PROPORTIONAL DEPTH	DEPTH OF FLOW (FT.)	FOR CIRCLE 1 FT. DIAMETER	FOR DIAMETER D
C – D	8000	56	1	600	4FT.	ult.	19.6	56	0.35	0.46	1.84	0.24	0.96
			1			5yr.							
			2										
			3										
D – E	6500	80	1			ult.							
			2										
			3										

FIG. 10.16(a) Calculation Chart—Columns 1 to 12.

13 (a)	13 (b)	14	15	16	17	18	19	20	21	22	23 (a)	23 (b)	24	25
DESIGN CHART No. 3	6 × 13 (a)	DESIGN CHART No. 1	DESIGN CHART No. 2	14 × 15	2 ÷ 16		BASIC	DESIGN CHART No. 4		20 – 21	16 ÷ 5	DESIGN CHART No. 5	1830 TIMES 23(b) × 13(b)	BASIC
MEAN DEPTH $H = \frac{A}{B}$ (ACROSS FLOW)						ASSUMED TEMPERATURE (°F)					VALUE OF K_2 (CLEAN WATER AT 68°F)			
FOR CIRCLE 1 FT. DIAMETER	FOR DIAMETER D	V_F = VELOCITY PIPE FULL (FT./SEC.)	$\frac{V}{V_F}$ FOR REQUIRED PROPORTIONAL DEPTH	V_S = VELOCITY AT REQUIRED PROPORTIONAL DEPTH (FT./SEC.)	TIME OF TRAVEL (SECONDS)	SEASON	TEMPERATURE	OXYGEN SATURATION CONCENTRATION (ppm)	UPSTREAM DISSOLVED OXYGEN CONCENTRATION (ppm)	OXYGEN SATURATION DEFICIENCY O_D	VALUE OF $V_S \times S_0$	VALUE OF K_2	VALUE OF $K_{O_2}D$ (CLEAN WATER)	$(BOD)_5$ (ppm)
0.36	1.44	4.5	0.77	3.46	2310	SUMMER	74	8.6	0	8.6	.0058	.0155	40.8	357

FIG. 10.16(b) Calculation Chart—Columns 13 to 25.

for 4 ft. diam. = 0·36 × 4 = 1·44 ft., Col. 6 × Col. 13(a) gives 13(b).

Columns 14 to 16: velocity for pipe flowing "just" full V_F = 4·5 ft/sec. (from Design Chart No. 1), and required proportional velocity, for $\frac{d}{D}$ = 0·46 (Camp's variation of "n"), $\frac{V}{V_F}$ = 0·77 (from Design Chart No. 2A); therefore sewage velocity at required proportional depth $V_S = \frac{V}{V_F} \times V_F = 0{\cdot}77 \times 4{\cdot}5 = 3{\cdot}46$ ft/sec., Col. 15 × Col. 14 = Col. 16.

Column 17: time of travel $= \frac{L}{V_S} = \frac{8000}{3{\cdot}46}$ = 2,310 secs., Col. 2 ÷ Col. 16 = Col. 17.

Columns 18, 19: season—summer, and temperature of sewage = 74°F, as calculated above, p. 126.

Calculation of dissolved oxygen relationships, including re-aeration, Columns 20 to 27, as follows:—

Columns 20 to 22: Oxygen saturation concentration for conditions T_C of 74°F and total barometric pressure assumed 30 in Hg., Design Chart No. 4 gives $(O)_{Sat}$ = 8·6 mg/l, and assumed zero dissolved oxygen concentration at C, $(O_{Sew})_C$ = O mg/l., then oxygen saturation deficiency, $O_D = (O)_{Sat} - (O)_{Sew\ 1}$ thence O_D = 8·6 mg/l, Col. 20 − Col. 21 = Col. 22.

Columns 23(a) to 26: value of $V_S \times S$, $3{\cdot}46 \times \frac{1}{600} = 0{\cdot}00577$ Col. 16 ÷ Col. 5 = Col. 23(a).

Value of K_{O_2}, for $\frac{A}{B}$ = 1·44, = 0·0155, from Design Chart No. 5, Col. 23(b).

Value of K_{O_2D} = 1830 × 0·0155 × 1·44 (see p. 24.) K_{O_2D} = 40·8, 1830 × Col. 23(b) × Col. 13(b) = Col. 24.

$[BOD_5]$ of Sewage = 357 mg/l, as calculated above, p. 126 Col. 25.

Proportion of clean water absorption rate, see p. 25, from Design Chart No. 6 for $[BOD_5]$ of 357 mg/l. and K_{O_2D} of 40·8, F_{dt} = 0·285, Col. 26.

Columns 27(a), (b), (c): value of K_{O_2S} at 68°F = $K_{O_2} \times F_{dt}$ = 0·0155 × 0·285, Col. 23(b) × Col. 26 = Col. 27(a), $K_{O_2S} = 4{\cdot}42 \times 10^{-3}$ mg/l per mg/l of O_D per minute.

Temperature correction factor for K_{O_2S} for temp. of 74°F = 1·05, from Design Chart No. 7 gives Col. 27(b), whence value of K_{O_2S} at temperature of 74°F = $4{\cdot}42 \times 10^{-3} \times 1{\cdot}05$, Col. 27(a) × Col. 27(b) = Col. 27(c), $K_{O_2S} = 4{\cdot}64 \times 10^{-3}$ mg/l per mg/l of O_D per minute.

Column 28: sulphate concentration in sewage, $[SO_4]$ = 65 mg/l, as calculated above, p. 126.

Basic rate of slime sulphide build-up, Columns 29(a) to (h): Shear stress Parameter, $\gamma \times A/P \times S = 62{\cdot}5 \times 0{\cdot}96 \times \frac{1}{600}$, 62·5 × Col. 12(b) × Col. 5 = Col. 29(a), τ_{op} = 0·099 lb/ft².

Prediction of Percentage Slimes, for τ_{op} = 0·099 and $\frac{d}{D}$ = 0·46, F_{sli} = about 15%, Col. 29(b), taken from Design Chart 10, see p. 124, Col. 29(b).

Slimes sheared off to above minimum flow line, see Figure 8.2(c), corresponding factor $F_1 = F_{sli} \times k$. Assuming k = 0·5, see p. 18, F_1 = 0·15 × 0·5, Col. 29(b) × k = 0·075, Col. 29(c).

Value of $1{\cdot}43 \times 10^{-7} \times [BOD_5]^{0.8} \times [SO_4]^{0.4}$, see p. 83. when $[BOD_5]$ = 357 mg/l and $[SO_4]$ = 65 mg/l, $1{\cdot}43 \times 10^{-7} \times [357]^{0.8} \times [65]^{0.4} = 83{\cdot}4 \times 10^{-6}$, Col. 29(d).

Value of $\frac{V}{A/P} = \frac{3{\cdot}46}{0{\cdot}96}$ = 3·60, Col. 16 ÷ Col. 12(b) = Col. 29(e).

Basic rate of sulphide build-up at 68°F due to generation in active slimes is $1{\cdot}43 \times 10^{-7}\ [BOD_5]^{0.8} \times [SO_4]^{0.4} \times F_1 \times \frac{V}{A/P}$, see equation [8·7], p. 85, $83{\cdot}4 \times 10^{-6} \times 0{\cdot}075 \times 3{\cdot}60$ Col. 29(d) × 29(e) × 29(c), $B_S = 22{\cdot}52 \times 10^{-6}$ mg/l per second, Col. 29(f).

Temperature correction factor for B_S, temperature of 74°F = 1·5, Design Chart No. 7, Col. 29(g).

Rate of build-up due to slimes for length C—D, $22{\cdot}52 \times 10^{-6} \times 1{\cdot}5$, Col. 29(f) × Col. 29(g), $(B_S)_{sli} = 33{\cdot}78 \times 10^{-6}$ mg/l/second, Col. 29(h).

Column 30: Rate of sulphide build-up in body of sewage for travelling time C—D 2310 secs., Col. 17, plus average age of sewage at manhole C assumed 2 hours, taking value of 0·08 ppm/hr from Design Chart No. 8 $(B_S)_{Sew.} = 22{\cdot}20 \times 10^{-6}$ ppm/second.

Columns 31(a), (b), (c): instantaneous rates of oxygen demand, for 68°F, $\frac{d}{dt}[BOD_0] = 6 \times 10^{-6} \times [BOD_5] = 6 \times 10^{-6} \times 357 = 2{\cdot}14 \times 10^{-3}$, Col. 31(a).

26	27			28	29								30
	(a)	(b)	(c)		(a)	(b)	(c)	(d)	(e)	(f)	(g)	(h)	
DESIGN CHART No. 6	23(b) × 26	DESIGN CHART No. 7(a)	27(a) × 27(b)	BASIC	DESIGN CHART No. 10	DESIGN CHART NO. 10	FIG. 8.2	DESIGN CHART NO. 9	16 ÷ 12 (b)	29 (d) × 29 (e) × 29 (c)	DESIGN CHART NO. 7(b)	29 (F) × 29 (G)	DESIGN CHART NO. 8 READ FOR APPROPRIATE AGE
							$= (B_S)_{SLI}$ = BASIC RATE OF SLIME SULPHIDE BUILD-UP (ppm/SEC)						BOD″ = BOD RATE (ppm/SEC)
FACTOR F_{dt}	VALUE OF K_{O_2S}, 68°F	TEMPERATURE CORRECTION FACTOR FOR K_{O_2S}	K_{O_2S} AT APPROPRIATE TEMPERATURE	SULPHATE CONCENTRATION $[SO_4]$	SHEAR STRESS PARAMETER τ_{OP}	PREDICTED PERCENTAGF SLIMES %	F_1 (slime factor)	$1.43 \times 10^{-7} \times [BOD_5]^{0.8} \times [SO_4]^{0.4}$	$V_S \times P_T \div A_S$	VALUE OF $(B_S)_{SLI}$ AT 68°F	TEMPERATURE CORRECTION FACTOR	$(B_S)_{SLI}$ AT APPROPRIATE TEMPERATURE	RATE OF SULPHIDE BUILD–UP IN BODY OF SEWAGE $(B_S)_{SEW}$
.285	4.42×10^{-3}	1.05	4.64×10^{-3}	65	.099	15	.075	83.4×10^{-6}	3.6	22.5×10^{-6}	1.5	33.8×10^{-6}	22.2×10^{-6}

FIG. 10.16(c) Calculation Chart—Columns 26 to 30.

Temperature correction factor for $\frac{d}{dt}$ [BOD$_0$], temp. of 74°F, = 1·15, from Design Chart No. 7, Col. 31(b), hence rate for sewage temperature of 74°F = 2·14 × 10^{-3} × 1·15, Col. 31(a) × Col. 31(b), $\frac{d}{dt}$ [BOD$_0$] = 2·46 × 10^{-3} ppm/sec. for length C—D, Col. 31(c).

Column 32: Oxygen demand;

$= \frac{d}{dt}[BOD_0] + 2(B_S)_{sli} + 2(B_S)_{sew}$

= Col. 31(c) + 2 Col. 29(h) + 2 Col. 30
$= 2{\cdot}46 \times 10^{-3} + 2\,(33{\cdot}78 \times 10^{-6} + 22{\cdot}2 \times 10^{-6})$
$= 2{\cdot}57 \times 10^{-3}$

Column 33: 60 × Col. 32

Column 34: Rate of re-aeration, $O_D \times K_{O_2S}$;
= Col. 22 × 27(c)
$= 8{\cdot}6 \times 4{\cdot}64 \times 10^{-3}$
$= 39{\cdot}90 \times 10^{-3}$

Column 35: Proportional loss of sulphide by oxidation; or
= Col. 34 ÷ Col. 33

$= \frac{39{\cdot}90 \times 10^{-3}}{154{\cdot}20 \times 10^{-3}}$

= 0·259

Column 36: Proportion of sulphide remaining (build-up)
= 1 — Col. 35
= 1 — 0·259
= 0·741

Column 37: Effective rate of build-up of sulphide sulphur mg/l/second;
$= (G_S)_{eff}$
$= G_S \times$ Col. 36
$= [(B_S)_{sli} + (B_S)_{sew}] \times$ Col. 36
$= [33{\cdot}78 \times 10^{-6} + 22{\cdot}20 \times 10^{-6}] \times 0{\cdot}741$
$= 41{\cdot}48 \times 10^{-6}$ mg/l/sec.

Total Sulphide Build-up, Columns 38 to 40. Equivalent sulphide discharge direct into sewage along C—D taken as negligible for this particular case, value of Col. 28 therefore = 0 mg./l.

Total rate of sulphide build-up therefore as for Col. 37, = 41·48 × 10^{-6} mg/l/sec., Col. 39.
Total build-up, along section C to D, = 41·48 × 10^{-6} × 2310, Col. 39 × Col. 17 = 0·096 mg/l (say 0·10 mg/l), Col. 40.
Sulphide concentration at upstream end C, as calculated above, p. 126, 0·39 mg/l, Col. 41.
Sulphide concentration in sewage at downstream end D, [S] = 0·39 + 0·10, Col. 41 + Col. 40 = 0·49 mg/l, Col. 42.

Since the computed total sulphide build-up at the downstream end of the sewer C—D is then considerably less than the limiting build-up of 1·5 mg/l (see p. 111), it may be taken that a 48 in. diameter pipe at a grade of 1 in 600, would be a satisfactory design for sulphide control so far as that section of the system is concerned. If the system is to be extended further beyond D, the effect of this level of total sulphide, at D, of 0·49 mg/l must be related to conditions downstream by computations in the same way, length by length.

COMPARISON OF CALCULATED AND OBSERVED VALUES FOR SULPHIDE BUILD-UP

The following case from an Australian capital city provides a direct comparison between calculated sulphide build-up and sulphide build-up values actually observed for a large main sewer.

The sewer cited is an 11 ft. diameter outfall sewer, some 16 miles long, serving one of Australia's largest cities. The relevant properties of the sewer are as listed below:—

Diameter	= 11 ft. 0 ins.
Length	= 84,300 ft.
Slope	= 1 in 2640
Manning roughness coefficient for conduit flowing full n_f	= 0.014

This sewer conveys a very high strength sewage (BOD_5 = 450-700 mg/l) containing a high propor-

31 (a)	31 (b)	31 (c)	32	33	34	35	36	37	38	39	40	41	42
$6 \times 10^{-6} \times$ COLUMN 25	DESIGN CHART NO. 7(c)	31(a) × 31(b)	31(c) + 2 × 29(h) + 2 × 30	60 × 32	22 × 27(c)	34 ÷ 33	UNITY − 35	[29(h) + 30] × 36		37	39 + 17		40 + 41
$\frac{d}{dt}[BOD_0]$ ppm/sec: RATE AT 68°F	TEMPERATURE CORRECTION FACTOR	RATE OF APPROPRIATE TEMPERATURE	$\frac{d}{dt}[BOD_0] + 2(B_S)_{SLI} + 2(B_S)_{SEW}$	$60\left(\frac{d}{dt}[BOD_0] + 2(B_S)_{SLI} + 2(B_S)_{SEW}\right)$	$O_D \times K_{O_2S}$	$\frac{O_D \times K_{O_2S}}{60 \times (\frac{d}{dt}[BOD_0] + 2(B_S)_{SLI} + 2(B_S)_{SEW})}$	$1 - \frac{O_D \times K_{O_2S}}{60 \times (\frac{d}{dt}[BOD_0] + 2(B_S)_{SLI} + 2(B_S)_{SEW})}$	RATE OF SULPHIDE BUILD-UP ppm/sec $G_S \times \left[1 - \frac{O_D \times K_{O_2S}}{60 \times (\frac{d}{dt}[BOD_0] + 2(B_S)_{SLI} + 2(B_S)_{SEW})}\right]$	SULPHIDE BOUGHT INTO SEWER ALONG WITH TRIBUTARY SEWAGE	TOTAL RATE OF SULPHIDE BUILD-UP ALONG SECTION ppm/sec	TOTAL BUILD-UP ALONG SECTION	[S] IN SEWAGE AT C	[S] AT D
2.14×10^{-3}	1.15	2.46×10^{-3}	2.57×10^{-3}	154×10^{-3}	39.9×10^{-3}	.259	.741	41.5×10^{-6}	0	41.5×10^{-6}	.10	.39	.49

FIG. 10.16(d) Calculation Chart—Columns 31 to 42.

tion of industrial wastes. Sewage temperatures vary from about 60°F to about 74°F with sulphate concentrations [SO_4] in the order of 90 mg/l. Over the period of observation mean dry weather flow was about 76 m.g.d.

The mean calculated sulphide build-up values, obtained by the above calculation procedures, and the actual mean observed values of sulphide build-up, at intervals over a 12 month period, were respectively 3.5 mg/l and 1.1 mg/l. (Emission was not allowed for in the computations.) Sampling times were selected so that there was no significant rainfall over the sewer catchment at the time, and for at least three weeks before.

This difference between mean observed and calculated build-up can be attributed partly to sulphide losses by emission, and also it was possible that some of the heavier forms of sulphides were precipitated and held in the sediments on the invert of the sewer, hence preventing their proper inclusion as part of the true build-up. However, considering the extensive length of this sewer (some 16 miles) the discrepancy, even without correction, is considered relatively not so large as might justify a general rejection of and lack of confidence in the foregoing formulations. Moreover, it should be remembered that these formulations are expected to be somewhat conservative for design purposes. The sulphide build-up values for this sewer, both as calculated and observed, suggest that the sewer is likely to suffer significant corrosion. This in fact has been the case, and it has been found necessary to take corrective action. This has included the installations of an extensive forced air ventilation system.

CHAPTER 11

DESIGN OF SEWAGE PUMPING SYSTEMS

Introductory

Pumping systems with pressure mains, or long inverted siphons can be an important source of sulphides; experience has shown that many serious sulphide problems originate in pumping systems. The problems usually manifest themselves in parts of the sewer system downstream of the pumps. When such systems discharge directly to treatment works or ocean outfalls the corrosion problem may be of less importance but on the other hand odour problems may become more important. In the case of discharge to a treatment works, treatment processes may be less efficient where the influent is septic.

Sewage pumping systems may be classified according to whether they are designed to pump continuously or only intermittently. These two general classifications will be considered separately because the design considerations, as they relate to sulphide problems, are somewhat different. Problems in arrangement and control of the pumping station also are different, and being part of the complex should be considered along with rising main layout and design.

Ideally, the sewage to be pumped should reach the pumping station with a normal redox potential, carrying at least 0.5 mg/l of dissolved oxygen. These circumstances should be consequential to sewer grades upstream being favourable for the oxygen balance either naturally because of topography, or because design was made according to the principles of Chapters 8 and 10. The overall design problem may be summarized briefly as follows:—

- The dissolved oxygen, following any accumulation and effective storage in the wet wells, should remain at or about the same level as in the incoming sewage before storage. This should apply particularly to normal dry weather conditions of temperature, humidity and other environmental factors which most favour the occurrence of sulphides in the system;

  (*Note*: In most sewerage systems, provision must be made not only for D.W.F. conditions, but for some proportion of stormwater infiltration in wet weather or prolonged "wet seasons". Such provision must be allowed for in pumping system designs also, but for *sulphide control* it is prolonged dry weather which is significant.)

- Pump capacities, rising main dimensions including overall length, route and grades, elevations, etc., must be determined in such a way as to obviate sulphide problems downstream of discharges into the receiving gravity sewer. At this stage economic factors become important. An appropriate balance is sought between:—

  (a) overall costs of the numerous alternatives,

  (b) capital and operating costs, including power charges, for the pumping system itself and

  (c) the incidental or consequential costs of remedial measures, including costs for additional ventilation or deodorization which may be required should sulphide and other problems result. This "choice" depends at least in part on the way wet weather conditions are provided for.

The following pages set out a procedure for the logical design of sewage pumping systems on this basis, but also having regard to the control of sulphide conditions throughout the associated gravity sewers and their appurtenances. The special case of discharge to treatment works directly is not given attention as such.

Sewage pumping systems may be designed with or without storage for sewage, as shown in Figures 11.1 or 11.2. In large systems which pump sewage continuously throughout the day and night it is customary to omit "wet well storage". Any tendency for the influent sewer to act as a balancing storage usually causes a serious increase in the intensity of the sulphide problem, and should be strictly forbidden either for design or as a development in operation.

Figure 11.1 illustrates this. Under wet weather conditions, ponding only becomes significant where detention periods are measurable in units of several hours, but need not and should not result in accumulation of silt or other deposits.

Design Features of Wet Well (Pumping Station Storage) Installations

Figure 11.2 shows a typical wet well installation and the processes affecting the sulphide problem which find a place in such a system, generally in diagrammatic form.

Although build-up of sulphide by wet wells commonly may be quantitatively unimportant, there may be nevertheless some unfavourable overall effect producing "staleness" of the sewage (pp. 34, 94), and this in turn may have an indirect disproportionate effect on the overall sulphide problem by increasing the subsequent build-up rate in the rising main. If the

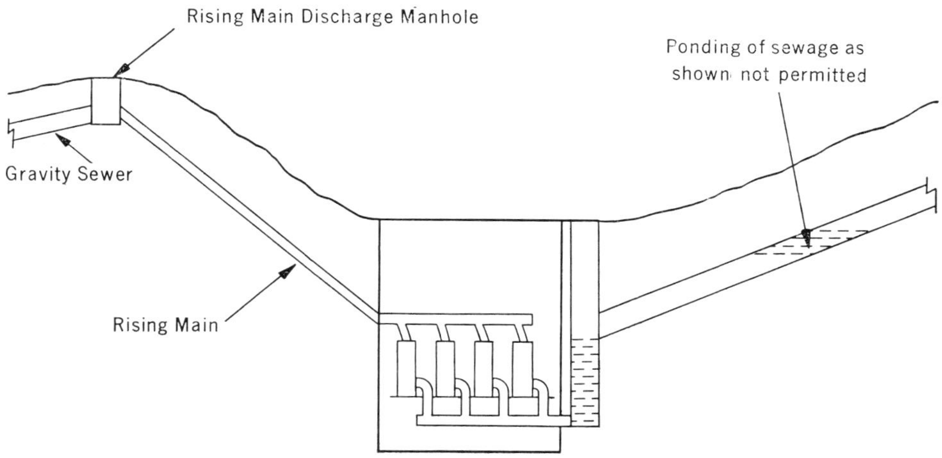

FIG. 11.1 Pumping System without Storage—continuous pumping.

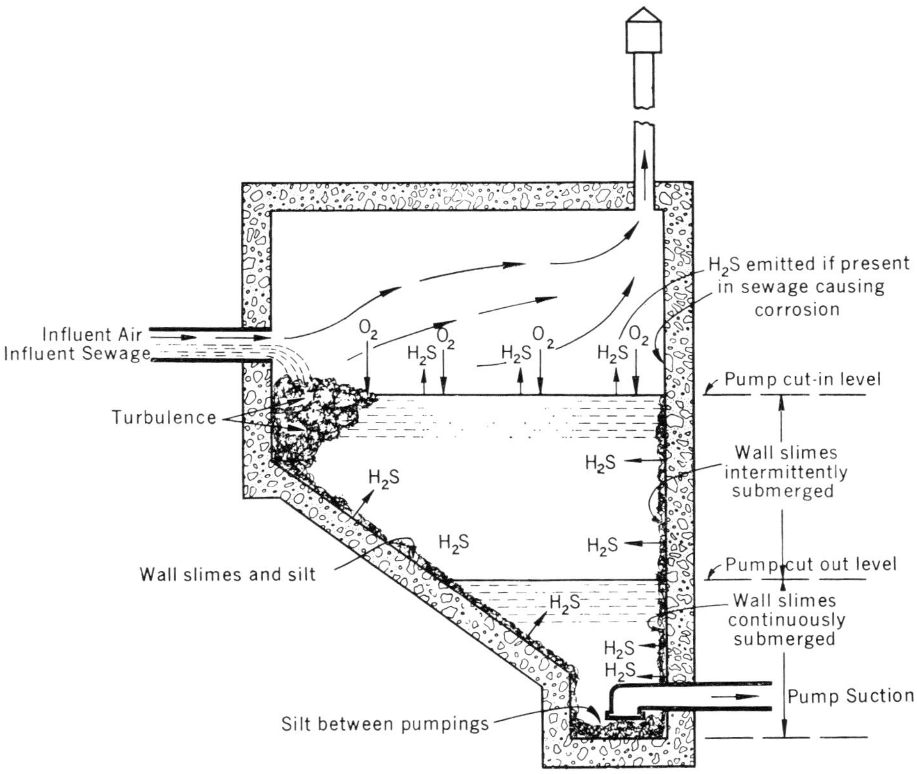

FIG. 11.2 Typical Wet-well Arrangement for Sewage Pumping.

sewage is still in a state of freshness when pumped from the wet well, then sulphide build-up will not occur at least in the inlet section of the rising main, due to the presence there of residual dissolved oxygen. The wet-well contribution to sulphide build-up could lie rather in the pumped sewage becoming stale, causing an unfavourable pre-conditioning of the sewage. Then maximum rates of build-up would commence nearer to the inlet end of the rising main than if a wet well had not been installed.

The comparatively very low rates of build-up which have been found to occur in most wet wells indicate that normal rates of generation in wall slimes do not apply to wet wells. However, these normal rates might well apply in cases of bad operation or design, or where regular cleaning is neglected.

The joint investigations carried out by the Australian Sewerage Authorities(*6*) included investigation into sulphide build-up in wet wells of eleven automatic intermittent systems. These investigations indicated that the sulphide build-up in sewage in wet wells between the time of its entry into and the time of leaving the wet well was much less *pro rata* than during its passage through the rising mains, and was, in general, negligible. However, one case was recorded where relatively high build-up occurred, the sewage having relatively high sulphate concentrations due to sea-water infiltration. The low sulphide build-up generally experienced appeared to be due to the presence of dissolved oxygen in the influent sewage, which would destroy some of the sulphide generated by the wall slimes on the submerged surfaces of the well. Probably also this influences the nature of the slime and its rate of sewage sulphide production, particularly since a large proportion of the slime area is intermittently exposed to the atmosphere for a substantial part of the time. Turbulent conditions in the wet well would be conducive to loss of some sulphide by emission to the air and also to absorption of some oxygen from the air.

Size of storage: The volume of wet well storage to be provided will depend upon the rate of sewage inflow and the time for which sewage is to be stored, that is to say upon the frequency with which the duty pump is assumed to operate. Modern starting equipment can now be obtained to provide for very frequent starting and stopping without undue maintenance problems, provided the required number of starts per hour is included in the purchase specifications.

The local conditions prevailing will no doubt influence the detailed design of a particular pumping station but the following suggestions are made as general guide lines for minimizing related sulphide problems.

The volume of wet well storage provided below the invert level of the influent sewer should be sufficient to hold the estimated ultimate peak wet weather sewage flow over a period of at least two minutes. Until full development of the area served takes place the cut-in and cut-out levels for automatic control of the pump should be adjusted from time to time so that the corresponding storage volume actually used is such that it will take not more than five minutes to fill at time of daily peak flow. However this volume also should be not less than about 1 to 1½ minutes equivalent minimum pump discharge rate. Sulphide build-up or oxygen depletion tests should be made to find the best adjustment of pump cut-in and cut-out levels feasible under all the circumstances of that time. Clearly, adequate provision must be made during design so that adjustments along these lines will be practicable.

In the event that these preferred arrangements are not practicable, the next most desirable condition would be one which ensured that the floor of the storage well was flushed clear automatically at the end of each filling-pumping cycle. This could be achieved by using inflowing sewage to flush the exposed floor, but in most cases only a fraction of floor area could be flushed in such a way unless the bottom of the storage were specially shaped. The only shape which might be flushed satisfactorily by inflowing sewage is the last of the types listed in Table 11.1 opposite. This type shows a relatively poor value for the ratio (area of sewage surface)/(area of wetted walls and floor), however, good flushing may largely offset the low ratio disadvantage.

Routine Maintenance and Cleaning: Present practice is to rely upon periodical manual cleaning (generally with hose jets) at intervals varying from a few

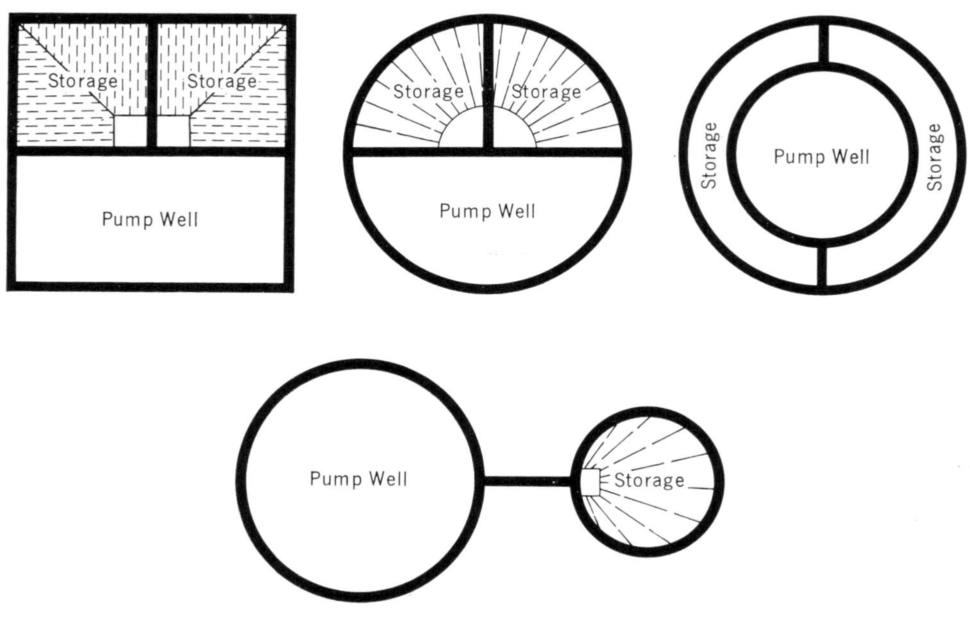

FIG. 11.3 Common Pumping Station Layouts.

days to a week or more. Until further knowledge is available as to the overall effect of the period between cleanings on the oxygen depletion or sulphide build-up through a pumping system, no proper assessment can be made of the effort warranted, at the design stage, to reduce siltation effects, etc., by such means.

Shape and other Structural Features: The shapes of wells in most common use in Australia are shown in Fig. 11.3.

Two wells which can be used singly or in combination are frequently provided. Such an arrangement permits one well to be used in the early stages of development or during periods of maintenance. Some of the characteristic geometry of these various arrangements is tabulated in the accompanying Table 11.1.

It would appear that in general the best shape of wet well is one which will provide a maximum of free sewage surface area for absorption of oxygen and a minimum of wall and floor slime and silt area for sulphide generation. That is to say the ratio of sewage surface area to submerged slime area of walls and floor should be as high as practicable. The word practicable is used here because the absolute upper limit of unity for the ratio referred to requires an impracticable combination of infinite area and negligible depth. For a circular well holding about 1000 cu. ft. at $2\frac{1}{2}$ ft. depth, this ratio is about 0.7 only.

Conditions for slime sulphide generation are much more unequal for slimes in flowing sewers. This is because the wall and floor slimes and floor silt below the pump cut-out level are continuously submerged by sewage, whereas all slimes above cut-out level are intermittently exposed to the atmosphere. Where the wet well is designed for short period storage these periods of exposure to the atmosphere will vary from a negligible time near cut-out point to practically the whole time near cut-in level. With relatively high pumping rates and substantial storage, conditions in wet wells with low ratios may well show rates of sulphide build-up comparable to the rates observed for sewers.

There is little quantitative knowledge regarding the variation in sulphide generating capacity of wall slimes where there is appreciable variation in the percentage of time such slimes are exposed to the air, but it seems reasonable to assume that the generating

TABLE 11.1. WET WELL SHAPE FACTORS

Shape Classification	For Total Stored Volume of 100 cu. ft. (625 gal.)				For Total Volume of 500 cu. ft. (3,250 gal.)				Comment
	Area of Sewage Surface sq. ft.	Dia. or side ft.	Max. Depth ft.	Value[1] of Ratio	Area of Sewage Surface sq. ft.	Dia. or side ft.	Max. Depth ft.	Value[1] of Ratio	
1. *Rectangular* (Vertical sides—flat floor) Ratio of sides 1:1	50 33 25	7.06 5.76 5.00	2 3 4	0.47 0.32 0.24	250 125 62	15.8 11.2 7.9	2 4 8	0.655 0.412 0.196	Marked tendency for deposition of solids. Manual cleaning practicable.
2. *Cylindrical* (Vertical sides—flat floor)	50 33	7.97 6.51	2 3	0.50 0.33					do.
3. *Inverted Pyramid* side slope 45° side slope 30°	71 102	8.42 10.10	4.22 2.82	0.709 0.866	208 300	14.4 17.3	7.2 5.0	0.709 0.866	Maintenance difficult.
4. *Half Inverted Pyramid* (one vertical wall) side slope 45° side slope 30°	55.7 80.5	7.45 8.96	5.29 3.68	0.522 0.695					Entry and cleaning easier than 3. (depends on dimensions).
5. *Inverted Conical* side slope 45° side slope 30°	65.5 94.0	8.09 9.70	4.57 3.16	0.709 0.866					"Self-scouring" but difficult to clean.
6. *Three Vertical Sides & Sloping Floor Floor Slope 45°* Width 4' " 6' " 8'				0.219 0.278 0.317					do.

Note 1: The "Value of Ratio" is the ratio of free sewage surface to potential slime area.

capacity will decrease as the percentage of time increases. As previously pointed out the floor slimes and silts and wall slimes below pump cut-out level are continuously submerged. It would seem therefore that high value for the ratio of (area of sewage surface)/(area of wetted wall and floor surface) is more important for the lower part of storage than for the upper. The values of the ratio depend upon wet well shape, and Table 11.1 gives some data reflecting the effect of shape. The various values in this table indicate clearly that shallow storages give higher values of the ratio, and require greater areas. Some advantages and disadvantages for wells of differing shapes may be seen from the table.

Another consideration which has some bearing on the selection of structural designs for wet wells is that of silt deposition. The ideal arrangement would be one in which no deposition occurred. This would require either constant and sufficient agitation to keep settleable solids in suspension, or the provision of self-cleansing slopes. However, floor slopes necessary to ensure self-cleansing are so great as to increase considerably both the overall depth and the cost of construction. Also maintenance access is more difficult and costly, and surface aeration is decreased to some extent.

Sewage Inlet Design: If the influent sewer is allowed to pond above the depth required for normal parallel flow of the incoming sewage, then the velocity will diminish and siltation will occur, and surface aeration will decrease also because of diminished velocity. If ponding were to result to the extent of completely filling the influent sewer then the section of sewer so filled becomes in effect a pressure main and the effective length of the rising main is increased by the length of the influent drain so filled. Designs should ensure that no ponding normally occurs in dry weather.

Ponding and the ensuing troubles can be eliminated by providing in the pumping station a suitable combination of variable and fixed capacity pumps automatically controlled in such a way that the pumping capacity of the station in operation at any time always matches the flow in the influent sewer(*59*). The types of sewer inlet in common use are shown in Fig. 11.4.

The considerations which enter into inlet design include the following:

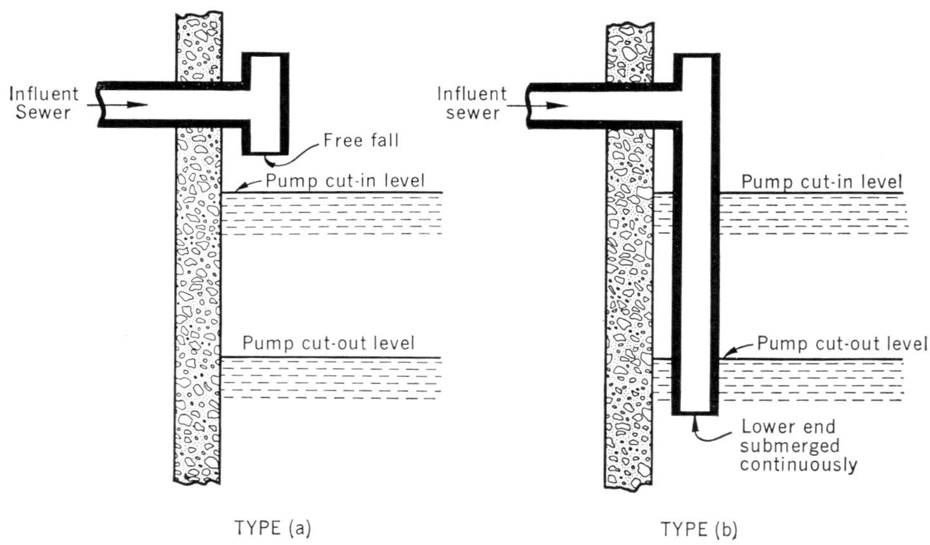

FIG. 11.4 Types of Inlets for Wet Wells.

- *Turbulence set up by dropping sewage*: Such turbulence either will cause increased emission of dissolved sulphides, etc., or increased absorption of oxygen from the air. If it be considered that negligible concentrations of sulphide, if any are to be expected in the sewage, a free drop type of inlet, Fig. 11.4(a), may be advantageous and help to keep the sewage fresh. If dissolved sulphides are considered likely to be present however then a submerged type of inlet, Fig. 11.4(b), would generally be preferable.

- *Cleaning of surface of the well floor by incoming sewage*: The extent to which this can be effected safely depends likewise on the dissolved sulphide content of the incoming sewage, unless sulphides be "fixed" by suitable chemical dosing (see p. 161).

- *Prevention as far as practicable of deposition of silt* on floor of well by encouraging turbulence just above floor surface. This would be unsafe also if the sewage contained too much dissolved sulphide. Both the foregoing require comparatively careful detailed hydraulic design.

Additional investigations of hydraulic models are needed to develop the best type and position of inlets for various circumstances. The use of the type of Fig. 11.4(b) is generally recommended.

● *Ventilation*: Reference to Fig. 11.2 suggests that unless effective ventilation is provided for the wet well, the flow of ventilating air via the influent sewer will be impeded, and perhaps stopped, leading to stagnant air conditions in both the sewer and the wet well itself; and if some dissolved sulphide be present the circumstances will be generally conducive to corrosion of concrete and ironwork above sewage level. The presence of sulphide corrosion in wet wells has in fact been noted frequently and to obviate these conditions adequate ventilation should be provided, see Chapter 12.

Recommended Procedure for the Design of Sewage Pumping Station Wet (Storage) Wells

Having regard to the foregoing, the following procedure is recommended as covering the requirements for wet well designs so far as their influence on sulphide problems is concerned:—

(i) Estimate storage volumes required for peak dry weather and wet weather flows during both early life (say first five years) and ultimate conditions.

(ii) Determine the pump cut-out level of sewage in the wet well for the proper mechanical and hydraulic functioning of the pump (or pumps).

(iii) That portion of the wet well required to hold one quarter of the ultimate required storage volume, together with the volume of storage provided below pump cut-out level, should be so shaped as to give a value of the ratio of (area of sewage surface)/(area of wetted walls and floor) of not less than 0.6.

(iv) Above the level of storage determined by the preceding requirement the walls of the wet well may be taken up vertically.

(v) The energy of the inflowing sewage and of outgoing sewage during pumping should be used to the best advantage to assist in cleaning the floor of the wet well between fillings, provided the inflowing sewage or sewage in wet wells does not contain more than about 0.25 mg/l of sulphide.

(vi) The layout and details of the pumping station ventilation system then should be worked out to best advantage.

(vii) Finally, consideration should be given to the control systems necessary to provide the pump flexibility referred to, covering the whole required range of pump capacities for rates of inflow at low and peak D.W.F. as well as for W.W.F. conditions.

Generally, the final decisions in this series of steps should be taken along with the corresponding design sequences for the rising main system, which are set out in the following section.

DESIGN OF PUMP DISCHARGE SYSTEMS

The principal consideration in designing such systems (rising or pressure mains and their discharge) for sulphide control is to minimize the overall build-up of sulphides in the sewage during its passage through the system. Sulphide corrosion problems associated with pumping systems generally do not occur in the rising main, but if they arise or are aggravated by sulphide build-up in the pumping system, it is downstream in the receiving sewers or treatment works that problems will occur. The rising main problem therefore cannot be considered in isolation as an entirely separate part of a sewerage system, and its design accordingly commences in a review of the whole associated low and high level gravity sewerage system as well as the whole storage and pump discharge system which connects them.

The rising main problem nevertheless is related mainly to sulphide generating rates. However important these rates may be for downstream gravity sewers, this downstream gravity flow system has no significant effect on them. The upstream low-level sewers, on the other hand, together with the interconnecting wet wells, have always a very important influence on sulphide build-up downstream by pre-conditioning the sewage entering the pumping system, whereby sulphide build-up either may be greatly lessened or retarded because the sewage is fresh, or may be encouraged because it is stale or septic. The effect of the pumping system continues along the downstream gravity sewers into which the sewage is pumped, especially where the proportion of sulphide is high.

Sulphide Generation in Rising Mains: An empirical formula for sulphide build-up in rising mains was presented in Chapter 8 as follows:

For continuous pumping, equation [3.8], p. 15

$$G_s = 32{\cdot}2 \times 10^{-6} \times V_s \times [BOD_5]^{0{\cdot}8} \times [SO_4]^{0{\cdot}4} \times 1{\cdot}139^{(T°C-20)} \qquad [11.1]$$

where G_s is the rate of sulphide generation, lb. S per 1000 sq. ft. of wall slime per hour.

For continuous pumping, this reduces to

$$(B_s) = 0{\cdot}573 \times 10^{-6} \times L/D \times [BOD_5]^{0{\cdot}8} \times [SO_4]^{0{\cdot}4} \times 1{\cdot}139^{(T°C-20)} \qquad [11.2]$$

The corresponding empirical equation for intermittent pumping takes the following form

$$(B_s)_{Av.} = 0{\cdot}406 \times 10^{-6} \times L/D \times \frac{1{\cdot}64^{\bar{V}_s}}{\bar{V}_s} \times [BOD_5]^{0.8} \times (SO_4)^{0.4} \times 1{\cdot}139^{(T°C-20)} \qquad [11.3]$$

Here (B_s) is the total sulphide sulphur build-up during continuous pumping through a circular rising main, diameter D ft. Length L ft., measured in mg/l.

$(B_s)_{Av.}$ and $\bar{V}_s$ are related as above, so that $(B_s)_{Av.}$ may be defined as the effective average rate of sulphide sulphur build-up in mg/l under conditions of intermittent pumping through a circular rising main, diameter D, length L, at an effective average velocity $\bar{V}_s$.

V_s is the velocity of sewage, ft./sec., for continuous pumping.
$\bar{V}_s$ is the effective average velocity for intermittent pumping and $= \frac{Q_s}{txA}$ where Q_s is the total volume of sewage pumped during time t, the pumping cycle period through rising main of cross-section A, ft./sec.
$[BOD_5]$ is the 5-day Biochemical Oxygen Demand of sewage pumped, $mgO_2/1$.
(SO_4) is the Concentration of sulphate in sewage $(mg/1\ SO_4)$.
T°C is the sewage temperature °C.

Examination of these equations indicates that the rate of generation of sulphides is less for intermittent than for continuous pumping, hour for hour. However, since it is generally to be assumed that the same total volumes of sewage are pumped from hour to hour; and that satisfactory equipment may be designed to handle sewage flows on either continuous or intermittent pumping basis; a velocity of not less than 2 ft./sec. during any actual pumping will be necessary to keep rising main clear; and the rising main is the same length for both types of pumping but of proportionately greater diameter, the more intermittent the pumping; then it will be found that the intermittent system produces no significant difference in build-up compared with a continuous system.

It is of interest to note that as the percentage of the time for which the duty pump (or pumps) operates increases to 100%, the intermittent pumping system becomes in effect continuous. Under this condition values of (B_s) and $(B_s)_{Av.}$ as given by equations [11.3] and [11.4] should approach equality. That is to say, since all other terms are the same,

$$\frac{1{\cdot}64^{\bar{V}_s}}{\bar{V}_s} \text{ should approximate } 1{\cdot}412$$

Owing to the empirical nature of the equations however, exact equality occurs only when V_s equals either 1.5 or 3.0 ft./sec. At velocities between these values or just outside their range the departure from equality does not exceed 8%, which is well within the overall precision to be expected.

As stated above, the downstream sewers come now into question. For continuous systems, the total build-up $G_s \times L \times \pi \times D$ sulphide sulphur per hour is equal to $G_s \times L \times \pi \times D \div 3600$ per second. Although the average quantity will be much the same for an intermittent pumping system, taken over the day, during say 12 minutes actual pumping per hour the rate of discharge of sulphide to the receiving sewers downstream then will be about 5 times greater. If for continuous pumping the contribution of sulphide sulphur amounted to 0.5 mg/l average for all the free flowing sewage downstream, then the particular 12 minutes actual pumping would contribute $2\frac{1}{2}$ mg/l; although this be for only 20% of the time, this could be serious. It might prove even more serious in some cases because of the high velocities during discharge.

To facilitate calculations, Fig. 11.5 gives values of $\frac{1{\cdot}64^{\bar{V}_s}}{\bar{V}_s}$ for various values. Design Chart No. 10 facilitates computation of equations [11.2] and [11.3].

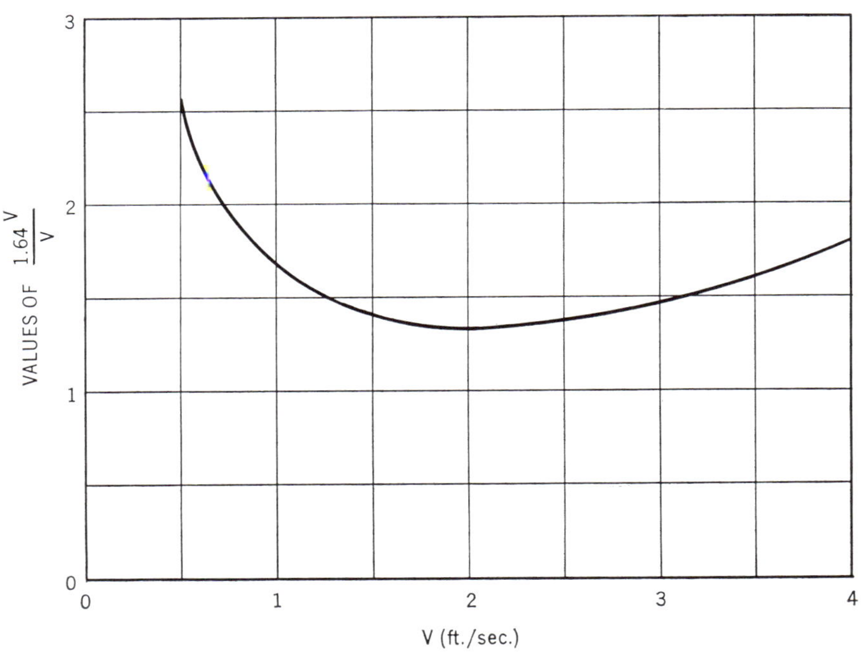

FIG. 11.5 Values of $\frac{1.64^V}{V}$ or $\left(\frac{1{\cdot}64^{\bar{V}_s}}{\bar{V}_s}\right)$

Before applying these equations directly to a design problem, however, there is another effect of the sewage flow to bring into account, viz. the shearing effect of sewage turbulence on the wall slimes. When the shear stress reaches about 0.08 p.s.f., the slimes are practically scoured away and sulphide generation virtually ceases.

Bringing in a factor φ, such that when $\tau_{om} \geqslant 0{\cdot}08$ p.s.f., $\varphi = 0$, and when $\tau_{om} < 0{\cdot}08$, $\varphi = 1$, and rewriting eqn. [11.2] accordingly, we have for continuous pumping,

$$(B_s) = 0{\cdot}573 \times 10^{-6} \times \frac{L}{D} \times [BOD_5]^{0.8} \times (SO_4)^{0.4} \times 1{\cdot}139^{(T^{\circ}C-20)} \times \varphi \qquad [11.4]$$

taking $\tau_{om} = 62{\cdot}32 \times S \times \frac{D}{4}$ (Eqn. 3.2) and taking $\tau_{om} = 0{\cdot}08$ p.s.f. then $S = \frac{0{\cdot}30}{62{\cdot}32D}$

Taking Manning's Equation $V_S = \frac{1{\cdot}486}{n} \times r^{\frac{2}{3}} \times S^{\frac{1}{2}}$

$$Q_S = \pi \times \frac{D^2}{4} \times \frac{1{\cdot}486}{n} \times r^{\frac{2}{3}} \times S^{\frac{1}{2}}$$

Substituting for S from above, we have *for slime control*

$$Q_S = \pi \times \frac{D^2}{4} \times \frac{1{\cdot}486}{n} \times \left(\frac{D}{4}\right)^{\frac{2}{3}} \times \left(\frac{0{\cdot}30}{62{\cdot}32D}\right)^{\frac{1}{2}}$$

$$= \frac{2{\cdot}9 \times 10^{-2} \times D^{2.167}}{n} \qquad [11.5]$$

Equation [11.5] therefore can be used directly for the design of rising mains calculated to restrain the development of sulphide-producing slimes. If slime growth also be controlled upstream, there is unlikely to be any anaerobic seed in the flowing sewage, and hence no sulphide build-up, unless from trade waste discharges.

Procedure for the Design of Rising Mains

With the foregoing in mind, a rational procedure for the design of rising mains has been drawn up as follows, written out specifically for continuous pumping systems. A procedure for intermittent pumping can be adapted directly from it.

Step 1.—Quantitative design data to be established include the rates of flow of sewage to be pumped from time to time, also the range of values for $[BOD_5]$ and (SO_4). Analyses of supply waters, per capita water consumptions, and industrial waste particulars also should be available, and other relevant information such as sewage temperatures and likely changes in conditions from time to time during the day.

Step 2.—With pumping station site and rising main discharge manhole located as thought best from considerations other than that of sulphide build-up, determine the length (L) of the corresponding rising main. If at all practicable the selected route should be such as will permit a continuous upward gradient from pump invert to that of the rising main outlet, allowing for air injection if found desirable later. The overall length (L) should be kept as short as practicable, thus restricting build-up to a minimum whatever the design arrangements finally arrived at.

Step 3.—Next determine the required range of the discharges (especially peak wet weather flows) and the numbers and sizes of pumps including standby. This will determine the number of rising mains and ranges of pumping rate combinations and standby provisions, again taking full account of wet weather peaks.

Step 4.—At this stage the diameter or diameters for zero slimes can be calculated directly from equation [11.5]. For sulphide control, the overall length may then lose its significance, unless there is substantial generation in the body of the sewage. With high enough wall shear, the length would be limited only by considerations of frictional losses and power costs, etc. The *lowest* velocity allowed in any case should be such as to avoid deposition and chokages in dry weather, and the *maximum* must avoid mechanical damage, erosion, etc., to the main.

Step 5.—Velocities, build-up of sulphides, etc., should now be checked for all D.W.F. operating conditions using equations [11.2] or [11.3]. If any check shows significant build-up, the velocities should be increased if practicable by re-designing for a smaller rising main diameter.

Step 6.—Using the methods of Chapters 9 and 10, estimate the average sulphide concentrations in the sewage flowing into the pumping station. Adding any estimated average build-up of sulphide during passage through the rising main system then gives the required estimate of the average sulphide concentration in the sewage discharged from the upper end of the rising main. Both the early and the later life stages of the system should be investigated.

Design of Intermittent Pumping Systems: As in the foregoing such systems may require one or more rising mains carrying discharges from one or more pumps. Similar design procedures, stepwise as above, should first be followed. From the daily discharge Q_S in cusecs, the daytime average should be reckoned (using appropriate comparisons from similar municipalities) normally about $Q_S \times 1\frac{1}{2}$. This discharge rate then should be examined as a potential discharge into the downstream sewer already envisaged, both from simple hydraulic and sulphide balance viewpoint, and from this the upper D.W.F. limits can be practically established. The remaining features of the flow patterns, viz., minimum D.W.F. and corresponding required storage, and so on, as well as W.W.F. possibilities, then follow.

Much as for the continuous pumping system, a flow level close to the peak D.W.F. could be used to provide continuous control of slimes by entering this equivalent rate into equation [11.5], thus ensuring complete removal of wall slimes each day, as a result of which build-up effectively ceases. Wet weather flows then may be provided for as suits best.

Rising Main Discharge Arrangements: Special attention is required to the techniques of introducing the pump discharges into the downstream gravity system. Even though the average sulphide be maintained within acceptable levels, experience has shown that there may be occasions on which, for some period, sulphide concentrations at the rising main discharge point will be sufficiently high to cause some local corrosion or odour nuisance. The design of the junction arrangements where a rising main discharges to a gravity sewer therefore needs special care, the considerations to be taken into account including the following.

The discharge ends of rising mains should run full and remain full continuously. If this is impracticable, provision should be made for run-full conditions for a minimum of six hours continuously at least once per week. If this is not done, some corrosion of the crown is likely to occur.

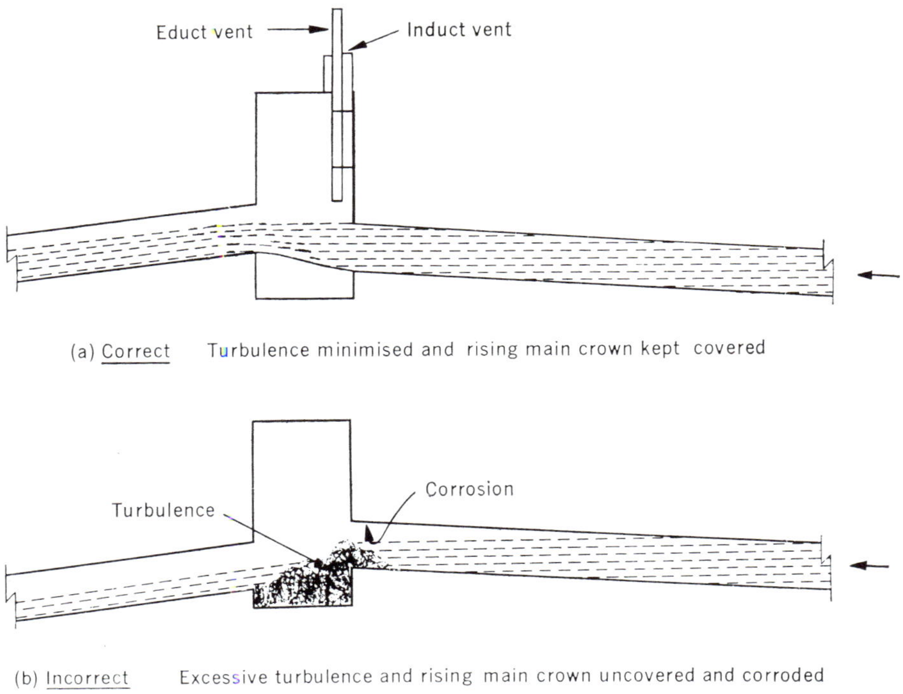

FIG. 11.6 Conditions at discharge end of rising mains.

Turbulence at the junction should be kept to a minimum. This is a particular case of junction design referred to in Chapter 9. Local ventilation may be desirable but likewise may lead to local odour nuisance especially if the rising main discharge point is in a built-up area. Figs. 11.6(a) and (b) show respectively the incorrect and correct principles for the design of rising main-gravity junctions.

CORRECTIVE OR REMEDIAL MEASURES FOR SULPHIDE PROBLEMS ARISING FROM PUMPING SYSTEMS

If the average sulphide concentrations as estimated are considered too high for acceptance, having regard to the proposals for bringing-in and discharging the sewage through the pumps to the gravity system, then the effectiveness and costs of alternative measure for sulphide control must be considered. These alternatives should be examined in turn as follows:—

(i) Change of site of pumping station and/or rising main discharge point to shorten the length of the rising main. The amount of such shortening necessary to reduce the sulphide concentrations to the acceptable value should be checked using equations [11.2] and [11.3], build-up of sulphide concentrations in the rising mains being nearly enough proportional to the length. If the full necessary control is impracticable or obviously uneconomical, then other alternatives should be examined.

(ii) Injection of compressed air into the pump system. This technique is described in detail below.

(iii) Chlorination, shock dosing with lime or other chemical treatments, also referred to below.

(iv) Resistant sewer construction downstream of rising main discharge point (see pp. 71 et seq.).

Before proceeding, a decision is necessary as to which of these three latter measures is most appropriate under all the circumstances of the particular case.

If (iv) offers the best choice, it must be provided for "*de novo*". For items (ii) or (iii) it is desirable to have provided in the designs for the introduction of air or chemicals required, including control equipment, but later modifications required are not difficult to install. It is, in fact, usual to defer detailed designs, purchase and installation of the equipment etc. until the station is actually operating, so that the extent of the sulphide problem can first be assessed more closely by actual measurements of problem conditions.

Grading of Rising Mains: Special care is needed in the laying of rising or pressure mains, especially when they are laid to flat grades, to ensure that the *crown* of the pipe is laid on a continuous upward grade in the direction of flow. Air pockets may form in individual pipes or sections of the sewer laid unevenly and severe local corrosion may follow.

Control of Sulphide Build-up by Injection of Compressed Air: Because of the relatively high rates of slime area to sewage volume and because there is no free surface to provide aeration to replace the oxygen consumed, there is commonly relatively high sulphide build-up in pressure mains and experience has shown pressure mains to be quite generally the origin of sulphide problems and odour nuisances.

In 1942, Lemcke in California suggested forcing compressed air into a rising main as a measure to control sulphides. This was tried with success for the particular case involved and has since been tested for other cases, sometimes with complete success and sometimes with partial success only.

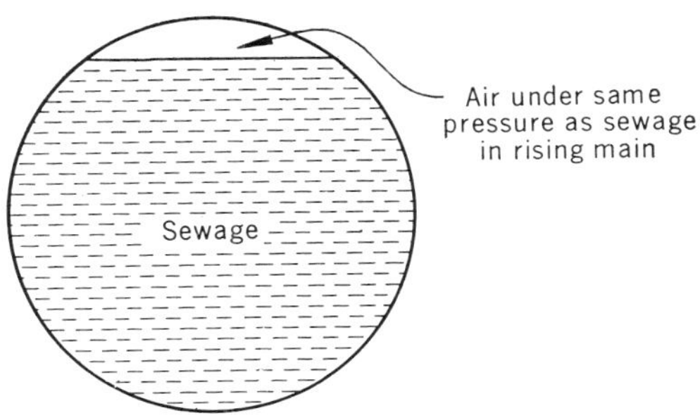

FIG. 11.7 Aeration of pressure mains.

The principles involved so far as they are at present known(*8*) can be explained as follows. Any air entering with the sewage accumulates as a moving layer at the top of the pipe as in Figure 11.7, under the same pressure as the sewage in the rising main. If injected, the air must be at a pressure appreciably greater than atmospheric and will transfer some oxygen to the sewage at a rate greater than would occur at the surface of sewage in a gravity sewer (see pp. 23 et seq.). The rate will be proportional to the absolute (total) pressure in the main, which remains above atmospheric but decreases progressively towards the outlet junction with the gravity system. However, because in a pipe operating near full flow the level is unstable, it seems likely that the air stream will travel up rising mains as a series of large bubbles, which greatly favours surface turbulence and enhances the rate of transfer of oxygen from air to sewage.

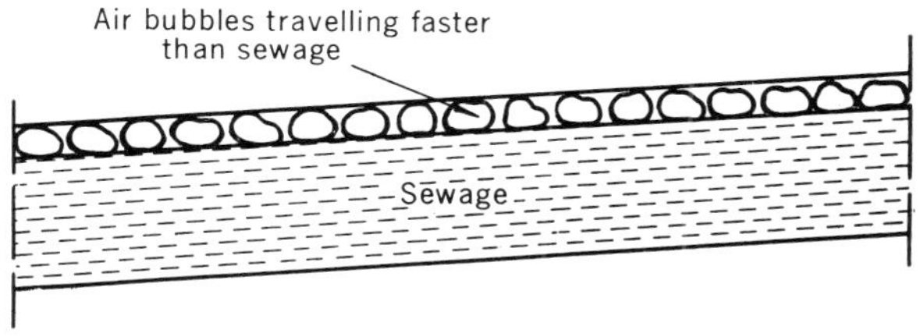

FIG. 11.8 Movement of air in pressure (rising) mains.

Conditions will of course change along the length of the rising main. As pressures decrease towards the upper end, hence the effective volume of the air will increase and therefore either size or velocity of bubbles will also increase. At the same time the oxygen content will tend to decrease. If the air introduced is effective in keeping the sewage aerobic the wall slimes will certainly generate less sulphide. When compressed air is injected into a rising main with active slimes however it may require one to two weeks before the full benefits can be assessed and the most suitable rate of air supply settled upon. In general, some trial and error must be allowed for.

The quantitative aspects of this method of control of problems in pumping systems require further practical studies before firm rules for design can be laid down. Further information is given in Chapter 13, p. 159.

Experience suggests that the final discharge of the air which travels along the rising main usually causes odour problems unless the air is deodorized. Such odour problems, in smaller scale installations, have been controlled by using soil beds for deodorization(*42*).

It also appears that the air picks up hydrogen sulphide from the sewage, and corrosion may take place along the top of the rising main. This point needs further elucidation, but it should be possible to control any such tendency by cutting off the air supply and allowing the rising main to run full for short periods once or twice weekly.

Control by Chemical Treatments: This is essentially the same for pumping systems as for sewerage systems generally, and is covered in some detail in Chapter 13.

CHAPTER 12

DESIGN OF SEWERAGE VENTILATION SYSTEMS

Introductory

In the design of ventilating systems for sewers, sewerage pumping stations, etc., three aspects must be kept always in mind by the engineer. The first aspect, with which this book is concerned primarily, is sulphide control. However, successful control depends on continuity of operation, whatever the system adopted, and significant odour nuisance finally will not be tolerated, and this is therefore a second aspect. Safe working conditions likewise are essential for proper operation, making the third aspect. This is more directly related to the second aspect than to the first, but safety matters extend much further than H_2S or odour control. Explosive conditions sometimes develop in sewers, and non-respirable atmospheres likewise, representing important extensions for the ventilation problem. Oxygen depletion or carbon dioxide accumulation have proved to be characteristic dangers which may arise wherever sewage or trade wastes accumulate, as in wet wells of sewage pumping stations. Deodorization may be an important ancillary feature.

These various matters are discussed in Chapter 6 in some detail, and are broadly reviewed in this Chapter, providing a background for the development of practical design procedures. This includes the assembly of appropriate mathematical formulae needed for or generally helpful to these procedures.

Natural and Artificial Ventilation

Natural and artificial ventilation were treated separately in Chapter 6. Although well-designed *natural ventilation* often may be sufficient on average to control not only corrosion but also safety and odour problems, such systems have some significant shortcomings. First, natural ventilation is quantitatively erratic and unreliable, and tends to be least reliable under climatic conditions and in seasons where and when sulphide conditions are generally worst. Second, it operates commonly practically as a plenum system over the major part of the sewerage system, again especially when sewer odour levels are highest. Finally, the rate of flow of ventilating air normally reaches a maximum equivalent to only about one change of air per hour or maybe less, which generally is inadequate for protection of personnel against sudden substantial discharges of volatile wastes which might contain toxic gases, or from the sudden releases of chokages carrying entrained vapours and gases. Therefore supplementary, albeit temporary, additional ventilation arrangements usually are required before entry to larger sewers or other structures can be permitted, and such temporary arrangements may be relatively costly.

Most of the objections to natural ventilation are eliminated by the provision of well-designed *mechanical ventilation* installations. Practical mechanical ventilation systems are relatively more flexible, with working ranges of up to 5 changes or more per hour consistently, although finally they may not be justified always from the viewpoint of economy.

One of the purposes of mechanical ventilation can be to control sulphide corrosion by reducing moisture levels in the concrete of the sewer walls; but this is a useful practical technique only where the atmospheric air taken into the sewer has relative humidity levels below say 70%. Another aim is control of the sewer air H_2S concentration levels by dilution, but increased ventilation air flow also results in increased total H_2S evolution, calculated as total weight evolved. Experience with one sewer for example showed that H_2S levels averaging 6 to 10 p.p.m. H_2S volume per volume with about 10,000 c.f.m. of air flow was reduced, at best, to about 4 to 6 p.p.m. with air flow increased to 30,000 c.f.m. Thus all the circumstances must be studied carefully before the engineer judges whether or not the installations required for mechanical ventilation are worth the costs.

General Design Aspects

Provision of Vents and Vent Stacks: Investigations of discharges through large openings at about 50 ft. to 100 ft. above ground level, on relatively high ground, have indicated that when the wind was blowing generally *along* the direction of the sewer, exhaust velocities from the stack reached levels up to about 0.16 times wind speed; whereas with the wind direction *across* the line of the sewer, upcast velocities were about 0.1 times wind speed, as summarised in the following expressions.

For direction of wind *downstream* in the direction of the sewage flow along sewer length,

$$Q_{DS} = V_{WD} \times A_{Sa} \times 14 \text{ c.f.m.} \qquad [12.1]$$

Wind direction *across* the sewer length,

$$Q_{AS} = V_{WA} \times A_{Sa} \times 9 \text{ c.f.m.} \qquad [12.2]$$

Where Q_{DS}, Q_{AS} are the total quantities of ventilating airflow for the respective wind directions

V_{WD} and V_{WA} are the respective wind speeds, m.p.h.

A_{Sa} is the cross-section of the sewer airway.

V_{WD} applies for the South-West to North-West quarters when the sewer is flowing from West to East; V_{WA} corresponding to the two quarters respectively North-West to North-East and South-West to South-East. Equations [12.1] and [12.2] must not be applied rigidly but are to be regarded only as rough guides. No similar relationship was found for the wind from the quarter South-East to North-East, because circumstances favourable for such easterly winds on the Australian Eastern Coast generally involve temperature relationships which outweigh the effects of wind speed and direction.

For reticulation sewers corresponding data are not available. For new systems generally boundary traps should be omitted, except for special areas involving special odour problems, and the total free ventilation area where practicable should be of the order of the total cross-sectional area of each property's common outlet drain.

In areas already connected to the reticulation sewer through boundary traps, boundary trap vents should be inspected regularly and defective fittings replaced. The replacements either should consist of fittings with two opposite-facing flap-protected openings or of cowled inlet vents with guide vanes to ensure one vent opening faces up-wind. Also in such areas, reticulation carriers of 12 in. diameter or more should be provided with *inlet* vent stacks at the rate of not less than one stack for every 1500 ft. of sewer length, and *outlet* vents also. The intake vents should be located downwind of the outlets, using the most frequent direction sector for the local area. Outlet vents should be located generally on high ground, using specially constructed separated pipelines to connect the vent, of about the same diameter as the sewer.

Larger sewers should be considered as special cases reckoned stage-wise along the sewer, allowing for inflow connections. *Calculations of head losses* may be simplified to the following relationships.

For turbulence and entry losses of inlet openings

$$H l_1 = 1{\cdot}1 \times 10^{-7} \times (V_A)^2 \text{ in. w.g.} \qquad [12.3]$$

For turbulence and entry losses due to bends, protuberances, beams, etc., which normally are less regulated, a better value may be

$$H l_2 = 2{\cdot}2 \times 10^{-7} \times (V_A)^2 \text{ in. w.g.} \qquad [12.4]$$

For friction head along sewer, requiring corrections along lines of [12.3] and [12.4] for effects of intakes and other such interference with the air-flow net, the following is of the right order:

$$\Delta p_{WA} = H_f = 4{\cdot}2 \times 10^{-10} \times L \times (V_A)^2 \times \frac{P_A}{A_A} \text{ in. w.g.} \qquad [12.5]$$

Where Δp_{WA} = friction head absorbed in travelling L. feet along the sewer, in. w.g.

P_A = perimeter of airway, feet

V_A = rate of ventilating air flow, f.p.m.

A_A = cross-sectional area of airway, sq. ft.

The following is an example of calculations for sizes of intake vent openings into a main sewer.

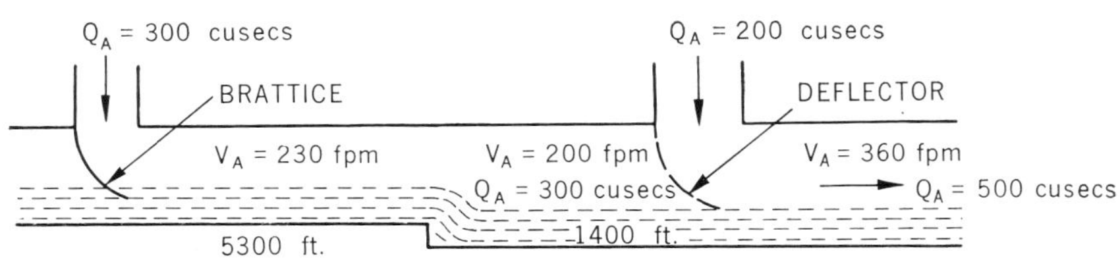

Say section of main sewer at brattice = 77 sq. ft., and airway perimeter 29 ft. Then as first try make upstream inlet vent area 77 sq. ft. Design requires say 300 cusecs upstream, 530 cusecs at downstream intake (total).

At upstream intake, $\frac{300}{77}$ = 3·9 ft. per sec.
= 230 f.p.m.

Velocity head = 0·0033 in. w.g.

Head loss into sewer, Eqn. [12.3] $= 1{\cdot}1 \times 10^{-7} \times 230^2$ in. w.g.
= 0·0058 in. w.g. (0·41 ft. head of air)

Say L_{1-2} is 3300 + 2700 = 6000 ft.
then head loss in sewer due to friction,

$$H_f = 4{\cdot}2 \times 10^{-10} \times [(5300 \times 230^2 \times \tfrac{29}{77}) + (400 \times \tfrac{29}{88} \times 200^2)] = 0{\cdot}066$$

Head loss at change in section, say, as per eqn. [12.4]

$$= 2{\cdot}2 \times 10^{-7} \times 230^2 = 0{\cdot}016 \text{ in. w.g.}$$

Head loss due to further obstructions, etc., not shown, say 0·02 in. w.g.

Total head required in sewer upstream from second intake $= 0{\cdot}02 + 0{\cdot}016 + 0{\cdot}066 + 0{\cdot}0058 + 0{\cdot}0033$ in. w.g.
$= 0{\cdot}11$ in. w.g.

Velocity required in 230 cusec intake is 350 f.p.m.

Velocity head $= 0{\cdot}0075$ in. w.g.

Head loss at entry, $H_1 = 1{\cdot}1 \times 10^{-7} \times 350^2$ in. w.g.
$= 0{\cdot}013$ in. w.g.

Total head required at downstream end

$= 0.11 + 0.0075 + 0.013$ in. w.g.
$= 0.12$ in. w.g.

These calculations apply equally for natural and mechanical ventilation systems but of course natural ventilation air flows are variable and only relatively low head losses, say less than 0.005 in. w.g. can be allocated for intake losses and turbulence at bends etc. Velocities approaching 1000 f.p.m. (0.02 in. w.g.) are about at the upper limit possible without mechanical assistance. It will be appreciated that this can be achieved only by designing openings and preventing protuberances etc. in such a way as to ensure substantially streamlined flow.

The head produced in high vent stacks is a function of stack height and actual density differences between atmospheric and sewer air, which in turn depend on both temperatures and humidities. The general formula may be written as follows:—

$$p_{Sb} - p_{St} = \Delta p_S = h_S \times (\rho_{OA} - \rho_{SA}) \times 4 \times 10^{-2} \text{ in. w.} \qquad [12.6]$$

where p_{Sb}, p_{St} are pressures respectively in the sewer and at the top of the ventstack, in. w.g.
Δp_S is the suction produced by the stack, in w.g.
h_S is total stack height measured above the soffit of the sewer.
ρ_{OA}, ρ_{SA} are the densities of the atmospheric and sewer air respectively lb. cu. ft.

Taking sewer air at 66·5°F and 100% R.H. and the free atmosphere at 52°F, 70% R.H., we have for a total effective stack height of 100 ft.,

$\Delta p_S = 1{\cdot}4 \times 10^{-2} \times 100 \times (0{\cdot}077 - 0{\cdot}0738)$
$= 1{\cdot}4 \times 0{\cdot}0039$
$= 0{\cdot}005$ in. w.g.
$= 0{\cdot}43$ ft. of air of density 0·075 lb./c. ft.

This may be taken as the natural draught per 100 ft. effective height of stack measured from sewer soffit.

Taking overall losses for intake and egress at say 3 ft. head of air, this natural draught is sufficient to produce $V_{SA} = 60 \times 8\sqrt{0.20}$ = say 200 f.p.m. air flow along the sewer. High stacks specially located along the sewer have produced under dry summer conditions measured sewer air flows of several times this order, see for example Table 6.1, p. 51.

Sulphide control aspects

Sulphide control aspects of ventilation are discussed in pp. 57 et seq., which should be read with other parts of Chapter 6 and with that part of Chapter 5 (pp. 34 et. seq.) which deals with H_2S emission from sewage. Overall control involves all aspects of the control of sulphide concentration build-up, in turn bringing in considerations of oxygen balance, the design aspects of which are covered in Chapters 8-11. Ventilation designs in this connection relate directly to the control of H_2S emission and to the control of sewer wall moisture levels. Equations [5.14] and [6.3] provide the basis for actual calculations required. Note that oxygen concentrations almost always are close to the normal 20.9%.

H_2S emission can be controlled within limits for any particular sewage sulphide conditions mainly by changing the relative movements of sewage and sewer air, remembering that it is the situation at the actual sewage-sewer air interface which is of direct concern. Air movements away from this zone of interplay are less important, remembering always however that turbulent conditions at the sewer air-sewer wall interface tend to increase the rates of transfer of gaseous H_2S to the wall moisture with consequent increases in acid formation and corrosion rates.

Referring to the diagram on p. 57 and equations [6.3] and [6.4], and having regard to the foregoing summary, it is clear that engineers should reckon in most cases that it will not be practicable to hold the interfacial sewer air flow to the same velocity as that of the sewage surface, although examination of the emission equations referred to, mainly equations [5.14] and [6.3], indicate that this would hold the total amount of H_2S emitted to about the minimum under the overall circumstances, and likewise the rate of corrosion. Provided the interface is streamlined this introduces the factor of sewer air H_2S concentrations, which show a direct correlation with corrosion activity.

Any attempt to control sewer air flow certainly will be more likely to succeed if mechanical rather than natural ventilation be applied, but it is pointed out that estimates such as those shown in Figure 12.1, opposite, can be developed for natural draught conditions basing designs solely on sewage-drag effects.

Control of Wall Moisture: Is discussed and a design example given in Fig. 6.9, pp. 58, 59. In practice, for most areas, drying out of sewer walls to prevent corrosive bacterial activity is not always practicable because where or whenever the relative humidity of the sewer air exceeds about 80-85%, wall moisture levels will be sufficient for some level of bacterial activity to be maintained, more and more as the R.H. approaches 100%. Probably 70% R.H. is about the upper limit for useful applications of drying out procedures. Designers therefore need to take account of typical atmospheric humidity levels and frequencies of occurrence, and corresponding sewer temperatures, before a proper decision can be made whether or not drying out is a practicable and economical measure.

Pages 57 et seq. provide the basis of designs for moisture control using equations [6.3], [6.4] and [6.5] with Design Chart No. 6, p. 121. Figures 6.8 to 6.11, and 12.1 are relevant also.

The procedure for designing a mechanical ventilation system for control of wall moisture involves the following steps:—

1. Assemble the necessary information respecting the ranges of sewage temperatures and the temperatures and humidities of the ventilating air, which will be those of the ambient atmosphere. This information should cover at least average winter and summer conditions, and should be investigated separately accordingly because in many locations the air introduced for ventilation will be colder than the sewage in winter and warmer in summer. Data respecting sewer dimensions also are required, and plans should be available showing junctions, and indicating general features inside and outside the sewer.

2. Divide the length of the sewer to be ventilated into convenient sections of, say, 200 feet to 300 feet each.

3. Make trial assumptions of the induct air volumes, etc. and the directions of air flow. Generally, the cheapest effective scheme will be one in which the velocity of the airstream relative to the sewage is kept low; it is suggested therefore that the first trial assumption be for an air velocity of 10% more than the mean sewage velocity and in the same direction.

4. Using equation [6.4], pp. 59, 60, determine for winter conditions from the information of step (1) the approximate air temperatures (dry bulb) at the various section points selected as per step (2).

5. Using equation [6.3] and making a trial assumption of the wet bulb temperature, calculate the partial pressure of the water vapour at the first section point below the induct. From the calculated air temperature and the above-mentioned partial pressure of water vapour, determine the wet bulb temperature (using tables, also weather data, Chapter 8). If this does not agree with the assumed value, make further trials, using the calculated value, until agreement between the assumed and calculated values is reached. From the calculated wet and dry bulb temperatures calculate the relative humidity of the sewer atmosphere, using the relationship, p. 59.

$$(\text{R.H.}) = \frac{p_{wa}}{P_{STa}}$$

6. Using the same trial and error procedure, calculate the relative humidity at each of the successive section points.

7. Repeat Steps (4) to (6) using data for summer conditions.

8. Plot the estimated relative humidities for winter and summer on the longitudinal section of the sewer as in Fig. 12.1

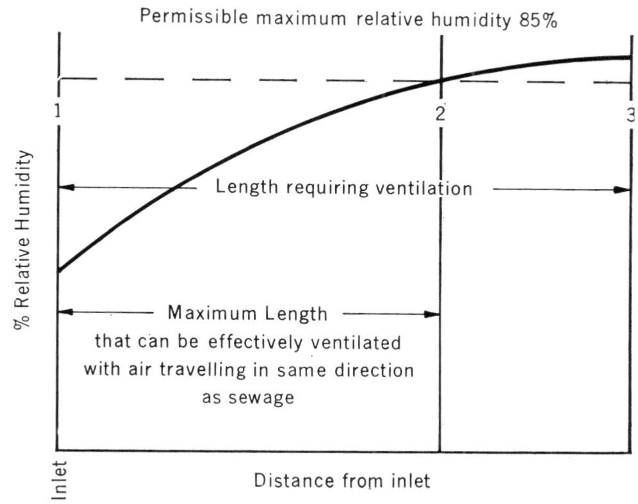

FIG. 12.1 Plot of Ventilating Air Humidities.

9. If the calculated relative humidity at the last section point is greater than 80 to 85% it indicates that the section of sewer to be ventilated is too long for one fan installation. However if the plot of calculated relative humidities indicates that the length of sewer to be ventilated is not greatly in excess of the selected length (say less than 50% more) then it still may be acceptable as a first try to ventilate the required length satisfactorily using a single fan and with most air travelling in the same direction as the sewage. For example a single exhaust fan located at point 2 on Figure 12.1 and drawing air from two directions, respectively from point 1 to point 2 and from point 3 to point 2, could be designed to maintain relative humidities within the permissible limits over the full length. The respective quantities of ventilating air required for the two sections 1 to 2 and 3 to 2 must first be computed having due regard to the fact that for Section 3 to 2 the air is flowing in the opposite direction to the sewage and therefore at a higher relative rate than for Section 1 to 2.

For such a system to function properly requires that induct vents at points 1 and 3 be provided with suitable means of adjusting the vent areas separately because finally it will be necessary to measure the flows and make adjustments by trial until the correct appropriate distribution of the ventilating air requirements for the two Sections 1 to 2 and 3 to 2 has been achieved. The air quantities entering the system at inducts 1 and 3 usually would be measured by using anemometers to determine the average velocities of air passing through the inlet openings. Systematic changes of the area of one or both induct vent openings and measurements of air quantities then will enable the required balance to be obtained.

10. Finally, consider the feasibility and comparative economy of mechanical ventilation as a suitable remedial measure having due regard to any associated odour problem.

Ventilation and odour

Ventilation and odour problems: Are discussed above in pp. 52 et seq., but the subject is also reviewed briefly here. It is pointed out again that odours usually are characterised by measuring H_2S concentration levels because it is recognised that although H_2S itself may not be the particular problem, odour levels and H_2S concentrations commonly are closely correlated. The principles governing the emission of odorous constituents are fundamentally the same as those governing H_2S emission, and those equations listed above, p. 145, will reveal the probable effect of any changes on the rate of accumulation of odours.

A technique for measuring odour levels referred to above is that of the Threshold Dilution Value (TDV) (Ratio, p. 55). This value provides the best available measure of the level to which the odorous air must be diluted in order that the odour be no longer detectable. It is suggested as a rough guide that in most cases the TDV will be given nearly enough by the value $(H_2S)_{Sa} \times$ (200 to 300). Some of the necessary dilution may be obtained as shown in Figure 6.6, p. 53, but Threshold Dilution Values as high as 300-fold are not likely always to be achieved at least in the immediate vicinity of the sewer air discharge (especially as with Fig. 6.7). Studies of sewer air dilution ratios are lacking, but mathematical models have been developed particularly for the dispersion of sulphur dioxide downstream of power station vent stacks, for example, under conditions applying to divergent plumes such as are shown in Figure 6.6.

Sutton, Bosanquet, Pearson and others have developed such models(*60*). Equations [12.7] and [12.8] due to Bosanquet and Pearson, are typical.

$$(Cg)_{max} = \frac{4 \times 10^6 \times Q_{ST}}{\sqrt{2\pi} \times e^2 \times V_W \times H^2_{SE}} \times \frac{k_x}{k_y} = \frac{2{\cdot}15 \times 10^5 \times Q_{ST}}{V_W \times H^2_{SE}} \times \frac{k_x}{k_y} \quad [12.7]$$

$$x_{Cgm} = \frac{H_{SE}}{2k_x} \quad [12.8]$$

Where $(Cg)_{max}$ is the maximum concentration of the stack gas downwind at ground level, p.p.m. vol./vol. and x_{Cgm} is the distance from the stack at which $(Cg)_{max}$ occurs.

These values are derived from the general expression,

$$Cg = \frac{Q_{ST} \times 1{\cdot}1 \times 10^4}{\sqrt{2\pi} \times k_x \times k_y \times V_W \times x^2} \times e^{\left(\frac{-H_{SE}}{k_x \times x} - \frac{y^2}{2k_y^2 \times y^2}\right)}$$

Where Cg is the groundlevel concentration of the stack gas, p.p.m.

Q_{ST} is the total rate of emission from the stack, assumed to be constant, c.f.m.

k_x, k_y are atmospheric eddy diffusion parameters, respectively, for the vertical and horizontal directions correlated with turbulence conditions. Thus k_x varies from 0·02 for calm to 0·10 (and as high as say 0·5) for more and more turbulent, with 0·05 for average conditions. Corresponding values for k_y are 0·04, 0·16 (0·65) and 0·08. The ratio $\frac{k_x}{k_y}$ ranges from about 0.5 to 0.63, commonly closer to 0·63.

V_W is the wind speed, miles per hour.

V_G is the exit gas velocity

x, y are respectively, the distance measured horizontally downwind from the stack and the displacement laterally therefrom, ft.

H_S is the actual height of the stack above ground.

H_{SE} is the total effective stack height above ground reckoned according to the following, ft.,

$$H_{SE} = H_S + (\Delta H_S)_{Vmax}$$

$[(\Delta H_S)_V]_x$, $[(\Delta H_S)_V]_{max}$ are the *additional* heights above the top of the stack to which the plume of the gaseous discharge rises due to exit velocity respectively at a distance x and at the maximum height of the plume axis.

$$(\Delta H_S)_{Vmax} = \frac{4{\cdot}77}{\left(1 + 38 \times \frac{V_W}{V_G}\right)} \times \frac{88\sqrt{Q_{ST} \times V_G}}{V_W} \quad [12.9]$$

$$\left[(\Delta H_S)_V\right]_X = \left[(\Delta H_S)_V\right]_{max} \times \left\{1 - 0{\cdot}8 \times \frac{[(\Delta H_S)_V]_{max}}{x}\right\} \quad [12.10]$$

TABLE 12.1 WIND PATTERN AND ODOUR DISPERSAL

	General description of wind behaviour pattern	Approx. distance where max. odour level occurs
AB	frequent changes of wind direction ranging over 75° of compass	H_{SE} x 3 to 5
C	changes of wind direction confined mainly within about 30° of compass	H_{SE} x (say) 10 to 15
D	changes of wind confined to less than about 10° to 15° of compass	H_{SE} x 50 or more

Adapted from Lowry(*60*).

More complicated expressions take account of different patterns of atmospheric conditions. Table 12.1 relates particular patterns of wind behaviour to odour dispersal by way of example:

Note: that these odour levels occurring near the ground will not be the same but will vary to some degree inversely with distance from the stack. It might be noted here also that k_x values of 0.02, 0.10 and 0.050, (eqn. [12.8]) give distances respectively of 50 times, 10 times and twice the effective stack height H_{SE}, which are of the same order as suggested by Lowry.

The engineer can develop some control of dilution-dispersion by taking account of the following:—

(i) Increasing the velocity of the ventilating air leaving the stack, increases the effective stack height roughly according to the relationships of equations [12.9] and [12.10].

(ii) Since increased air flow in the sewer tends to increase the total amount of odour evolved such increase in the velocity of the exit gases can be obtained by drawing in additional fresh air near or at the base of the stack, rather than from along the sewer itself; then with one step the effluent H_2S concentrations are reduced and the upcast velocity increased.

A series of observations of a large outfall sewer vent stack gave the following data:—

TABLE 12.2 VENTILATION AND H_2S CONCENTRATIONS

	Air Flow Up Stack c.f.m.		Additional Air From Ground Level c.f.m.	opening	Total Head Across Fan	Mean H_2S p.p.m.
1. Natural Draught	Mostly 5,000-10,000		—		—	45
2. Mechanical Ventilation	From Sewer 45,000		0	0	0.8″	14
3. Mechanical Ventilation	From Sewer 25,000	Total 46,000	21,000	9½ s. ft.	0.4″	10

Control of Odour Nuisance and Deodorisation: The discussions in Chapter 6 indicate that dispersal by dilution cannot be relied upon to control odour nuisance under all climatic conditions especially under circumstances of atmospheric temperature inversion, see p. 53, Fig. 6.7. The engineer must examine local weather frequency data and decide from this the probable frequencies and duration of nuisance periods, and hence determine to what extent he should provide for direct odour nuisance control. Broadly the engineer will distinguish respectively between temporary measures and installations which subsequently are "permanent", the selection for each particular case depending on such questions as economics and expedience. However, they are determined not only from safe working and public risk aspects which always should be paramount, but from other technical considerations respecting reliability and so on, also

Safe working and safety control aspects: Are discussed in Chapter 6 under the heading of Toxicity Factors in sewage ventilation pp. 56, 57 and in Chapter 13 pp. 153, 154. These discussions deal with wide ranging technical problems which largely are beyond the scope of this book and surely demand that the sewerage authority be advised regarding them by persons with specially developed knowledge and experience. The history of many authorities has demonstrated that in the main most of the hazards of working with sewerage systems arise from unpredictable coincidences. Sewerage systems always will be the repository of unwanted or waste substances of almost any type of material, however infrequently and erratically any one material might occur therein. In a general way, the likely amount of most such materials tends to be limited, but the limits are very wide as between one particular material and another, and also as between one particular time and another. There are a number of sections in this book in which special reference is made to trade or industrial waste discharges, emphasizing that it is proper also for designers and operators to ensure that in handling related problems they are reinforced with the best advice they can get. Safe working and safety require the same careful attention especially maybe in industrial areas; but even in purely domestic areas dangerous coincidences can and do arise.

It is always difficult to make economic assessments and particular comparisons in connection with such aspects and the engineer's judgement in general will outweigh purely formal assessments. Explosion-proofing of electric motors and provision of special anti-spark tools are not uncommonly written into specifications by some engineers as a matter of course. Other engineers, however, treat such problems by making individual judgements on some basis of permissible or acceptable levels of risk. The basic idea in respect of the design of ventilation should be determined by engineers in some way or ways which are related to the volume-time relationship of the structures and the local sources of the waste materials in question, taken in comparison to ventilation air supply rates and system air storage.

Attention is required to both routine operating conditions and to the different special conditions associated with essentially non-routine circumstances. In this sense, for example, regular visits, say, to inspect the superficial working of a pumping station may be regarded as simple routine, except perhaps for the actual manual operation of stop-start switch boards; but *entry* to the wet well, say, for cleaning purposes, although routine in the sense that it is done say regularly each week, should be always *non-routine*. The engineer therefore normally would provide the Sewage Pumping Station with two levels of ventilation accordingly, levels of say two to three changes of air per hour for the machinery well, including also the wet well, plus additional provisions, say a second fan to be actuated before an entry is made to the wet well, providing say five to six changes of air per hour therein. This fan then would be operated for at least 15 to 20 minutes prior to actual entry and working instructions would stipulate further that should the workers either inside or outside the wet well notice any detectable change in circumstances, all workmen should move at once to leave the atmosphere of the wet well. Two such levels of ventilation, respectively say two to three and five to six changes of air per hour, are recommended as general guide lines for *all* structures carrying "live sewage".

General experience with sewer ventilation

In the foregoing some of the differences between natural and mechanical or artificial ventilation have been mentioned. One principal difference of import for all aspects lies commonly in the situation that on the one hand positive, and on the other hand negative, barometric pressures prevail in sewers operating under natural, compared with mechanical or artificial, systems. It is possible also to ventilate sewers by applying mechanical pressures using so called plenum systems, but this is not common practice, mainly because of the general idea that suction fans largely prevent the escape of odours from sewers, whereas plenum systems tend to encourage the escape of odours to an uncontrolled extent. It is nevertheless possible to design plenum systems using a multiplicity of outlet vents sufficient to keep the odour level at any one vent from reaching nuisance levels.

Mostly sewers are laid out on a river-tributary system and as a result, in the case of suction fan systems, air enters not only through any special intake vents provided but also adventitiously through sub-main and other junctions tributary to the main sewer being ventilated. This limits the overall length of

TABLE 12.3—MECHANICAL VENTILATION OF MAIN SEWER

Comparison of Designed and Operating Conditions

Branch Inlet	Flow in Main at Section c.f.m. for 2 changes per hour	Flow from Submain c.f.m. for 2 changes per hour	Selected compromise air flows		Flow measurements after fan actually installed	
			From Sub-main c.f.m.	Progressive Total in Main c.f.m.	From Sub-main c.f.m.	Progressive Total in Main c.f.m.
Sub-main No. 1	1440	4080	2400	2400		
2	2640	660	2400	4800		
3 & 4	1320	2580/	1800/	6600/		
		2220	1200	7800		
5	2160	1440	2400	10200		
6	3720	720	3000	13200		
7	3540	780	3000	16200		
8 & 9	7380	1140/	4200/	20400/	–/2460	13400/
		600	1800	22200		15600
S.P.S. No. 1	2940	780	3000	25200	3900	19500
2 & 3	3540	1380/	1800/	27000/	5280/	24700/
		1620	1800	28800	3900	26800
	1320	420	1200	30000	720	30000
Totals	30,000	18,420	30,000			

sewer throughout which negative pressure conditions can be maintained by a single fan installation. Equation [12.5] provides the basis for calculating head loss due to simple friction along the sewer, while equations [12.3] and [12.4] allow adjustments for the effects of intake vents, beams, piers or other projections, bends, and so on. Other corrections should be made on a similar basis for the effects of non-streamlined sewage flow which may occur at junctions, drops, etc.

Some data in this connection were reported from investigations of a mechanical ventilating system installed by the London County Council in the 1930 decade. The sewers investigated included one principal low level sewer with two much smaller major tributaries. General patterns of air flow indicated that sewer pressure levels could be held always negative in sewers of about 4 ft. to 9 ft. diameter for distances up to about 8000 to 9000 ft. upstream of the working fan. Air flow in other sewers of about 2 ft. to 5 ft. diameter on the other hand could be controlled effectively for distances of only about 1200 to 1500 ft.

Table 12.3 gives data for a main sewer in Australia which varied also from about 5 ft. to 10 ft. equivalent diameter.

The scheme of Table 12.3 was based on a uniform increment of requirements progressively along the sewer to maintain a regular ventilation rate of two changes per hour for the upstream length of both main and tributary sewers. A complete series of air flow measurements proved not to be feasible at that time under any one given set of conditions, but such measurements as could be made indicated that whereas it is relatively easy to predetermine the total output from a fan installation it is not usually easy to achieve the designed pattern for all intakes along a given length. Numerous trial and error adjustments of intake dimensions would be required before an intake pattern might be approached for satisfactory compromise to be achieved. The overall length of the sewer of Table 12.3 from Submain No. 1 to the fan was of the order of six miles, and the distance from Submain No. 9 to the vent, about three miles. The fan suction head under the quoted test conditions was about $\frac{7}{8}$" overall.

In the case of the London County Council system quoted earlier, the sewers reduced upstream to much smaller sizes than in this latter example. Under such circumstances of relatively rapid decreases in sewer diameter, $1\frac{3}{4}$ to 2 miles probably is of the right order as a maximum length of sewer for any one fan ventilation installation, recognising that longer lengths also may be ventilated satisfactorily in cases where the circumstances are more favourable. Two changes per hour for 10,000 feet along any section requires a minimum average air velocity of 20,000/60 = 330 ft. per minute. This velocity is rather high for normal working comfort, and it may be concluded therefore that the lengths of sewer to be ventilated by any one fan installation should be restricted normally to not more than about 2 miles along the main sewer.

It is expected as a general rule that the maximum static suction head appropriate for collecting sewers is about $1\frac{1}{4}$", otherwise unwatering of boundary traps or even of household fittings is likely to result. Probably 1 in. is a safe maximum figure for working. This figure is to be related also to the quantity of air exhausted from the sewer, and the corresponding power charges. Roughly, power costs vary as Q_A^3 and Q_A is proportional to $(V_A)_{av}$ for any given section. For mechanical ventilation, reckoning $(V_A)_{av}$ as also proportional to $\sqrt{H}$, we have power costs $P_{ow} \alpha H^{3/2}$. The engineer thus faces the task of assessing overall costs as an alternative to installing more fans with corresponding regulating intake and so many discharge vent structures; and balancing operating and capital charges against the advantages or disadvantages of lower air velocity with more potentially odorous discharges, more or less comfortable working conditions, greater or less hazard-buffering capacity, and so on. Moisture control, if looked for, introduces further complications.

Ventilation Design for Low Level Sewerage Areas

Low level sewerage areas may be viewed as comprising three different parts, respectively the reticulation system, the sewerage pumping system, and the pressure mains system.

The low level sewers in general do not necessarily differ significantly in this connexion from normal high level sewers. There may be a tendency for these low level sewers to be laid at shallower slopes but this generally should be avoided by engineers, and the sewers themselves for all practical purposes should be designed as for gravity sewers, using the mechanical models and data of Chapter 8 according to the procedures of Chapters 9 and 10. However, low level sewers probably always terminate at practically the lowest contour levels of the areas served, and this means usually that greater than normal care is demanded of engineers in laying out and shaping the airways required from place to place along the sewers. This applies particularly in respect of natural ventilation. The possibility of free ventilation from the above-roof house vents down into the sewers will be largely obviated if the airway transitions are not designed so as to absolutely ensure the full benefits of normal sewage drag, providing easy transitions for the original inflow of air, almost negligible, to sewer air flow rates which should reach 70 to 100 f.p.m. say even at the upstream end of the carrier sewers. Suitable sequences of air intakes, again mostly from house vents, together with discharge vent arrangements, see pp. 48 et seq., should be provided progressively for the sewers downstream.

The sewage pumping station structure itself should

be provided with a dual ventilation system. One cowled vent assembly should take the air from the inlet sewer as directly upwards and as high as practicable, again ensuring smooth aerodynamic transitions throughout. The other system would serve the sewage pumping station only. If the inflowing sewage is to be screened, whether using open or chambered screens, the sewer should be sealed from the screen chamber preferably by water seal, otherwise using a brattice system which does not produce dead-end effects or cause any significant loss of sewage velocity upstream. The sewage pumping station system then comprises inlets, dry well, and wet well if any. As stated, the inlet structures should be isolated from the upstream sewers and separate provision made for their ventilation, the inlet ventilation installation normally being integrated with the ventilation system for the wet wells. Thus, if a screen chamber be incorporated in the layout, air should come into the wet well with the sewage leaving the screens, the airway being such as will suit the wet well layout, allowing for any sewage level changes, and play its part as wet well ventilating air before entering the outlet vent system. Careful design might well be based on natural ventilation for routine work and operation, but in line with the foregoing, a mechanical system provided as an alternative available for use as and when required. The dry or pump well of the sewage pumping station also requires some provision for ventilation. In cases where sulphidic or other odours are not likely to prove a nuisance, corrosion provision alone usually would require at least standard building regulation standards for structures above ground, but something more would be required for the deeper below-ground parts of the structure. This could take the form of carefully designed ducting systems which carry air to the bottom level on the one hand and on the other provide outlet vents above roof level. Such designs should take full advantage of heat output, etc. which might permit routine operation entirely as a natural system with alternative mechanical ventilation for non-routine working.

The rising main system also demands consideration. If the rising main be aerated the air discharged with the pumped sewage should not be allowed to join the ventilating air of the downstream gravity system, but should be led directly out of the receiving manhole either to the open air or to a deodorizing system (pp. 63, 64). Likewise there should be provision at that place also for intake of air into the downstream sewer. Designs should take account of possible rapid changes as a result of breathing when pumping starts and likewise when it stops.

Design procedures for general sewerage ventilation

Having regard to all of the foregoing, including the more detailed discussion in Chapter 6 of the many practical aspects and features, the following generalized procedure is presented as a guide for the design of sewerage ventilation systems generally.

Step No. 1: The engineer must first decide the basic purpose or purposes of the ventilation system which is to be provided, see pp. 47, 57. The relevant considerations are discussed in detail in Chapter 6 and have been reviewed in the foregoing. Mainly the engineer will need to consider all aspects, although particular aspects may be paramount. However, even though for example the engineer judges that the specific purpose of the installation should be to provide for, say, wall-moisture control as the main basis of the design, odour control and safe working should never be overlooked. Likewise, if the main idea be say to control sewer air sulphide concentrations on the one hand, or say on the other hand to maintain negative pressure conditions in a main sewer, so also should the *overall* picture be kept always in mind.

It should be noted that the engineer's deliberations as to such purposes do not necessarily fix a single main purpose for the *whole* sewerage ventilation system which is being considered. The sewerage system in this respect is to be regarded as made up of a number of parts, the different parts perhaps showing different requirements. This first step therefore is carried out in two stages.

1 (*a*) consists of an overall review of the system actually under design, taking the general sewerage layout along with such information as may already have been collected by the engineer respecting details such as time of travel of sewage, and the character of the sewage, including such factors as the relative proportion and nature of industrial content and so on, taken in a general way. Again only in a broad way, the engineer may determine a series of zones or areas accordingly, preliminary to 1 (b).

1 (*b*) involves a second stage of similar considerations but requires a detailed break-down along athe above lines, commencing with a study of the zones or areas which have been broadly marked out in 1 (a). Separate plans should then be marked out afresh, marking out more closely defined areas, particularly those demanding some special attention. This would require, for example, further detailed examination of the preliminary designs (Chapters 9 and 10) of the sewers, with drops, chambers, etc. (Note that at this stage the whole operation should be regarded as preliminary in the sense that during these deliberations and at other stages of the following Steps 2-6, it will appear that some adjustments of the earlier preliminary designs are desirable or necessary).

Step No. 2: Becomes possible after completion of the above first stages and consists of making a first approximation at the particular areas for which, respectively, natural or artificial ventilation might be chosen, either directly or as first trials. In most cases for example natural ventilation will be the first trial for most if not all reticulation areas.

Step No. 3: Is the next stage wherein the details of ventilation air inlet-outlet arrangement should be settled.

3 (*a*): Presumably this will require, first, some actual design detailing work whereby preliminary working drawings for vents, etc., are made, adapted to the preliminary sewer designs referred to, in order that the final overall designs for inlets etc. are standardised at least to some extent.

3 (*b*) involves the many considerations which arise because of the particular local terrain, contours, residential and other occupational details and so on, relevant for the selection of air intake and sewer air discharge vent sitings.

Step No. 4: Then demands an actual determination of the forms and numbers of vents to be provided for. For odour control for example, this brings in the question of the dispersal of the sewer air at discharge locations. Thus the choice for outlets may be either relatively high stacks at relatively long intervals, or for example as an alternative, unobtrusive lower stacks spaced so that no significant odour problems are likely. It may be that high stacks connected to relatively long lengths of sewers which carry or are likely to carry septic sewage would be rejected as an alternative, using deodorisation installations instead, because of the residential densities.

4 (*a*) is the stage of detailed considerations and site inspections involving contour maps, etc. together with careful and detailed studies of climatic factors(*60*) and population forecasts.

4 (*b*) requires preliminary studies of odour control systems, including the principles of dilution-dispersion, and following this, the assembly of estimates of installation and operating plus maintenance costs.

Step No. 5: Covers the first stages of preparation of plans which correspond to the first General Arrangements, respectively Step 5-1 for natural ventilation systems, and Steps 5-2 and 5-3 for mechanical ventilation systems. Step 5-3 in particular covers systems designed specifically for wall-moisture control, and is set out in greater detail as an example.

The following series of steps, 5-1, 5-2 and 5-3, apply separately for natural and artificial systems, with the break-down of Step 5 given only for mechanical ventilation for moisture control, which was set out in a partly expanded form at pp. 58-9. This provides an illustration for the generally similar design sequences which are required for all of the foregoing Steps 1 to 4, as well as for Steps 5 and 6, for all cases. Thus the detailed sequences given for 5-3 should be used analogously as a model for the other cases where detailed itemised procedures are not given.

Step 5-1: *Natural Ventilation Systems*: The engineer's basic decision requirement before carrying out detailed working through a sewerage area, is the adoption of a ventilation rate standard; for example, either a changes per hour average criterion, or an "air-o-surs" criterion (p. 51) or other such basis.

5-1*a*. The chosen criterion is combined with average sewer air velocity reckonings, pp. 58-61, working downstream progressively through the whole system. Should a mixed natural and mechanical system have been selected, the ventilation rate criterion usually will be different for different parts of the system. The rate for the mechanically ventilated sections normally would be at least twice the average rate for the natural ventilation systems, for which equation [6.1], p. 47, determines V_A. Because natural ventilation is more or less erratic, it is suggested that it is necessary to select an intermediate standard for design purposes. Allowing that winter conditions are not so critical in most climatic areas as those in summer, the recommended basis for design, especially respecting odour nuisance and safe working, is to aim at favourable conditions being achieved for not less than 70% or say two-thirds of the summer period, but giving greatest weight to the evening hours. The hours taken may be from 4 p.m. to 10 p.m., alternatively from 6 p.m. to midnight according to local habits, with designs to suit the most humid three months of the average year.

Step 5-2: *Design of mechanical ventilation for general purposes,* that is, not specifically for moisture control: The procedure for mechanical ventilation systems not aimed specifically at moisture control, generally should follow the same lines as for natural ventilation. However, whereas natural systems may depend mainly or entirely on sewage drag, which applies uniformly and continuously in most cases, the motive force for intakes at design rates with mechanical systems is normally an electrically operated fan, and for each case there is a practical "length of sewer" limit. Equations [12.4] and [12.5] suggest how the probable frictional losses may be calculated say on the basis of per 1000 ft. of sewer, corrected for bends, projections, etc., as a theoretical value to be used as a guide. This requires perhaps a discount of at least 30% of the theoretical length, but an upper limit should normally be fixed respecting the sewer length ventilated and sewer air velocity. It is suggested that these upper limits be respectively $<$ 10,000 to 12,000 ft. of sewer, and $V_A \not> 350$ f.p.m.

Step 5-3: *Mechanical ventilation for moisture control*: The general procedure of pp. 57, 145 is summarized in the following, item numbers referring to the former presentation.

5-3 (a), (1), assemble design data and general information.

5-3 (b), (2), divide the sewer into appropriate sectional lengths.

5-3 (c), (3), make initial trials respecting air flow volumes and directions for these various systems.

5-3 (d), (4)-(7), calculate related intake air and outlet air characteristics respecting temperatures and humidities for each section.

Note that these two steps (c) and (d) involve "internal" trial and error repetition, possibly in two or three stages.

5-3 (e), (8) and (9), involves plotting and consideration of the implications and significance of the findings, and may result as a consequence in return again to Steps 5-3 (c) and 5-3 (d).

5-3 (f), (10), this penultimate step requires a detailed costing, having due regard to related aspects such as odour problems, before the final proposal of the foregoing steps are accepted for detailed design as in Step 6.

Step No. 6: Is the final step of proceeding to fully detailed designs for each section. Many details normally would be repetitive; for example, standardized type designs for all fixed or adjustable fresh air intakes. This standardization applies practically for all systems, whether natural or artificial, and for many specific items. It should be noted that the foregoing procedures are set down primarily for the design of permanent ventilating plant. The engineer also should consider the non-routine requirements for sewerage maintenance work, for extensions, renewals and new connections, and when the designs are finally presented they should include some recommendation respecting temporary arrangements for non-routine circumstances.

Finally, engineers are reminded again that the details of any sewer ventilating installations should follow the best practice applying to mechanical ventilation schemes generally, as described in text books on ventilation and air conditioning. Care should be taken that vents and ducts are of adequate size with properly streamlined entries, bends, and so on, so that power requirements are kept to a minimum.

CHAPTER 13

CORRECTIVE MEASURES FOR EXISTING SYSTEMS

Introductory

Generally, the techniques for controlling or preventing the development of sulphide corrosion in newly constructed systems by the appropriate design procedures set out in the foregoing chapters are not practicable for correcting the conditions found in existing systems. However, familiarity with all of the principles already discussed provides an excellent background for proper consideration and determination of the techniques which will be practicable and appropriate for the particular circumstances of the particular system in question.

Broadly, the techniques or procedures which may be practicable fall under three main headings as follows:—

(i) Control methods involving changes in operation or management of the sewerage systems,

(ii) Control methods based on chemical or biological treatment of the sewage (including aeration),

(iii) Control by practical methods, including structural alterations and provision of special ventilation.

Before proceeding with more detailed consideration of these matters, however, attention is drawn to an important aspect of maintenance and operation of sewerage systems; that of safe working and safety precautions.

Safe Working and Safety Precautions

The importance of safety precautions for personnel working in sewers and ancillary structures is indicated by the relative frequency with which serious accidents, sometimes involving fatalities, have occurred, from time to time, in many parts of the world. The precautions which must be observed by any person entering a sewer or sewerage structure can be specified so that all but the unpredictable accidents would be avoided. It is important, however, that such safety precautions are clear, concise and practicable and that ample and adequate equipment is available for use by the workmen. Otherwise, compliance with the instructions may prove too irksome and the very persons that they are intended to protect may tend to ignore them.

In developing a satisfactory set of instructions, the maintenance engineer should first consider the hazards involved and then base his instructions on the precautions that any reasonably prudent man should observe to ensure his personal safety in the situation being considered. The frequent exposure to hazardous situations during their daily tasks tends to lessen the awareness of some maintenance personnel to the inherent dangers which, although not obviously apparent in some instances, are nevertheless ever-present. These dangers comprise those which are common to normal construction works as well as those peculiar to sewerage works arising from the nature of the sewers and sewage. Of the dangers resulting from the presence of sewage only those due to hydrogen sulphide and other vapours and to possible deficiencies of oxygen in the sewer atmosphere, belong properly within the scope of this Manual. Any precautions prescribed for protection against hazards resulting from the presence of sulphide or from oxygen depletion must of course be related to and integrated with the precautions prescribed for protection against other hazards such as carbon monoxide, methane, petrol fumes, etc., which have much in common with those for hydrogen sulphide and oxygen depletion problems. The precautions necessary to give protection against hydrogen sulphide and oxygen deficiency hazards are listed in the following.

Sewer maintenance personnel should be acquainted with the general nature of all the hazards involved, the early symptoms produced, and the action to follow accidental occurrences. The most likely sources and sites of dangers should *be marked on plans on view* in each maintenance district (*revised periodically*). However, the staffs should realize and be reminded clearly and regularly that constant vigilance is the rule everywhere, not just near or in "trouble areas".

Special attention is required in the following cases:—

(i) Unvented air pockets including those near siphons or where a sewer becomes a pressure main or during ponding upstream of a pumping station or upstream of sewer blockages.

(ii) Localities where industrial wastes containing or producing high concentrations of sulphides are discharged to sewers. Such wastes include tanneries, boiling down works, glue works, chemical works, wool scouring works, cotton textile dyeing works, photographic processing works and others. When new industries discharging potentially dangerous wastes are connected to a sewerage system the sewer maintenance organisation should be notified immediately, the information added to the plan referred to in (i), and attention drawn thereto.

(iii) The vicinity of junctions where sulphide-containing wastes and acid-containing wastes may meet and mix. Sudden increase in acidity may cause a quick escape of hydrogen sulphide into the sewer air. This set of conditions has resulted in fatalities in Australia as well as overseas in spite of the fact that most sewerage authorities require neutralization of acid wastes.

(iv) Junction areas where rising mains discharge to gravity sewers.

(v) Local high turbulence regions, such as vertical drops, especially in sewers where the sewage contains high sulphide concentrations.

(vi) Covered tanks and wells.

As noted above, any sewer blockage should be regarded as a potential danger. If the sewage mixes with an acid industrial waste downstream or after clearing the block, the danger may be extreme.

The use of sewage and sewer air surveys covering carefully selected representative points in the system should be considered as a routine way to determine if and where additional points of special danger exist (see pp. 147 et seq.).

Maintenance gangs should be provided with the safety equipment now normal in most sewer maintenance organisations which may include safety harness and hoisting gear, masks and portable air blowers, self-contained breathing apparatus, etc. In addition to this standard equipment already mentioned, some effective device for detecting and roughly measuring hydrogen sulphide concentrations may be desirable. The most common devices are those which depend on the action of hydrogen sulphide which turns lead acetate paper brown when exposed to air containing the gas, BUT such tests must never be taken as proof that conditions are safe. They may be used to show that because proper precautions have been taken, conditions generally are safe, nevertheless, rapid changes must be allowed for.

It should be mandatory that one or more men be present at the surface access when personnel are working below ground. Adequate ventilation of any sewer or chamber should be ensured before anyone enters. In some cases this could perhaps be achieved by removing manhole covers upstream and downstream of the manhole to be entered, but in large or deep sewers, the use of portable mechanical air blowers is needed.

Before a person enters such a sewer, the sewer air should be tested after ventilation has operated for five minutes and tests repeated until all tests indicate that the air in the sewer is normal. This would mean at least that the O_2 and CO_2 concentrations were those of the open air, and H_2S concentrations were negligibly different from those of the open air. Tests might be required also to show the absence of explosive gases, or of CO, or some other particular substance relevant for the particular case.

Points of recognizable continuing danger revealed by surveys or tests should be provided with specially designed ventilation. If sulphide gases are present in toxic concentrations it is almost certain that damage to sewer structures is taking place, and special attention should then take account of corrosion of sewer components such as ladders, platforms, step-irons and so on.

Where toxic hazards are associated with particular industries, it may be useful to install holding tanks so that relatively high intermittent concentrations can be obviated, the aim being if feasible to maintain concentrations, for example, of toxic volatiles or of constituents such as those relating to unusually high and rapid oxygen demand, at a uniformly safe low level. If safety depends on the proper operation of float controls, valves, etc., or on proper operation of particular pre-treatment units, then not only should their proper performance, operation and maintenance be ensured, but a close inspection and, if needed, a test of these works should be made at the time of entry, and adequate supervision maintained while men are in the sewer.

Corrosion control by operational or managerial changes

These measures include both changes wholly within the operating organisation itself, and changes in respect of the usage of the sewers by the public and by industry.

The operation of a sewer system involves the consideration of many aspects if the system is to be effectively and economically maintained in a good and safe working order and condition free from nuisance. One of these aspects is the control of problems originating in the presence of sulphides in the sewers in undue amounts. Sulphides cause not only corrosion and damage to sewer materials but are often associated with odour nuisances to those living near sewers and the build-up of toxic conditions within the sewer which would be potentially dangerous for personnel. It is therefore essential that each person engaged in the operation and maintenance of sewers have a knowledge of these aspects appropriate to his area of operation and level of responsibility. Such knowledge should include the ability to recognize early symptoms of sulphide attack and to perceive danger areas for personnel.

Proper operational measures can often do much to reduce the intensity of sulphide problems and the following pages describe broadly some important measures which can be recommended as likely to be effective under most conditions. The detailed measures adopted by a particular sewerage authority will of course depend to some extent on the technical and financial resources available to that authority. Usually some degree of adaptation is required to suit local circumstances.

Sewage and corrosion surveys: Basic sewage and corrosion surveys are necessary to show where operational and maintenance procedures should be altered or remedial measures undertaken to prevent or reduce to a minimum the corrosion and the other problems which have arisen from the presence of sulphides.

Basic sewage surveys also serve to uncover dangerous conditions, even where sulphide attack is not yet detectable by superficial inspection of the structure. Such surveys must be repeated from time to time, since the volume and character of the sewage is always likely to change significantly due to development of the sewered areas or because of additional industrial waste discharges and other circumstances. In planning surveys of this type, it must be borne in mind that conditions vary also according to the season of the year, the day of the week, the hour of the day, the rainfall (as affecting dilution, dissolved oxygen content and velocity), and other factors (see Chapter 5). Conditions may vary even in a given cross-section of a sewer, and, in addition, "plugs" of particular wastes or intermittent pumping station discharges flowing down a sewer can cause rapid temporary changes in sewer composition. The sampling programme, including frequency of sampling, duration of the programme, etc., can only be determined after consideration of these variations and the type of information required. The tests to be made depend on the factors being investigated; at least the tests should include sulphide, pH, BOD, and sulphate determinations and temperature measurements. The results of such a survey should be plotted on a suitable diagram showing the layout of the sewerage system; the plotting should show where no sulphide was detected, as well as actual concentrations where sulphides were present. Fig. 13.1 shows one method of plotting such data.

Thus Fig. 13.1 presents the simplified plan of an outfall sewer, showing by the thickness of the hatched band to an appropriate scale, such information as the total sulphide concentration in mg/l, or alternatively the total pounds of total sulphide per day, increasing with passage along the sewer. The thickening of the bands thus might show the progressive sulphide build-up as the sewage flows towards the treatment works. BOD and other determinations may be plotted in a similar manner, presenting the "picture" of conditions clearly and distinctly.

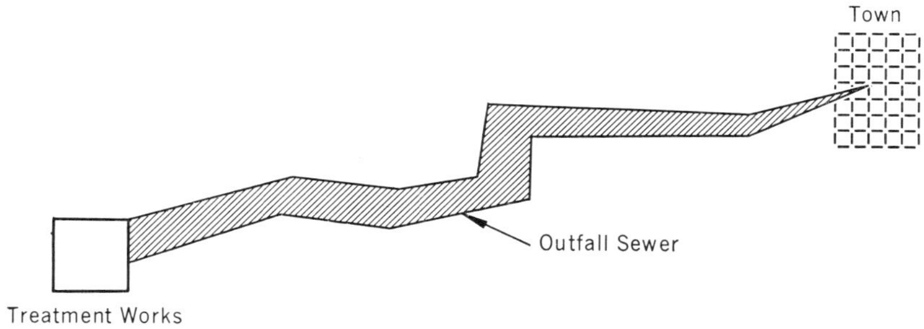

FIG. 13.1 Sulphide Survey Data Plot.

As well as the basic surveys, it may often be desirable to conduct a series of relatively localized tests to demonstrate the effectiveness of various alternative measures tried, in order to determine the best operational method of removing or reducing some local sulphide problem—to which attention may well have been drawn by the wider basic survey. The solving of local problems of this nature and indeed establishing any sewer improvement routines generally is best achieved by the co-operative efforts of both Operations and Laboratory staffs.

A corrosion survey is a special inspection of the system's above-sewage structures to determine the existing condition of sewers, manholes, metalwork, etc.; particular attention being directed to places where softening or roughening of the concrete surface has started or where the aggregate may be showing. Preferably, the examination should be on a quantitative basis. It is essential that competent observers be employed for these surveys and that observations and analyses be recorded properly. Regular inspections are necessary, since a sewer may not exhibit any conspicuous signs of attack for some years after it has been put into service, nevertheless within a short period, may show quite marked superficial corrosion. Even with regular visual inspections, incipient attack often is not detected; for instance, visual inspections of manholes may give a false sense of security, since local ventilation of a manhole may retard local attack, whilst along the sewer between manholes the corrosion may be quite severe. Visual detection of incipient corrosion actually requires considerable experience. Certain conditions of the concrete surfaces can be recognized as potentially aggressive, with pH say below 6, but the discovery of existing aggressive conditions prior to the appearance of visible signs of corrosion is usually dependent on careful sampling and detailed chemical analyses.

The use of polychrome type pH indicators painted or sprayed on concrete surfaces is sometimes of assistance in detecting potentially dangerous pH conditions.

The presence of sulphides in sewage as indicated by the sewage surveys also provides a guide as to where corrosion is likely to occur.

The visual inspection of sewers too small for a man to pass through renders corrosion surveys of such sewers difficult except at manholes. However, devices have been developed to assist in making visual inspections in such cases. The earliest of these devices were based on optical systems employing telescopes set up in manholes and an arrangement of lights and mirrors carried on a sledge moved along within the sewer. Later devices have incorporated closed circuit television systems in which small television cameras have been moved along within the sewer.

Control of Industrial Wastes: The characteristics of the domestic fraction of the sewage of a town or city depend both on characteristics of the water supply and those of the human and household wastes. The primary object of the sewerage system is to deal with this domestic sewage, and little or no control over the characteristics of this fraction is generally possible.

The industrial waste fraction of municipal sewages appears to be growing in magnitude and is a potential contributing source to sulphide problems. Most sewerage authorities have some legal powers with respect to the acceptance of industrial wastes into their sewers and may require such pre-treatment of these wastes as they consider necessary to make them acceptable for discharge into the sewer system. When pre-treatment prior to discharge is required, the problem becomes two-fold. Not only has the nature of the pre-treatment works and the acceptable standard of their performance to be determined but, once the pre-treatment is put into operation, it is also necessary to ensure that its standard of performance is sufficiently maintained from time to time throughout the different days and from year to year.

The technical problems associated with the control of industrial wastes often are very complex and it is recommended that the advice of a sewage chemist, or other trade waste technologist, be obtained when decisions are to be made on the admission of specific industrial wastes.

In general, each trade waste discharge must be considered in relation to the composition and volume of the sewage with which it will be mixed on entering the sewerage system. An alkaline waste may be beneficial in raising the pH of the sewage and reducing sulphide emission; a waste containing iron salts may help by precipitating some of the dissolved sulphide already present in the sewage; a waste may contain substances which inhibit bacterial activity. On the other hand, a waste with a high BOD is potentially harmful, especially if its sulphur content is also high. The addition of an acid waste to a sewage already containing sulphide may increase the sulphide emission rate significantly. Hot wastes, which raise the sewage temperature materially, will increase the biological activity in the sewage, and also reduce the solubilities of both oxygen and hydrogen sulphide, unfavourable for sulphide control.

The rate at which a waste is discharged to a sewer may determine its acceptability. Thus a relatively small volume of strong sulphide waste might be admissible if the discharge were so regulated as to be always small in proportion to the sewage flow, resulting perhaps in no significant difference for existing local or downstream sulphide conditions.

Wastes from the following industries are among those which experience indicates are likely to contribute to hydrogen sulphide problems:—Abattoirs, dye works, meat packing plants, photographic processing works, fellmongering and tanning and wool-scouring plants, glue and gelatine works and gas-making works, and many others.

The proper control of trade wastes requires properly organised and constant supervision or control of the wastes actually being discharged to sewers. Any harmful discharge conditions should be detected and warning given at once, whereupon appropriate action for adequate control should be taken promptly.

With greater emphasis now being placed on "Town Planning" it may become possible to restrict to some degree the locations at which industrial installations with large or potentially troublesome industrial wastes are established, and it is recommended that the effect of industrial wastes on sewerage problems be given due weight when Town Planning aspects are under consideration.

Control of Infiltration of Groundwater: In many cases, groundwater infiltration into a sewer may result merely in the dilution of any sulphide present in the sewage and may therefore have perhaps even some beneficial effect on sulphide problems. However, instances have been reported where infiltrating groundwater containing significant concentrations of sulphate or sulphide has resulted in increased sulphide content of the sewage. Such infiltration also may bring in undesirable concentrations of anaerobic bacteria, or produce low redox conditions for other reasons.

Quite frequently where sewers are laid close to the sea or estuarine waters or for other reasons, the infiltrating groundwater contains concentrations of sulphates sufficiently high to cause greatly increased sulphide build-up in the sewers. This may apply particularly in the early stages of development of a sewered area, before such time as the flow from the connected properties becomes sufficient to render the groundwater infiltration insignificant in respect to sulphate and sulphide concentrations.

If the infiltration of sulphate or sulphide-containing

groundwater is likely to cause a permanently serious problem, the possibility of staunching the groundwater inflow by internal or external grouting should be considered.

Groundwaters generally are characterized by relatively low dissolved oxygen contents, which can be another highly undesirable feature.

Controlled admission of water: Suitable surface waters or groundwaters have been admitted to sewers to augment the sewage flow during the early stages in the development of the areas served. This may reduce siltation, while another beneficial effect is dilution of the sewage; also if the water contains sufficient dissolved oxygen the sewage may be kept fresher. The freshening and diluting effects of stormwater in some sewers result in an interruption of sulphide generation during the storm flow, and the rate of generation remains negligible, or is frequently lowered significantly for at least a number of days afterwards. It is of interest to note that "combined" sewerage systems seldom suffer from serious sulphide corrosion. If a source of suitable fresh water is available, consideration should be given to its use to prevent or reduce sulphide build-up in a sewer during the early stages of the catchment which it serves.

The precautions to be observed when extraneous water is brought in for such purposes include the following. First the water must be of acceptable quality, with respect to pH, sulphide and sulphate in particular. Also silt must not be carried in with the groundwater, and the water admitted must not cause surcharge. It is desirable that the water normally contains adequate dissolved oxygen to maintain aerobic conditions downstream. Automatic control of the admission rate can be provided by a float-operated valve, which is closed when the subject sewer is already carrying stormwater in excess.

The pumping of sea water into a sewer has been suggested as a means of diluting and freshening the sewage. There is evidence to indicate that these benefits would not always offset the harmful effects of adding so much more sulphate and other salts to the sewage. The possible use of sea water should therefore be approached cautiously, and all relevant factors taken carefully into account.

Groundwaters also may carry ferrous iron, and their possible use on this account may warrant some consideration.

Operation of a Sewer as a Pressure Main: Corrosion control can be achieved by operating a sewer either continuously or intermittently, as a pressure main. The requirements may be summarised as follows:—

(i) Frequency and duration of intermittent filling must be adequate to control corrosion.

(ii) Where corrosion has occurred the effectiveness of treatment will be increased by first removing corrosion products by wire-brushing or other suitable means.

(iii) The sewer must be filled completely while operating as a pressure main.

(iv) Continuous siltation should be prevented or provision made for frequent periodical de-silting.

(v) The sewer must be free from house connections or branches in which ponding or surcharge may result.

(vi) If the sewer is filled continuously or for long periods, hydrogen sulphide attack on works downstream must be provided for. Means to obviate odour nuisance may also be required.

Experience suggests that in any case where substantial attack has taken place, subsequent control by intermittent drowning is unlikely to be successful unless treatment is repeated at weekly or at most fortnightly intervals.

Control of Pumping System Operation

A sewage survey will show whether a pumping station and rising main are responsible for the exhaustion of dissolved oxygen in the sewage or for a significant build-up of sulphide. There are measures which may be taken as part of normal operations aimed at reducing sulphide build-up to a minimum, so far as the pumping system may be the cause. The following should be read along with Chapter 11, which deals with the design of new pumping systems. The technique of compressed air injection is referred to in pp. 141 and 159.

Intermittent Pumping Systems—Operation of Pumps: In an existing intermittent pumping station, the only adjustment which the operator can make is the alteration of the levels in the wet well at which the pump (or pumps) cut-in and cut-out. This merely alters the length of time for which the duty pump operates following each start, causing an equivalent change in the length of the idle period; *the effective average* velocity $\overline{V}_S$ in the rising main being unaltered. Since investigations carried out by the Australian Sewerage Authorities indicated that the build-up of sulphide in p.p.m. in the rising main was dependent, to some extent, on this average velocity, little or no change in sulphide build-up through the rising main is to be expected due simply to the cut-in and cut-out level settings. However, these investigations also indicated that when the rising main was of normal length, the build-up of sulphide concentration in the sewage detained in the wet well was generally a small fraction only of the build-up through the whole system, but that the reduction of any dissolved oxygen present in the incoming sewage, in some cases even a slight build-up of sulphide in the wet well, preconditioned the sewage in some way which encouraged the build-up of sulphide in or downstream of the rising main, despite the apparent negative value of any change in cut-in and cut-out levels. This may apply particularly where the sewage is very strong and the rising main very short. Such improvements may seldom be sufficient to overcome the adverse effect of a pumping system, nevertheless they would help to simplify further remedial measures.

It is again emphasized that the best setting of cut-in and cut-out levels can be determined only by on-site testing. These tests should cover the actual measured sulphide build-up or dissolved oxygen decline through the pumping system at various settings. Wherever possible, each test should cover a full 24-hour weekday period and definitive tests should be made only after the particular setting of cut-in and cut-out levels had been in operation for at least two weeks.

The cut-in level in a wet well should always be such that no ponding of sewage upstream in the inflowing sewer occurs particularly under normal dry weather flow conditions.

Continuous Pumping Systems: Pumping stations in which one or more pumps operate continuously are usually employed only for handling large flows and generally operate without wet well storage. The adverse effect on sulphide build-up may arise from either the build-up in the rising mains, or from the ponding of sewage in influent mains because of low flexibility of station output capacity. This latter may be because this capacity can be varied only by steps about equal to the capacity of one additional pump. Ponding in influent sewers causes lower velocities, reduced surface aeration of sewage, greater area of slimes contributing to sulphide generation, and deposition of silt.

For an existing continuous pumping system, the generation equation shows that little can be done to alter the sulphide build-up through a rising main so far as pumping programmes are concerned. However, ponding may well be reduced by improved pumping programmes and hence some improvement should result, reducing sulphide build-up through the whole pumping system which in such a case also includes the lengths of the influent mains affected by ponding.

Other Remedial Measures: The preceding pages deal with simple adjustments to the pumping mechanisms. When the sewage survey shows a critical build-up of sulphide through the pumping system, other or additional measures for sulphide control are likely to become necessary. Some of these are discussed in the broader context of the system as a whole; but may be applied even more effectively to specific local problem areas.

Treatment of sewage to control sulphide problems

Treatments Using Chlorine: The addition of a sufficient quantity of chlorine to septic sewage destroys the sulphide already present. Higher doses also have some inhibitory effect on the sulphide-producing slimes. To maintain control in an outfall sewer over a length of several miles, repeated dosing has been found necessary. Experience at two Australian cities has shown that the first dose upstream preferably should fully satisfy the chlorine demand of the sewage. This should be followed, at intervals of from one to two miles, by further doses, each amounting to about one-third of the initial dose. For shorter term suppression of hydrogen sulphide generation as in a rising main, chlorine doses amounting to only portion of the full demand may be adequate. Slimes can be killed by dosing the sewage with sufficient chlorine for a sufficient time.

For existing systems it will be practicable to determine the required dosage of chlorine in a particular case by trial and error methods. As chlorine is usually expensive, and as excessive chlorine residuals in sewage may introduce further problems with respect to metallic corrosion and personnel safety, it is desirable that dosing rates should follow actual requirements fairly closely. Careful surveys of chlorine requirements from hour to hour should therefore be made before regular dosing is designed. Chlorination equipment can now be obtained which will follow any predetermined dosing programme.

Treatment by Mechanical Aeration of Sewage in Gravity Sewers: In some cases mechanical aeration may be an economical method using compressed air or surface agitation. In long sewers aeration stations may be needed at a number of sites located at appropriate intervals. A possible difficulty with this method is the prevention of odour nuisance and localised corrosion at the aeration stations, where the hydrogen sulphide emission rates may be high. However, if aeration treatment is commenced at a site upstream of any significant sulphide concentration in the sewage, there is then no reason why local sulphide emission should be troublesome because controlled sulphide conditions then will be maintained along the whole sewer. However, as with other alternative remedial measures of possible application to an existing system, it would be desirable to test the probable effectiveness of this remedy for each particular case. This can be done by using some effective means of aeration temporarily; such as, for example, the discharge of jets of sewage into the sewage stream, employing a portable pumping plant to produce the necessary jet pressure. If aeration were found to be effective, more efficient apparatus then could be permanently installed. It is important that any preliminary trial be continued for some weeks at least to allow full adjustment of slime growth, etc., to the new environmental conditions.

The amount of oxygen, in mg/l which it will be necessary to introduce and dissolve in the sewage so that it can travel a distance "L" ft. along the sewer without additional sulphides appearing in the sewage, will be given by the following general equation

$$\text{(oxygen absorbed through sewage surface)} = \left(\text{BOD rate} \times \frac{L}{V_S}\right) + \text{(oxygen demand of slimes over distance, L)} \quad [13.1]$$

Substituting appropriate values for the various terms, we have

$$\text{BOD rate} \times \frac{L}{V_S} = \frac{6 \times 10^{-6} \times L}{V_S} \times [BOD_5] \text{ from p. 31}$$

$$O_2 \text{ demand of slimes} = 1{\cdot}43 \times 10^{-7} \times V_S \times F_1 \times P_T \times [BOD_5]^{0.8} \times (SO_4)^{0.4} \times 1{\cdot}139^{(t^\circ C-20)}$$

$$O_2 \text{ absorbed by sewage} = \frac{K_{O_2 s} \times O_D \times L}{60 \times V_S}$$

For calculations, Design Charts 4 and 5 are used.

The rate at which oxygen passes into solution will depend upon the oxygen saturation deficit applying. The rate will decline to a negligible value as the deficit approaches a zero value. Therefore it usually is not economical to aim at saturating sewage with dissolved oxygen either initially or at any of the aeration stations.

It should also be pointed out that only some of the oxygen in the aeration air will be dissolved in the sewage. Most of it enters the sewer air just downstream of the aeration station where entrained air bubbles leave the sewage. Unless full sulphide control is achieved, these air bubbles will contain sulphide and serious corrosion risk may result.

Injection of Compressed Air into Rising Mains: One of the most successful methods applied to pumping systems is the injection of compressed air into rising mains. It has been used with success in a number of instances in America(*8*), where the earliest application was made about thirty years ago. The essential principles of the method are shown in Figs. 11.7 and 11.8, p. 141. It would appear that much of the success achieved is due to the relatively high pressure of the air in the main, which permits an increased rate of oxygen absorption proportional to the pressure, and to the peculiar form of turbulence produced at the air-sewage interface.

Pomeroy(*8*) states that injection of compressed air into a force main or other filled pipe is a useful method of sulphide control, usually accomplishing either a large reduction or complete elimination of sulphide generation. Whereas the quantitative aspects of the method are not yet fully worked out, Pomeroy suggests that air requirements are usually about one cubic foot per minute for each inch of pipe diameter, with adjustments for abnormal lengths, pressures or sewage strengths. Where the method is under consideration for application to an existing rising main, trials can be made using portable air compressing plant, testing for sulphide conditions before and after air injection is applied. The trials should be conducted for a sufficiently long period to allow the slimes to stabilise under each new set of environmental conditions. Tests should be aimed at determining the most desirable operating conditions having regard to all the circumstances of sulphide conditions, odour nuisance, overall costs and so on.

There is need for some caution in adopting this method of sulphide control until more is known of the corrosion risk to the top of the pressure sewer, although there can be little risk if control of sulphides is complete. The method may also require the effective treatment of residual injected air at its discharge point if nuisance is to be avoided. Pomeroy has used earth filtration for deodorization of this vented air(*40*).

It has been reported also that the injection of substantial quantities of compressed air may cause sufficient agitation to dislodge slime and, for slime control, such treatments might be repeated at weekly or shorter intervals. There is as yet insufficient evidence to indicate whether in larger pipes the slime affected would be only along the top of the pipe or around the full periphery.

Treatment Using Nitrates: Oxygen introduced into sewage in the form of nitrates can satisfy the biological oxygen demand in preference to sulphates and so prevent or reduce the biological production of sulphides, and if a low-cost source of nitrates were available, this remedy may be economical.

Treatment by Partial Purification: The rate of oxygen demand of sewage can be reduced by standard methods of treatment such as sedimentation combined with trickling filters, or possibly using partial activated sludge processes. It would be expected that effluents from such treatment processes could be conveyed along a sewer with greatly reduced risk of sulphide problems arising since the oxygen absorbed at the sewage surface could more easily cope with the oxygen demand in the flowing sewage and from the slimes on the sewer walls. To arrive at the most economical application of this method for a given sewer, it is necessary to determine what level of BOD can be tolerated in the existing sewers without sulphide problems and hence what design of plant can most economically achieve the degree of partial treatment necessary.

An accurate estimation of the BOD level in question presents difficulties but an assessment of this may be made by the use of equation [10.2]. Practical data may be obtained by making sulphide and BOD tests at intervals following periods of rainfall, at different flow stages, to detect the first re-appearance of sulphide build-up. The order of the partially treated effluent BOD_5 level required is normally found to be somewhat upwards of 70 mg/l, actual values depending upon grade and size of sewer and proportional depths of dry weather flow.

The purpose of the partial treatment is to raise the redox of the sewage so that in the chain of processes involved, there is no resultant concentration of H_2S in the flowing sewage. Trickling filters, by appropriate loadings of sewage measured say as equivalent lb. of BOD_5 per cubic yard of filter media, can be designed to produce any desired degree of purification between full stabilisation and something just better than settled sewage. A similar range in quality of effluent may be obtained using "partial treatment—activated sludge" systems. Most of the work so far done in this particular field refers to trickling filter installations and this type of installation is taken as the basis of the following discussion.

To arrive at a preliminary plant design sufficient for estimating the costs of alternative solutions of a particular sulphide problem, it is commonly necessary to construct and operate a pilot plant using the sewage concerned for a reasonable period. The pilot treatment necessary involves sedimentation and sludge digestion and disposal as well as the trickling filters. However, before embarking on such a pilot plant, a preliminary estimate of the required capacity of the various units could be obtained from previous work by investigators in this field. Should this preliminary estimate show that partial treatment is an economic possibility, the question of a pilot plant then would be given further consideration.

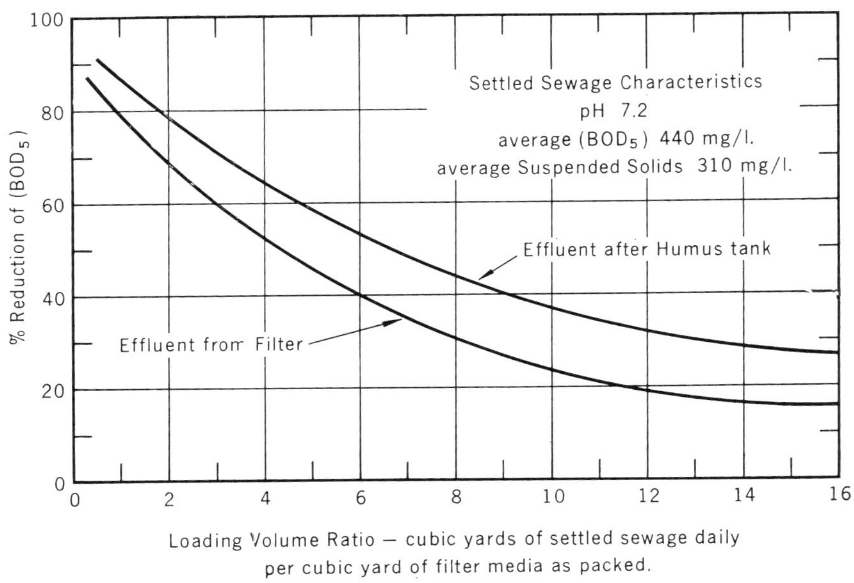

FIG. 13.2 Purification by Trickling Filtration (partial treatment).

The whole subject of partial purification in relation to sulphide problems is complex and readers particularly interested in this subject are referred to a report on investigations carried out by Parker(*61*). The salient features of these investigations are summarized in the following.

For the sewage described in Fig. 13.2, the required degree of partial purification could be accomplished by various arrangements of filters and detention tanks, of course, with ancillary arrangements. Pumping plant with its associated power costs was normally part of this.

The plant arrangements investigated included:—

(a) single stage filters:—

(i) with primary and secondary settling tanks,

(ii) with recirculation and settling tanks as in (i) above;

(b) two stage filters:—

(i) with primary, secondary and tertiary settling tanks,

(ii) with recirculation on second stage and with settling tanks as in (a).

For these various plant arrangements, the relationships between filter loadings in pounds of 5-day BOD per day per cubic yard of filter media and the percentage reduction in BOD are shown in Fig. 13.2. From this and the daily volume of sewage flow, an approximate idea of the volume of filter media required to produce an effluent with the required BOD can be computed. Filter depths up to 10ft. do not appear materially to influence the overall filter performance for partial purification, but do affect pumping costs. When sewage composition departs materially from that of the sewage employed by Parker, filter performance also may be affected. For such a case, a better approximation may be obtained by use of the relationship of pounds of BOD removed to pounds of BOD applied, and such a relationship may be obtained by suitable manipulation of the information of Fig. 13.2.

The pilot investigations referred to were based on settling periods of one hour. Thus, with single stage filter arrangements the total volume of settling tanks would be one-twelfth, and with two stage filters, one-eighth, of daily dry weather flow volume.

The pumping capacity required to lift effluent from the filter outlets to settling tanks inlets would be equal to daily flow and twice daily flow for single and two stage filters respectively. Where wet weather flows greatly exceed dry weather flows, a large proportion of the wet weather flows can by-pass the plant since sulphides are not then usually a problem.

Treatments Using Lime: Treatments using lime may be applied (i) for pH control of H_2S emission, or (ii) for direct control of slime growths.

(i) *pH control of H_2S emission:* The amount of sulphide emitted to the sewer atmosphere depends on the concentration of molecular sulphide present in the sewage, and the fraction of molecular sulphide present depends on the pH of the sewage (pp. 34, 35). By increasing the pH to about 8.5, emission is virtually stopped, and this can be achieved by adding sufficient alkali. The most economical alkali to use generally is lime. The dosage usually is high, and in fact considerably higher than may be indicated by laboratory experiments. For example, when laboratory tests indicate the need for a dosage rate around 100 p.p.m. of lime to achieve the desired pH, it may be found in practice that 200 p.p.m. or more is required to maintain this pH along a mile of sewer.

It should be pointed out that this treatment does not normally reduce the total dissolved sulphide, but only the undissociated molecular fraction of sulphide. Therefore, build-up of total dissolved sulphide downstream could produce again the conditions which are being treated upstream.

(ii) *Control of slime activity:* It has been found that if very high dosages of lime are used, even during relatively short periods, the "shock" effect on the slimes causes them to cease normal sulphide generation for periods ranging from a few days to a week or more. This has led to the introduction in America(*67*) of "shock treatment" with lime. In this treatment, dosing may be at intervals which range from three days to a week or longer, according to the season. As we are here dealing with existing systems, the actual amount and frequency of dosage should be determined by trials using sulphide determinations as a measure of effectiveness.

Treatments using metallic salt dosing: This is a control measure, the primary purpose of such treatment being to convert any dissolved sulphide present to the insoluble form. Iron, zinc, or copper salts have been used most commonly.

(i) *Iron Salts:* Iron sulphides may take any of the forms FeS, Fe_3S_4 or FeS_2. Therefore, according to the form of the sulphide produced, the theoretical iron requirements will vary from 0.87 to 1.74 lb. per lb. of sulphur reckoned as H_2S. In practice it is common to provide for a dosing rate greater than the calculated rate. This allows for a residual of iron to cover any further sulphide which may be added downstream. General experience has been that even with this provision, the whole of the dissolved sulphide is not converted to insoluble forms, nevertheless dosing with iron salts can be used to reduce comparatively high dissolved sulphide concentrations to less than 1 mg/l. Simultaneous dosing with lime sometimes helps to improve precipitation conditions. The comparative economy of the method depends upon the cost of the iron salts, which often consist of ferrous or ferric sulphate or chloride, produced on site. Occasionally, in certain localities, the availability of industrial wastes or by-products containing iron salts, or the raw materials to produce such salts cheaply, may make this means of control attractive, the Scott-Darcey process being such an example. Provided other problems are avoided, the discharge to the sewers of industrial wastes containing iron or other metallic salts may be encouraged as a potential sulphide control measure likewise.

(ii) *Zinc Salts:* Zinc sulphide takes the form of ZnS and therefore the theoretical zinc requirements are 2.04 lb. of zinc per lb. of sulphur as H_2S. While there is no reason why this method also should not prove successful, the economy of its use depends even more on the price of the zinc salt used. However, the possible use of industrial wastes from zinc manufacture and other industries using zinc may make this method attractive also.

(iii) *Copper Salts:* The position regarding the possible use of copper compounds is much as for zinc. Copper compounds probably also would tend to show more marked bacteriostatic effects.

Sewer cleaning and flushing

Solids accumulate in sewers as debris caught on projections, as silt deposits on the invert and as slime growths on the walls. Within these masses, conditions are favourable for the generation of hydrogen sulphide and also their slow fragmentation may influence generation downstream. Furthermore, these accumulations delay the passage of sewage through the system, and this also encourages sulphide build-up. The increasing use of rubber ring type joints may lessen the risk of accumulation of debris which occasionally occurred on cement mortar projecting into pipes at badly made joints.

To assist in determining the frequency of cleansing necessary to maintain a particular length of sewer in the best conditions for controlling sulphide generation rates, the sulphide concentrations should be checked at the downstream end of the length of sewer just before cleansing commences and at intervals after its completion, repeating the trial several times. It is usually feasible to work out a systematic schedule of cleansing operations, grouping the sewers according to the required frequency of cleansing. These frequencies may range from once every few weeks to once every year or even longer. Further information regarding preventative sewer maintenance may be sought in the literature.

It is recommended that a positive policy be adopted of keeping sewers clean by regular maintenance, rather than waiting until blockages or other troubles occur. This involves regular cleaning and/or flushing where solids accumulate, particular attention being given to flat sewers carrying only a small fraction of their designed ultimate flow.

The following example is from the records of an Australian Sewerage Authority. During the early life of a sewered area it was noted that sulphide concentrations in a sub-main sewer appeared to be unduly high and it was considered that this condition could have arisen from deposits in the contributing reticulation sewers caused by low flows as, at this time, only a small fraction of the houses had been connected to the sewers. It was decided to study the effect of cleansing of reticulation sewers on sulphide concentrations in the sub-main. Fig. 13.3 shows the effect of cleansing a portion of the total reticulation sewers involved and the duration of the effect.

Silt and other deposited debris from sewers may be

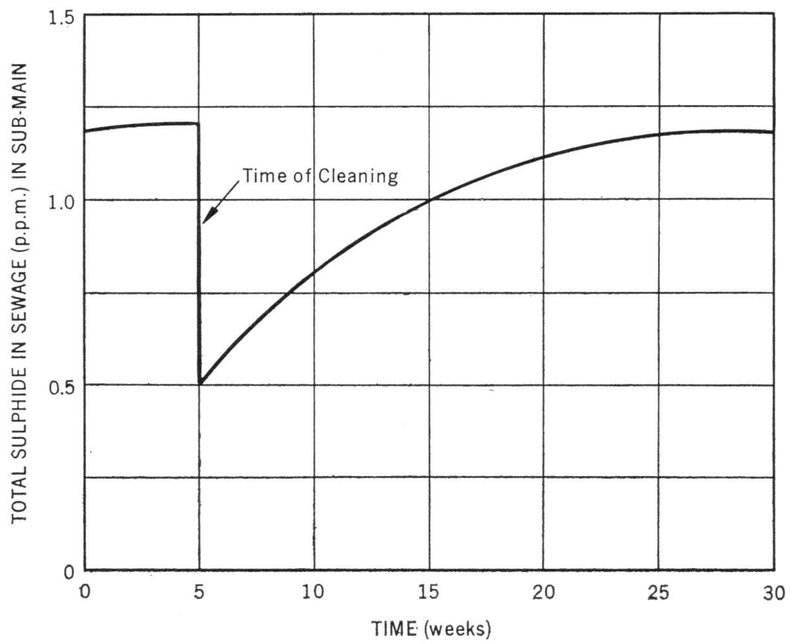

FIG. 13.3 Effect of sewer-cleansing on Sulphide build-up.

removed either by flushing or by the use of the various cleaning devices developed for the purpose. The effective and economical use of the various methods and equipment now available depends largely on the experience and skill of the personnel using them.

The flushing of sewers may be carried out either by using additional water from water supply hydrants, water tankers or convenient streams; or by making use of the sewage itself, which normally requires the use of penstocks provided at the time of construction of the sewer. For flushing, the penstock is closed for a period to accumulate sufficient sewage in the upstream sewer, which then is suddenly released and sends a "flush" down the sewer. The lengths for which such a flush is effective can only be determined by experience. This technique for cleansing has some added advantage in that sewer surfaces above normal dry weather sewage level also are wetted and their pH may be corrected for a period. This helps to reduce the corrosive activity which may have developed on surfaces above sewage level. However, some additional silting may result at the upstream end of sewage ponded behind the penstock, and this could be something of a disadvantage.

Mechanical procedures: In addition to the use of hand rods, the following mechanical equipment for sewer cleaning has proved effective:—

Flexible steel rods with cutting tools, etc., propelled and rotated by a power unit at street level, as depicted in Fig. 13.4.

Using a Rubber Squeegee or Ball slightly smaller than the pipe diameter, largely carried through the sewer by the sewage flow (see Fig. 13.5).

Using a Sewer Plough or Hoe, this system makes use of a light truck or sledge with a scraper blade-squeegee which also serves as a dam, to raise the

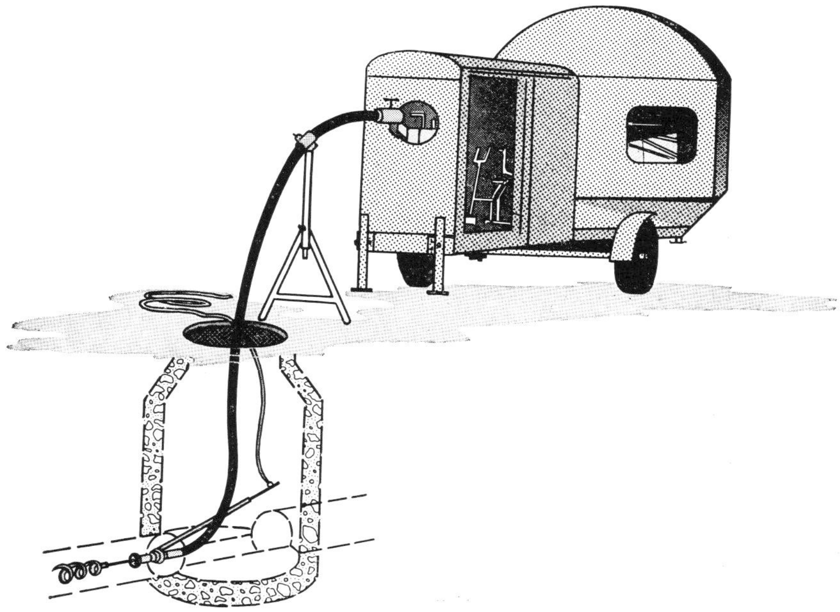

FIG. 13.4 Mobile power system for operating flexible tools.

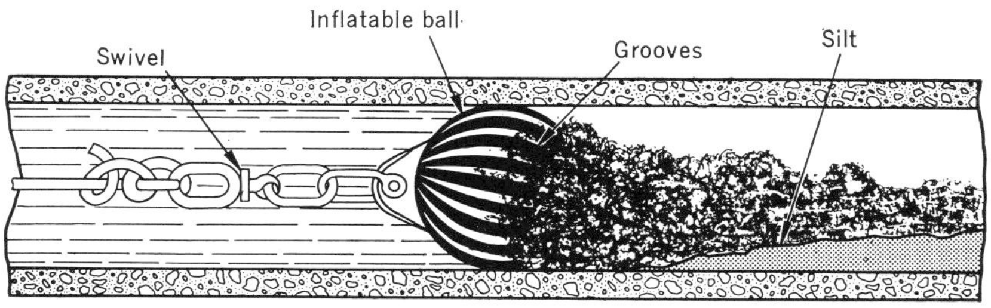

FIG. 13.5 Rubber ball-type squeegee.

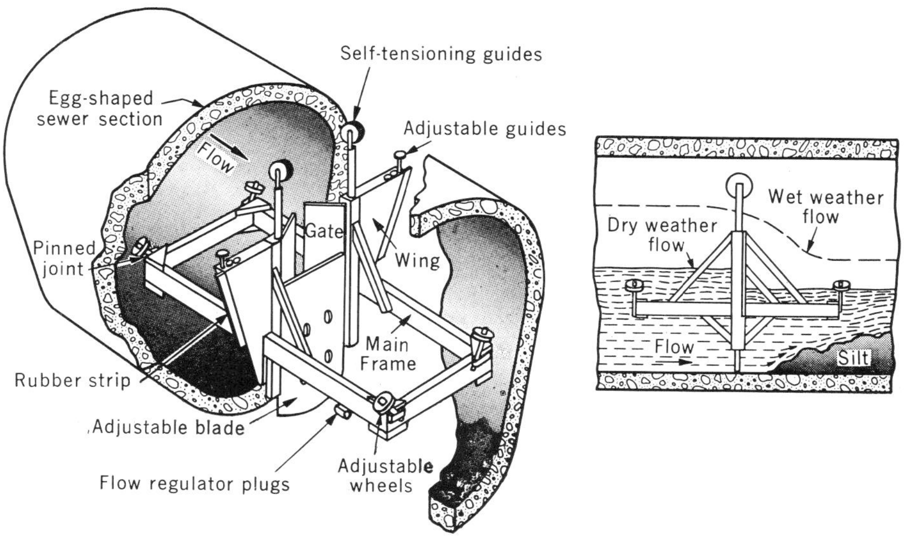

FIG. 13.6 Sewer "plough" or "hoe" system.

sewage level by a foot or so. This is propelled through the sewer by the head of sewage behind it, and scrapes the lower part of the sewer as well as scouring the inverts by jets of sewage directed downwards (Fig. 13.6).

Sewer cleaning machines, trucks located at street level, carrying power-driven winches, for hauling drag buckets to and fro through the sewer between manholes. (Fig. 13.7).

FIG. 13.7 **Mechanical winching or "bucket" system.**

Cleansing of wet wells: Any accumulation of solids on the floor of the wet well should be removed periodically using regular cleansing schedules based on experience and testing of the particular station involved.

Cleansing of rising mains: Silt accumulating in an oversized rising main should be removed by frequent enough flushings with higher rates of flow. Thus, a stand-by pump may be brought into service with the regular duty pump from time to time to flush a rising main too large for the normal flows of the time. Where more than one rising main is available, not more than the necessary number of mains should be in use at any one time. Mains not in use preferably should be emptied or filled with mains water.

Corrosion in wet wells: A number of cases of sulphide corrosion of the upper walls of wet wells above sewage level have been reported. This corrosion can usually be controlled by installation of a proper ventilation system, although a well designed natural ventilation system sometimes will suffice. The mechanical ventilation system might be based on 2 to 5 changes of air per hour, also providing for regular hosing down of the walls. Sometimes the problem is aggravated by the turbulence created by a free drop from entry pipe to surface of sewage in well. In such cases the use of a vertical drop pipe may reduce corrosion of the walls (see p. 136).

Repairing corroded structures

In the case of reinforced concrete structures, repairs should be undertaken long before corrosion has penetrated to the reinforcement. Repairs can be made by resurfacing the sewer periphery with gunite, cement rendering, bricks or tiles.

Before any resurfacing is done, the corroded concrete must be cut back to the solid and all soft material removed from between the projecting coarse aggregate; an air or water jet, an air-water jet, or sand-blasting, are effective methods for doing this. Since sulphuric acid and sulphate normally penetrate the concrete beyond the zone of obvious surface deterioration, it is recommended that some of the apparently sound concrete, perhaps a further ¼in., be removed also. Relining of sewers down to a diameter of 48 inches has been successfully carried out. Australian experience has shown that high-alumina cements (e.g. Ciment Fondu) will bond satisfactorily with underlying Portland cement concrete.

In some cases a structure can be saved by external blanketing with concrete. A large circular sewer may be repaired by installing tubular liners, although these reduce its capacity and allowance should be made for local ponding in the sewer upstream of the lined section.

In small sewers, particularly where excavation for replacement would be costly, drawing plastic tubing through the existing sewer pipe has been undertaken where the original sewer still has sufficient structural strength to support the external loads.

LIST OF REFERENCES

AUTHOR INDEX

SUBJECT INDEX

REFERENCES

1 POMEROY, R. *Hydrogen Sulphide in Sewage.* Sewage Works Journal 13, No. 3 (May 1941), 485-505.

2 SUMMER, W. *Methods of Air Deodorization,* Elsevier Publishing Company, 1963.

3 BAUMGARTNER. *Effect of temperature and seeding on hydrogen sulphide formation in sewage,* Sewage Works Journal, 6 (1934), 399-412.

4 POMEROY, R. and BOWLUS, F. D. *Progress Report on Sulphide control research,* Sewage Works Journal, 18, No. 4 (July 1946).

5 VIEHL, K. *About the reasons for hydrogen generation in sewage,* Gesundheits Ingenieur Vol. 2 (1947) pp. 41-44. Original in German, translation available in library of Melbourne and Metropolitan Board of Works.

6 TECHNOLOGICAL STANDING COMMITTEE OF THE PRINCIPAL AUSTRALIAN WATER AND SEWERAGE AUTHORITIES ON HYDROGEN SULPHIDE CORROSION, *Report on co-operative investigations into hydrogen sulphide build-up in sewage pumping systems, inverted siphons and sedimentation tanks* (1959).

7 PARKER, C. D. et al., *Concrete Sewer Corrosion by Hydrogen Sulphide,* Melbourne and Metropolitan Board of Works Technical Paper No. A.8, Part 2, Generation of Sulphides in Sewers.

8 POMEROY, R. *Generation and control of sulphides in filled pipes,* Sewage and Industrial Wastes, 31, No. 9 (1959), 1082-1095.

9 KINNEEN, C. W., *Hydrogen Sulphide Generation in pressure mains,* Thesis submitted to the University of Melbourne, Australia (1956).

10 REID and TIEN-CHENG YANG, *Sewer Slimes cause Odours,* Waste Engineering, Sept. 1959.

11 ROBERTSON, W. J., *Controlling Hydrogen Sulphide Corrosion in Sewers,* Australian Civ. Engineering 5.6.69, 41.

12 REPLOGLE, J. A. and CHOW, V. T., *Tractive Force Distribution in Open Channels,* Proc. Am. Soc. Civ. Engrs. 92 (Hy2) (March 1966), 169.

13 Studies mainly by officers of the Melbourne and Metropolitan Board of Works, incorporated appropriately in the text and formulations of this book but without specific reference.

14 FERGUSON-WOOD, D. A., *Marine Microbial Ecology* (1965), Chapman and Hall, London.

15 DAVY, W. J., *Influence of Velocity on Sulphide Generation in Sewers,* Sewage and Industrial Wastes, Sept. 1950.

16 IMHOFF, K., MÜLLER, W. J. and THISTLETHWAYTE, D. K. B., *Disposal of Sewage and Other Water-borne Wastes,* 2nd ed. (1971), Butterworths, London.

17 ADENEY, W. E., *The principles and practice of the dilution method of sewage disposal,* Cambridge University Press, 1928.

18 WHITMAN, W. G., Chem. & Met. Eng. 29 No. 4, July 23rd, 1923; also LEWIS and WHITMAN, Ind. Eng. Chem., 16 (1924), 1215.

19 DOBBINS, W. E., *B.O.D. and Oxygen Relationship in Streams,* Proc. Am. Soc. Civ. Engrs. 90 (SA3) (1964), 53 (SA6), 131.

20 CHURCHILL, M. A., ELMORE, H. L. and BUCKINGHAM, R. A., *The Prediction of Stream Re-aeration Rates,* Proc. Am. Soc. Civ. Engrs. 88 (SA4) (1962), 1; cf. GUNNERSON, C. G. and BAILEY, T. E. 89 (SA4) (1963), 95; 90 (SA1) (1964), 175.

21 GUSTAFFSON, B. and WESTBERG, N., *Oxygen Consumption and Re-aeration in Sewers,* Proc. 2nd Int. Conf. Wat. Pollut. Res. Tokyo 1 (1964) 221.

22 KRENKEL and ORLOB, Turbulent Diffusion and the Reaeration Coefficient, Proc. Am. Soc. Civ. Engrs., Journal of the Sanitary Engineering Division, 88 (SA2) March 1962, Part 1, Paper No. 3079.

23 OWENS, M., EDWARDS, R. W. and GIBBS, J. W., *Some re-aeration studies in streams,* Int. J. Air Pollut. 8 (1964) 469.

24 GAMESON, *Weirs and the Aeration of Rivers,* T. Inst. Wat. Engrs., Oct. 1957, 477-490.

25 OGDEN, GIBBS & GAMESON, *Some factors affecting the aeration of flowing saline water,* Water Pollution Research Laboratory Paper published in Water and Waste Treatment Journal, Sept./Oct. 1959.

26 LANGMUIR, I., *Surface Chemistry,* Chemical Reviews Vol. 13, No. 2, Oct. 1933, 156-165, 186.

27 RAND, *Surface reaeration of water as affected by domestic sewage,* Public Works Dec. 1957, 93-94.

28 DAVIS, D. S., Monograph for *Solubility of Oxygen in Water,* Water and Sewage Works, Oct. 1957.

29 SIERP, F. and FRÄNSEMEIER, F., *Fortschritte in der Kenntnis des biochemischen Sauerstoffbedarfs,* Tech. Gemeindebl. 34 (1931), 233-313.

30 MONTGOMERY, H. A. C., *Determination of Biochemical Oxygen Demand by Respirometric Methods,* Wat. Res. 1 (1967), 631.

31 THERIAULT, E. J., *The Oxygen Demand of Polluted Waters,* Public Health Bulletin No. 173, United States Public Health Service, U.S. Government Printing Office, 1927.

32 GOTAAS, H. B., *Effect of temperature on the biochemical oxidation of sewage,* Sewage Works Journal 22 (1948), 441; also *The Effect of Seawater,* 21 (1949), 818.

33 HUNTER, J. V. and HEUKELEKIAN, H., *The Composition of Domestic Sewage Fractions,* J. Wat. Pollut. Control Fed. 37 (1965), 1142. See also FELDMAN, L. A. and McGARVEY, F. X., *The Moon-base Problem,* Effl. & Wat. Treat. Jnl. 5 (1965), 209. Note that tradewaste discharges may alter sewage composition substantially.

34 MITCHELL, I. G., *Concrete Sewer Corrosion by Hydrogen Sulphide,* Melbourne and Metropolitan Board of Works Technical Paper No. A.8—Part 3, Influence of Oxygen Solution on Sulphide Content.

35 POWELL and GRIFFITHS, Trans. Inst. Chem. Eng. 13 (1935), 175.

36 MASSEY, B. S., *Mechanics of Fluids* (1968), Van Nostrand, London.

37 GAGE, S. D. M., *Methods for determining the origin, prevalence and effect of obnoxious odors,* Amer. Soc. Civ. Engrs. 51 (1925), 820.

37a STERN, A. C. (Ed.), *Air Pollution* (Vols. 1 to 3) 2nd ed. (1968), Acad. N.Y.

38 POMEROY, R. and BOWLUS, F. D., *Stay alive in that Sewer,* American City, (1959), May, 123.

39 BEER, A. D. D., *Influence of Sewer Air Conditions on*

Corrosion (of Sewer Concrete), Melbourne and Metropolitan Board of Works, Technical Paper No. A8, Part 7.

40 POMEROY, R. D., *Deodoring of gas streams by the use of microbiological growths,* U.S. Pat. 2,793,096 of 21.5.57.

41 THISTLETHWAYTE, D. K. B. and GOLEB, E. E., *The Composition of Sewer Air,* to be presented as part of the proceedings, 6th International Conference on Water Pollution Research, Israel, June 1972, and with HARDWICK, B., *The Deodorization of Sewer Air by Chemical and Biological Techniques,* Conference, Chemical Engineering in the Service of Mankind, Paris, Sept. 1972, Societe de Chimie Industrielle.

42 CARLSON, D. A. and LEISER, C. P., *Soil Beds for the Control of Sewage Odours,* J. Wat. Pollut. Control Fed. 38 (1966) 829.

43 PARKER, C. D., *Isolation of a species of bacterium associated with the corrosion of concrete exposed to atmospheres containing hydrogen sulphide,* Aust. J. Exptl. Biol. & Med. Science 23 (1945) 81, 91.

44 PARKER, C. D., *Species of bacteria associated with the corrosion of concrete,* Nature 159 (1947), 439.

45 PARKER, C. D., *Hydrogen Sulphide Corrosion of Sewers,* Melbourne and Metropolitan Board of Works Technical Paper No. A8, (1965), Part 6, Bacteriology of Corrosion Process.

46 MORRIS, S. S., *The deterioration of concrete in contact with sewage,* J. Inst. Civ. Eng. 13 (1940), 337.

47 STUTTERHEIM, N., VAN AARDT, J. H. P., *The Corrosion of Concrete Sewers and Some Possible Remedies,* Sth. African Industrial Chemist 7 (Nov. 1953).

48 MILLER, D. G. and MANSON, P. W., *Laboratory and field tests of concrete exposed to the action of sulphate waters,* U.S. Dept. of Agriculture Tech. Bull. No. 358, June 1933.

49 HAWKEN, R. W. H., *Economy of Purchase,* Revision dated 1941 of Paper No. 4, Proceedings of the Institution of Engineers, Australia, Vol. 6 (1925).

50 POMEROY, R. D., *Sanitary Sewer Design for Hydrogen Sulphide Control,* Public Works Oct. 1970, 93.

51 BINNIE, A. M. and HOOKINGS, G. A., *Laboratory Experiments on Whirlpools,* Proc. Roy. Soc., Ser. A, 194 (1948), 398.

52 ACKERS, P. and CRUMP, E. S., *The Vortex Drop,* J. Inst. Civ. Eng. Paper No. 6430, Aug. 1960, 433.

53 LELIAVSKY, S., *Fluvial Hydraulics* (1955), Constable.

54 KALINSKI and HSIA, Univ. of Iowa Engineering Studies Series, Bull. No. 29.

55 DURAND and CONDOLIOS, Le Houille Blanch, Dec. 1953, 815.

56 VALLENTINE, H. R., *Transportation of Solids in Pipelines,* Commonwealth Engr. 1.4.55, 349.

57 NING CHIEN, *The present status of research on sediment transport,* Trans. Amer. Soc. Civ. Engrs. Paper No. 2824, 121 (1956), 833.

58 SWAB, B. H., *Effect of Hydrogen Sulphide on Concrete Structures,* Proc. Amer. Soc. Civ. Engrs., San. Div. Sept. 1961, 1.

59 DAVY, W. J. and BIRD, A. W., *Functional Design of the Brooklyn Sewerage Project,* Civ. Eng. Transactions of the Institution of Engineers (Australia) Sept. 1961.

60 MAGILL, P. L., HOLDEN, F. R. and ACKLEY, C., *Air Pollution Handbook* (1956), McGraw-Hill N.Y., Chapter 5.

61 PARKER, C. D., Report on partial treatment investigations not yet published by the Melbourne and Metropolitan Board of Works.

62 POMEROY, R. D., *Control of Hydrogen Sulphide Generation in Sewers,* Water & Sewage Works 103 (March 1956), 133.

AUTHOR INDEX

SUBJECT INDEX